RILEM 最新技术发展报告 TC 224-AAM

碱 激 发 材 料
Alkali Activated Materials

John L. Provis ［英］　Jannie S. J. van Deventer ［澳］　主　编

刘　泽　彭桂云　王栋民　王群英　主　译

张大旺　陈仕国　副主译

中国建材工业出版社

图书在版编目（CIP）数据

碱激发材料 / 刘泽等编译；（英）约翰·普罗维斯
(John L. Provis)，（澳）詹妮·范德文特
(Jannie S. J. van Deventer) 主编 . —北京：中国建
材工业出版社，2019.1（2020.1 重印）

书名原文：Alkali Activated Materials：State-of-
the-Art Report，RILEM TC 224-AAM

ISBN 978-7-5160-2434-8

Ⅰ. ①碱… Ⅱ. ①刘… ②约… ③詹… Ⅲ. ①建筑材
料-胶凝材料 Ⅳ. ①TU526

中国版本图书馆 CIP 数据核字（2018）第 228356 号

First published in English under the title
Alkali Activated Materials：State-of-the-Art Report，RILEM TC 224-AAM
edited by John Provis and Jannie van Deventer，edition：1
Copyright© RILEM，2014 *
This edition has been translated and published under licence from Springer Nature B. V..
Springer Nature B. V. takes no responsibility and shall not be made liable for the
accuracy of the translation.

本书中文简体字版专有出版权由 Springer Nature B. V. 授予中国建材工业出版社。未经出版
者预先书面许可，不得以任何方式复制或抄袭本书的任何部分。

碱激发材料

John L. Provis ［英］ Jannie S. J. van Deventer ［澳］ 主　编
刘　泽　彭桂云　王栋民　王群英　主　译
张大旺　陈仕国　副主译

出版发行：**中国建材工业出版社**
地　　址：北京市海淀区三里河路 1 号
邮　　编：100044
经　　销：全国各地新华书店
印　　刷：北京雁林吉兆印刷有限公司
开　　本：787mm×1092mm　　1/16
印　　张：20.5
字　　数：490 千字
版　　次：2019 年 1 月第 1 版
印　　次：2020 年 1 月第 2 次
定　　价：98.00 元
著作权合同登记图字：01-2018-6527 号

本社网址：www.jccbs.com，微信公众号：zgjcgycbs
请选用正版图书，采购、销售盗版图书属违法行为
版权专有，盗版必究。本社法律顾问：北京天驰君泰律师事务所，张杰律师
举报信箱：zhangjie@tiantailaw.com　　举报电话：(010) 68343948
本书如有印装质量问题，由我社市场营销部负责调换，联系电话：(010) 88386906

RILEM 最新技术发展报告　第 13 卷

　　国际材料与结构研究实验联合会（the International Union of Laboratories and Experts in Construction Materials, Systems and Structure, RILEM）成立于 1947 年，是旨在推动建筑科学、技术和工业化发展的非政府科技联盟，实现该领域研究与应用间的技术转化。RILEM 关注建筑材料及其在建筑和市政工程材料中的应用，涵盖了从建筑建造、使用和材料回收的全过程。RILEM 的更多信息和前期出版物都可以在 www.RILEM.net 上查到。RILEM 最新技术发展报告（STAR）是由技术委员会（TC）撰写的。这些报告是 RILEM 最重要的对外公开资料——报告该领域最前沿的科学与工程资讯。TC 的工作是 RILEM 重要职能之一。TC 的成员都是本领域的专家，在他们的空余时间分享他们的专业知识。所以，广大科技界同仁都受益于 RILEM 的相关活动。RILEM 的目标就是要把该领域的最新资讯广泛传播给科学界。因此，RILEM 非常重视 STAR 技术委员会的发展报告，并且尽可能鼓励他们出版。发展报告以及类似报告的提出为实际标准规范的提高提供科学基础。只有具备坚实的科学基础，建筑工程实战才能更高效、更经济。RILEM 也期望这份报告内容能广泛用于科学界。

　　了解更多内容请登录 http://www.springer.com/series/8780。

编者（外）

John L. Provis
Department of Materials Science and Engineering
University of Sheffield
Sheffield, South Yorkshire, UK

Jannie S. J. van Deventer
Zeobond Group, Docklands, Australia
Department of Chemical and Biomolecular Engineering
University of Melbourne
Melbourne, Australia

ISSN 2213-204X ISSN 2213-2031 (electronic)
ISBN 978-94-007-7671-5 ISBN 978-94-007-7672-2 (eBook)
DOI 10.1007/978-94-007-7672-2
Springer Dordrecht Heidelberg New York London
Library of Congress Control Number: 2013954333

©RILEM 2014

　　本作品受版权保护。发行商保留一切权利，无论是全部还是部分材料，具体涉及翻译、重印、使用插图、朗诵、广播、复制缩微胶片或其他物理方式的权利，以及传输或信息储存和检索、电子转印、计算机软件或通过现在已知或后来发展的类似或者不同方法。与评论或学术分析及材料相关的摘录豁免此法定保留，这些专门用于计算机系统的输入和执行，供买方作品专用。本出版物或其部分的复制仅可在本地"出版者著作权法"条款规定下进行，在当前版本中，使用许可必须在 Springer 获得。使用权限可以通过版权清除中心的 RightsLink 获得。违反相关著作权法的行为可能受到起诉。

　　本出版物中对使用的一般描述性名称、注册名称、商标、服务标志等没有作出声明，即使没有具体声明，这些名称也免除相关的保护性法律法规，因此一般可免费使用。

　　尽管认为在出版时本书中的建议和信息都是真实正确的，但作者、编辑及发行人员对出现的错误和删漏都不承担任何的法律责任。出版商对本文出现的材料不作任何说明或暗示的保证。

　　本出版物印在无酸纸上。

　　Springer 是 Springer Science+Business Media 的一部分（www.springer.com）。

中文版前言

碱激发材料（AAM）的概念自20世纪初为人所知，至今已超过一百年，在前苏联、中国、澳大利亚等国有几十年的工程应用示范，但一直没有在世界范围内大规模推广。随着水泥基胶凝材料原材料的大量开采使用，从20世纪90年代，国际上对碱激发材料的基础研究又开始蓬勃发展，特别是在中国，改革开放和工业化发展不断深入，产生出大量铝硅酸盐工业固体废弃物，活性铝硅酸盐固废（如粉煤灰、冶金渣、煤系高岭土、赤泥等）制备碱激发材料的研究不断扩展和深入。工业固废制备碱激发材料的基础理论、技术、标准和产业化推广亟需一本专业的指导教材。

国际材料与结构研究实验联合会（International Union of Laboratories and Experts in Construction Materials, Systems and Structure, RILEM）在此背景下，于2006年发起成立了碱激发材料技术委员会，并由英国John L. Provis教授和澳大利亚Jannie S. J. van Deventer教授组织编写《碱激发材料》一书。该书是针对碱激发材料发布的最新研究报告，具有广泛的国际影响力，得到业界同仁的一致认可和赏识。

为此，在获得出版公司（Springer）授权许可下，我们一年前组织了行业力量对本书进行中文版的翻译工作。经过一年多紧锣密鼓的组织翻译和校对，终于将此书审阅完成。

中文版的组织翻译工作，主要由中国矿业大学（北京）混凝土与环境材料研究院王栋民老师和刘泽老师及其研究生，以及华电电力科学研究院彭桂云院长和王群英主任及其团队成员完成。其中第1章由王栋民、刘泽主译，第2章由张大旺、房奎圳主译，第3章由于杰、崔勇主译，第4章由任才富、刘少卿主译，第5章由刘泽、王琳琳主译，第6章由彭桂云、苏彤主译，第7章由王群英、芮雅峰主译，第8章由周瑜、张明娟主译，第9章由王栋民、张川川主译，第10章由张雷、陈仕国主译，第11章由刘泽、程国东主译，第12章由亢一星、陈仕国主译，第13章由王栋民、刘泽主译。之后，刘泽老师和张大旺博士对全书进行了校对和审定，王栋民老师和彭桂云院长再次进行专业校对，最终由刘泽老师全面审阅和最终定稿，始今交付正式出版。

《碱激发材料》从组织翻译、校对到最终审阅、出版印刷，众多参与者付出巨大辛劳，在整个出版过程中，得到了各方面专家和领导的支持与帮助。在此，对所有参与本书出版工作的朋友和业界人士致以最真诚的敬意和感谢！

毋庸讳言，由于水平和精力有限，翻译、校对和审阅工作仍有遗漏和瑕疵，敬请读者批评指正！

刘　泽
2018年8月

序 言

作为一种硅酸盐水泥的替代品,碱激发材料(AAM)的概念早在1908年之前就已为人所知。另外,碱激发材料在实际应用过程中的耐久性已经过比利时、芬兰、苏联和中国以及近期在澳大利亚几十年的验证。尽管如此,AAM的基础研究从20世纪90年代才在国际上蓬勃发展起来,并且大多数的研究都集中在碱激发材料的微观结构,缺少使用寿命、耐久性和工程应用性能的研究。近年来,碱激发材料的研究被视作一项具有应用潜力的研究方向,但在大宗工程应用中还不能替代硅酸盐水泥。

碱激发材料技术因诸多原因,导致其不能像硅酸盐水泥一样受到市场青睐。首先,石灰石几乎遍布各地,硅酸盐水泥能够就近市场生产。相比之下,碱激发材料的原料像粉煤灰、冶金矿渣等不能随处可得,或者其供应链在当地市场还没有建立起来。第二,从20世纪70年代以来,外加剂的快速发展极大地提高了现代硅酸盐水泥混凝土的和易性或硬化特性。与现代混凝土相比,碱激发材料缺乏适应性良好的外加剂的问题,已成为一个重要挑战。第三,硅酸盐水泥拥有超过150年的广泛使用记录,而AAM技术的应用范围有限,如苏联早期的示范项目并没有得到推广。第四,不同区域的普通混凝土以及粉煤灰和高炉矿渣等作辅助胶凝材料的混凝土标准均以硅酸盐水泥为依托,导致了AAM混凝土市场应用时并不符合现有标准,造成决策者尤其是咨询工程师和投资方将会承担更大的风险和责任。第五,与硅酸盐水泥混凝土标准相关的一系列耐久性测试方法虽然不完善,但已被市场接受,这主要由硅酸盐水泥混凝土的大量应用跟踪记录作支撑。由于缺少长期的应用跟踪记录,以及实验室耐久性测试在预测使用寿命时的不确定性,使得AAM进入市场面临挑战。第六,现有的硅酸盐水泥混凝土工程设计准则都是建立在一系列不同环境条件下混凝土微观结构行为的隐含假设上,这些假设不一定适用于AAM,修订AAM设计标准还需更多的研究。

显然,AAM在市场应用方面面临着巨大挑战。AAM的发展和商业化应用取得的重大进步,承担着过去几十年许多人的愿景和希望。然而,近年来AAM受到广泛关注的主要原因是:与硅酸盐水泥相比,其CO_2排放大幅减少。CO_2潜在交易市场以及当前或未来CO_2减排法规的出台,为AAM在精心选择的商业项目中应用提供了动力。大规模AAM示范项目推动了进一步的基础研究,也为了解AAM耐久性和工程性能提供依据。与此同时,某些区域对非硅酸盐水泥基材料的认可度提高取得了巨大的进步,例如在结构建筑方面的应用。因此,世界各地对AAM的信心随着与关键研究进展相关联的商业进步而增长。现在已有坚实的科学体系支撑着AAM技术,并且该领域的论文数量呈指数增长,而且更多文章出现在高水平的期刊上。

非常感谢西班牙马德里爱德华多·托罗亚建筑科学研究所Angel Palomo教授的建议,他建议在2006年年底,建立一个AAM领域的RILEM技术委员会(TC)。Palomo教授也是该技术委员会的首位秘书长。衷心感谢他对本委员会和AAM领域的贡献。2007年

RILEM 批准成立 TC 224-AAM，这是一种新型建筑材料走上正规化道路的重要一步。值得一提的是，TC 224-AAM 是该领域第一个国际委员会，其成员涵盖混凝土材料的各个方向，都是 AAM 方面的权威专家。技术委员会成立 6 年来，会员人数略有变化，最后 36 名成员写出了代表 15 个国家的发展报告。

根据 RILEM 的职能范围，TC 的活动主要专注于 AAM 在建筑材料方面的应用，包括混凝土、砂浆和灌浆料，不包括 AAM 作为高温陶瓷型材料的应用。TC 有三个主要的目标：(1) 审查 AAM 化学和材料科学的研究发展报告以及 AAM 混凝土的服役性能；(2) 评估现行的标准体系，搭建一个适用于不同区域的 AAM 标准体系；(3) 评估 AAM 耐久性的物理和化学测试方法，制定出合适的测试方法并纳入到新标准的框架。显然从一开始，对可用于 AAM 的各种原材料和激发条件来说，现有规范标准并不合适。AAM 本身就是一个复杂的领域，需要有一个新的 AAM 混凝土性能测试标准体系，这还需进一步的工作。

TC 基本符合上述三个目标。同时，TC 充分认识到，对 AAM 的理解我们还有很大差距，因此在以下领域需要做更多的工作：(1) 当低钙和高钙原料组合使用时，需进一步理解凝胶相形成的机制，指导实际混料和养护条件下形成不易干燥开裂的稳定凝胶；(2) 在调节混凝土的凝结时间、流变性、工作性以及养护条件方面，化学外加剂的开发还很缺失；(3) 尽管 AAM 中的碱-骨料反应比水泥混凝土中的影响较小，但对 AAM 混凝土中的膨胀问题依然知之甚少；(4) AAM 混凝土在化学和物理侵蚀条件下表现良好，但还没有足够的工作使实验室测试数据与服役性能关联；(5) 现有的测试方法似乎不适用于 AAM 混凝土的碳化测试；(6) AAM 混凝土的徐变行为缺乏深入的研究，这对结构设计是一个瓶颈。希望该发展报告能够增强 AAM 作为建筑材料的信心，促进对其进一步的研究，推动 AAM 的商业化，进而服务于 AAM 相关建筑协会或学会。

TC 224-AAM 由以下三个工作组 (WG) 组成：(1) WG 1，由 John Provis 教授（先前就职于澳大利亚墨尔本大学，现就职于英国谢菲尔德大学）领导，组织撰写了 AAM 研究和工业化应用的相关内容；(2) WG 2，由瑞士 Holcim 的 Lesley Ko 博士领导，组织撰写了不同区域的现有标准，制定了 AAM 性能标准的框架；(3) WG 3，由 Ing. Anja Buchwald 博士（先前就职于德国魏玛包豪斯大学，2009 年以来就职于荷兰 ASCEM）领导，组织撰写了 AAM 测试方法的开发，旨在为性能标准提供框架。三个工作组的工作有所重叠，技术委员会在 2011 年写出发展报告的草稿，工作组通过技术委员会联席会议合并所有内容，成员间通过电子邮件对报告内容做了充分的讨论及修改。衷心感谢 John Provis 教授、Lesley 博士和 Ing. Anja Buchwald 博士推动工作组的相关工作。也要感谢所有技术委员会成员对工作组工作的贡献、指导和建议，这些工作不一定能完全反映在本报告中。

TC 224-AAM 成立大会于 2007 年 9 月 5 日在比利时 Ghent 举行。WG2 于 2008 年 3 月 27 至 28 日在瑞士苏黎世和 10 月 2 日至 3 日在西班牙 València 举办了专题研讨会。TC 224-AAM 第二次联席会议于 2008 年 6 月 9 日在捷克共和国 Brno 举行，连同 6 月 10 日至 12 日在 Brno 举办的第三届非传统水泥和混凝土国际研讨会，大多数技术委员会成员都参加了该会议。2009 年 5 月 28 日，技术委员会在乌克兰基辅举办了关于 AAM 的研讨会，在此期间提交了 15 篇技术论文。这次研讨会之后是 2009 年 5 月 29 日的第三次技术委员会联席会议。

第四次技术委员会会议，连同作为第七届水泥和混凝土国际会议一部分的第一届国际化学活性材料进展大会（CAM'2010）于 2010 年 5 月 9 日在中国山东济南举办。CAM'2010 年会议记录由史才军和沈晓冬编辑，并由 RILEM 出版刊物 S. A. R. L 于 2010 年在 Bagneux 出版作为第 72 卷论文集。许多技术委员会成员为 2010 年的 CAM 作出了贡献，其中包括 6 场主旨演讲和 23 篇学术论文。第五次 TC 224-AAM 联席会议于 2011 年 7 月 1 日在西班牙马德里举办，结合第十三届国际水泥化学大会（XIII ICCC），其中许多技术委员会成员参加了此次会议。2011 年 7 月 2 日在马德里大会前由 RILEM 和第十三届 ICCC 共同主办 AAM 培训课程，60 位代表出席。第六次，也是最后一次 TC 224-AAM 联席会议于 2011 年 9 月 7 日在中国香港特别行政区举行，作为 RILEM 交流周的一部分。技术委员会主席和秘书长向 RILEM 交流周的与会代表就 TC 224-AAM 的工作作了详细汇报。特别感谢众多 TC 224-AAM 成员担任技术委员会和工作组会议的会议主席，对他们的专业、慷慨、热情、好客、友谊表示赞许。作为技术委员会的一员，我们一起努力工作，共享了许多乐趣和美好时光，希望每个人都能像我一样珍惜。2012 年全年及 2013 年上半年我们都致力于完成这份发展报告。衷心感谢编写本报告 13 章的所有作者，感谢他们在撰写文稿时的努力工作、奉献精神和合作精神。许多作者花了大量的时间和精力来起草，希望本报告能够成为在 AAM 领域很有价值的参考出版物。应当注意的是，每章作者在一般情况下是按字母表顺序排列；对于那些大部分由一位作者撰写的章节，该作者首先列出，而其他人按字母表顺序排列。特别要向澳大利亚珀斯科廷大学的 Arie van Riessen 教授和瑞士 Empa 的 Frank Winnefeld 博士表示感谢，他们慷慨地为各个章节的草稿做了审阅和修订。相信所有的技术委员会成员，尤其是作者们，很高兴与我一起特别感谢 John Provis 教授为 TC 224-AAM 运行付出的巨大努力和贡献。作为技术委员会秘书长，大部分时间里他确保会议顺利进行，会议记录一丝不苟，保证 RILEM 办公室与技术委员会的有效沟通。此外，Provis 教授作为共同主编远远超出了其责任范围，以确保这份发展报告成为我们都引以为傲的文件。我代表技术委员会成员向他无私的奉献表示深深的感谢。

Prof. Jannie S. J. van Deventer（墨尔本 Zeobond 有限公司及澳大利亚墨尔本大学）
RILEM 技术委员会 TC 224-AAM 主席
作者：John L. Provis[1,2] and Jannie S. J. van Deventer[2,3]
发展报告主题："碱激发材料：最新技术发展报告，RILEM TC 224-AAM"

---

1 谢菲尔德大学材料科学与工程系，谢菲尔德 Mappin 大街 Robert Hadfiled 爵士大楼 S1 3JD，英国。
2 墨尔本大学化学与生物分子工程系，维多利亚州 3010，澳大利亚。
3 Zeobond 有限公司，维多利亚州 8012 码头区 23450 号邮箱，澳大利亚。

技术委员会成员名单

K. Abora, UK

I. Beleña, Spain

S. Bernal, Australia/UK

V. Bilek, Czech Republic

D. Brice, Australia

A. Buchwald, Germany/Netherlands

J. Deja, Poland

A. Dunster, UK

P. Duxson, Australia

A. Fernández-Jiménez, Spain

E. Gartner, France

J. Gourley, Australia

S. Hanehara, Japan

E. Kamseu, Italy/Cameroon

E. Kavalerova, Ukraine

L. Ko, Switzerland

P. Krivenko, Ukraine

J. Malolepszy, Poland

P. Nixon, UK

L. Ordoñez, Spain

M. Palacios, Spain/Switzerland

A. Palomo, Spain

Z. Pan, China

J. Provis, Australia/UK

F. Puertas, Spain

D. Roy, USA

K. Sagoe-Crentsil, Australia

R. San Nicolas, Australia

J. Sanjayan, Australia

C. Shi, China

J. Stark, Germany

A. Tagnit-Hamou, Canada

J. Van Deventer, Australia

A. Van Riessen, Australia

B. Varela, USA

F. Winnefeld, Switzerland

目　录

第 1 章　绪论和研究范围 ······ 1
 1.1　报告结构 ······ 1
 1.2　背景 ······ 1
 1.3　激发剂与固体原料复合体系 ······ 3
 1.4　术语说明 ······ 4
 参考文献 ······ 5

第 2 章　历史背景与概述 ······ 8
 2.1　碱激发材料发展的必然性 ······ 8
 2.2　碱激发历史 ······ 14
 2.3　碱激发溶液化学 ······ 15
 2.4　碱激发技术的全球发展 ······ 20
 2.5　碱激发材料在北欧、东欧和中欧地区的发展 ······ 20
 2.6　碱激发材料在中国的研究状况 ······ 21
 2.7　西欧与北美碱激发水泥的发展 ······ 23
 2.8　大洋洲碱激发水泥的发展 ······ 25
 2.9　拉丁美洲碱激发水泥的发展 ······ 26
 2.10　印度碱激发水泥的发展 ······ 26
 2.11　小结 ······ 26
 参考文献 ······ 26

第 3 章　胶凝材料化学——高钙碱激发胶凝材料 ······ 50
 3.1　介绍 ······ 50
 3.2　矿渣基碱激发胶凝材料的结构与化学特性 ······ 50
 3.3　矿渣体系的激发剂 ······ 52
 3.4　孔溶液化学 ······ 59
 3.5　BFS 性能的影响 ······ 60
 3.6　C-S-H 中的碱（包括碱激发硅酸盐水泥） ······ 61
 3.7　非高炉矿渣前驱体 ······ 63
 3.8　小结 ······ 65
 参考文献 ······ 65

第 4 章　胶凝材料化学——低钙碱激发胶凝材料 ······ 78
 4.1　低钙碱激发胶凝材料 ······ 78
 4.2　胶凝材料的结构特性和激发剂的影响 ······ 79
 4.3　粉煤灰及其与碱性溶液的相互作用 ······ 83
 4.4　天然矿产资源 ······ 85

4.5 火山灰和其他天然火山灰 … 87
  4.6 低钙冶金渣 … 88
  4.7 合成体系 … 89
  4.8 小结 … 89
  参考文献 … 89

第5章 胶凝材料化学——复合体系和中钙体系 … 107
  5.1 引言 … 107
  5.2 复合胶凝体系中的共生凝胶 … 107
  5.3 中钙体系中的激发剂 … 108
  5.4 单一原料 … 109
  5.5 复合原料 … 110
  5.6 小结 … 115
  参考文献 … 115

第6章 外加剂 … 124
  6.1 碱激发体系的外加剂定义 … 124
  6.2 引气剂 … 125
  6.3 促凝剂和缓凝剂 … 125
  6.4 减水剂和高效减水剂 … 127
  6.5 减缩剂 … 128
  6.6 小结 … 129
  参考文献 … 129

第7章 碱激发混凝土——配合比设计标准和早期性能 … 133
  7.1 简介 … 133
  7.2 TC 224-AAM标准编制规划——基于性能实施方案 … 134
  7.3 碱激发胶凝材料的标准 … 135
  7.4 碱激发混凝土的标准 … 137
  7.5 与AAM有关的现行标准调研 … 140
  7.6 原材料分析 … 141
  7.7 养护的重要性 … 142
  7.8 小结 … 142
  参考文献 … 143

第8章 耐久性和测试——化学基体裂化过程 … 149
  8.1 引言 … 149
  8.2 抗硫酸盐测试 … 149
  8.3 碱-骨料反应 … 154
  8.4 浸出试验 … 162
  8.5 耐酸性 … 168
  8.6 耐碱性 … 171
  8.7 海水侵蚀 … 172

8.8	软水侵蚀	172
8.9	生物诱导侵蚀	172
8.10	小结	172
	参考文献	173

第9章 耐久性和测试——劣化与物质传输 182

9.1	简介	182
9.2	渗透性和孔隙率	182
9.3	界面过渡区	189
9.4	氯化物	190
9.5	钢筋锈蚀直接分析	193
9.6	碳化	197
9.7	风化	204
9.8	小结	205
	参考文献	206

第10章 耐久性和测试——物理性能 226

10.1	引言	226
10.2	力学性能测试	226
10.3	收缩和开裂	231
10.4	徐变	235
10.5	冻融和抗冻性	236
10.6	小结	239
	参考文献	240

第11章 建筑与市政基础设施的示范项目和应用 251

11.1	引言	251
11.2	碱激发 BFS 混凝土构造	251
11.3	混凝土路面	254
11.4	地下及沟槽结构	255
11.5	比利时布鲁塞尔 Le purdociment 公司	256
11.6	Pyrament 水泥及相关产品（法国的 Cordi-Géopolymère/美国的 Lone star）	256
11.7	芬兰开发的 F 混凝土以及其他材料	257
11.8	ASCEM 水泥（荷兰）	258
11.9	E-Crete 水泥	263
11.10	不同地区的铁路轨枕	267
11.11	小结	267
	参考文献	267

第12章 碱激发材料的其他潜在应用 272

| 12.1 | 引言 | 272 |
| 12.2 | 轻质碱激发材料 | 272 |

12.3	油井水泥	274
12.4	高温性能	274
12.5	废弃物的稳定化/固化	281
12.6	纤维增韧	287
12.7	小结	290
参考文献		290

第13章 碱激发技术的总结与展望 ... 307

13.1	技术委员会的成果总结	307
13.2	碱激发的未来	308
13.3	结语	311

第 1 章 绪论和研究范围

John L. Provis

1.1 报告结构

本报告由 RILEM 碱激发材料技术委员会（TC 224-AAM）编写。技术委员会的目标有三个：分析碱激发技术的工艺现状；就目前对碱激发材料的了解，为国家标准发布提供建议；开发出合适的测试方法并纳入到推荐标准当中。TC 224-AAM 成立于 2007 年，是首个碱激发领域的国际技术委员会。技术委员会的重点在于建筑（混凝土、砂浆、灌浆和相关的材料）相关方面的应用，因而不包括碱激发胶凝材料作为低成本陶瓷材料在高温领域的应用。

该发展报告由五个主要部分组成，内容如下：

（1）第 1 章和第 2 章包含了世界不同地区碱激发技术发展的历史概况，为理解该领域特有的发展模式提供了基础。

（2）第 3、4、5 和 6 章重点讨论和分析了碱激发化学，以及在不同碱激发体系中结合相的性质：

① 高钙碱激发体系，特别是那些以冶金渣为基础的体系；

② 低钙碱激发体系，主要涉及碱铝硅酸盐，包括现在广为人知的"地质聚合物"；

③ 关于中间组成区的讨论，可以通过混合钙基和铝硅酸盐基的原料，也可以使用一些专有的原料；

④ 在开发具有所需性能的碱激发胶凝材料和混凝土时，化学外加剂在其中的作用。

（3）第 7、8、9 和 10 章包含 TC 224-AAM 开展工作的技术核心，解决耐久性和工程性质、标准依据和测试方法等问题。

（4）第 11 章和第 12 章概括了碱激发技术的一些应用（历史的和正在进行中的）以及实施碱激发技术的潜在领域。

（5）第 13 章总结报告，概述了碱激发领域在研究、开发和标准化中未来的重大需求。

1.2 背景

碱性原料和铝质及硅质固体原料反应，形成能与硬化硅酸盐水泥相比的固体材料，这是 1908[1] 年由德国著名的水泥化学家及工程师 Kühl（图 1-1a）首次提出的，其中玻璃质炉渣与碱金属硫酸盐或碳酸盐组合，加或者不加碱土氧化物或氢氧化物作为"激发材料"，文献指出"完全等同于优良硅酸盐水泥"的性能。Purdon 在 1940 年的一篇重要期刊文献中（图 1-1b）详细地测试了超过 30 种不同的高炉矿渣，分别由 NaOH 溶液、NaOH 和 $Ca(OH)_2$ 组合及不同的钠盐激发，并且得到了可与硅酸盐水泥比拟的强度发展以及最终强度指标，为胶凝材料的发展提供科学基础。他也指出，与相同抗压强度的硅酸盐水泥相比，碱-矿渣水泥有更高的抗拉强度和抗折强度，硬化凝胶相具有低的溶解性和放热。Purdon 还提出这种混凝土制备方法对预拌混凝土和预制混凝土制品来说是理想的，其中激发剂的用量可以精确控制。然而，激发条件对需水量的敏感性，以及高浓度腐蚀性溶液的使用被认为是潜在的问题。随后 70 年的实验表明，Purdon 对这些问题的顾虑是正确的——例如 Wang 等人[4]在 1995 年列出了相似的困难，但也表明每个人都可以通过正确运用科学知识来解决手头的问题。

图 1-1　关于矿渣反应活性的早期出版物

(a) 1908 年 Kühl 的专利；(b) 1940 年 Purdon 发表的"碱对高炉矿渣的作用"；
(c) 1959 年 Glukhovsky 出版的图书《Gruntosilikaty》

关于碱激发技术的研究，早期主要集中在西欧地区，随后研究热潮逐步向东部扩展。苏联和中国的水泥短缺导致了对其他替代材料的需求；碱激发技术在这两个区域都有发展，通过利用现有原材料，尤其是冶金矿渣，作为克服水泥短缺问题的方式之一。特别是苏联的工作由 Glukhovsky[5]（图 1-1c）发起，位于基辅的研究所现在以其命名，主要关注冶金矿渣的碱碳酸盐激发。

在 Purdon 的工作之后，直到 20 世纪 80 年代，碱活性的研究在西方世界非常有限，正如 Roy[6]在一篇评论中发表的时间表所强调的那样。Davidovits 在法国工作，他从 20 世纪 80 年代初开始申请了大量适用于小众应用的铝硅酸盐基配方专利[7]，并且首次对这类材料使用"地质聚合物"的名称[8]。美国陆军在 1985 年发布了一项报告，讨论了碱激发技术在军事方面的潜在价值，尤其是作为一种混凝土高速公路的修补材料[9]。这份报告是与当时以

Pyrament 为名出售碱活性胶凝材料的生产商一起编写的。

1.3 激发剂与固体原料复合体系

自从 20 世纪 90 年代以来，碱激发的研究在世界各地急剧增长，目前有超 100 个活跃的研究中心（学术和商业）遍布在每个有人居住的大陆上进行详细的研发活动。这项工作的主要目的是基于特定区域的原材料进行优异性能的材料的开发。文献中提供的大量技术资料，报告了从特殊原材料和碱激发剂组合中得到的基本物理和/或微观结构性质。除了详细介绍这些成果，表 1-1 也总结了过去几十年出版的关于不同原料对不同碱来源激发适应性的总体结果。

表 1-1 不同固体原料与激发剂组合的总结（激发剂表现为可行/优异）

激发剂 固体原料	MOH	$M_2O \cdot rSiO_2$	M_2CO_3	M_2SO_4	其他
高炉矿渣	可行 §3.2，§3.3.1	优异 §3.2，§3.3.2	合适 §3.3.3	可行 §3.3.4	
粉煤灰	优异 §4.2.1，§4.3，§5.4.1	优异 §4.2.2，4.3，§5.4.1	差—加入水泥/熟料变得可行 §5.5.5	只能添加水泥/熟料 §5.5.4	$NaAlO_2$—可行 §4.2.3
煅烧黏土	可行 §4.2.1，§4.4.1，§4.4.2	优异 §4.2.2，§4.4.1，§4.4.2	差	只能添加水泥/熟料 §5.5.5	
天然火山灰质材料及火山灰	可行/优异 §4.5	优异 §4.5			
骨架铝硅酸盐	可行 §4.4.3	可行 §4.4.3	只能添加水泥/熟料	只能添加水泥/熟料	
合成玻璃质原料	可行/优异（取决于玻璃组成） §4.7	优异 §4.7			
钢渣		优异 §3.7.1			
磷渣		优异 §3.7.2			
镍铁渣		优异 §3.7.3，§4.6			
铜渣		可行（磨细矿渣不确定） §3.7.3			
赤泥		可行（添加矿渣更好） §5.5.4			

续表

激发剂 固体原料	MOH	$M_2O \cdot rSiO_2$	M_2CO_3	M_2SO_4	其他
底灰及市政固废焚烧灰		可行 §3.7.4			

本报告中详细分析的这些不同组合的节编号也有给出。

注意：

1. 分类如下所示：

优异：通过使用这种激发剂可以得到高性能（高强度，耐久）的胶凝材料和混凝土。

合适：性能一般稍微低于最优激发剂得到的，但依然能够得到较好的结果。

可行：制备有价值的碱激发胶凝材料是可能的，但就强度发展、耐久性及/或工作性而言有显著的缺点。

差：强度发展对大多数的应用来说通常不够；另外这里提到的体系（需要大量地添加硅酸盐水泥熟料）在这种分类中会降级。

2. M 代表碱金属阳离子；在高 pH 值条件下但没有碱金属化合物（例如石灰-火山灰水泥）的激发不在本报告的范围。

3. 不同原料的混合在此表中没有明确描述。

4. 所列的 $M_2O \cdot rSiO_2$ 描述了所有范围的碱金属硅酸盐化合物，不考虑模数（r）。

5. 空白单元格表示此体系没有在公开科学文献中描述过。

1.4 术语说明

作为发展报告的一部分，一些对术语方面的建议也很重要，因为这是碱激发领域的一个争议点。这里的讨论一般来说遵循 Van Deventer 等人[10]的介绍；应当指出的是，技术委员会并不完全同意这里提出的所有观点。

对相似材料的描述存在许多的名称，包括"矿物聚合物""无机聚合物""无机聚合物玻璃""碱胶凝陶瓷""碱灰材料""土质水泥""土质硅酸盐""SKJ 胶凝材料""F-混凝土""水化陶瓷""沸石水泥""沸石陶瓷"等许多其他的名称。衍生出的不同名称（本质上描述相同的材料）主要会对那些不太熟悉该领域的研究人员造成影响，他们不仅会严重混淆哪些术语指哪些具体材料，在学术搜索引擎上进行简单关键词搜索时还会找不到相关的重要研究成果。在本报告的背景下，"碱激发材料（AAM）"和"地质聚合物"至少是值得使用的。

（1）碱激发材料（AAM）最广泛的分类，包括任何本质上是由碱金属原料（固体或溶液）和固体硅酸盐粉体反应得到的凝胶体系[11,12]。这里的固体可以是硅酸钙（如在碱激发中常用的熟料），或富含铝硅酸盐的原料，如冶金矿渣、天然火山灰、粉煤灰或底渣。使用的碱来源可包括碱氢氧化物、硅酸盐、碳酸盐、硫酸盐、铝酸盐或氧化物——本质上任何可以提供碱金属阳离子的可溶物质，以提升反应混合物的 pH 值并加速固体原料的溶解。应注意的是，在陶瓷产品中的酸性磷酸盐化学[13]并未涵盖在此定义中；尽管在描述磷酸盐陶瓷化学成键时偶尔会使用"地质聚合物"[14]，但这只会使问题更加混乱。这份报告也特别包括石灰石-火山灰体系和其他提升 pH 值的碱土化合物材料。可能在某些粉煤灰（通常是 C 类）和水[15]之间，或（在很长一段时间内）在矿渣和水[16]之间发生的反应也不在这一范围。为了加以比较和完整阐述，报告中简要讨论了这些大类材料中的每种材料。

（2）地质聚合物在许多情况下都被视作是 AAM 的一类，其胶凝相基本使用纯铝硅酸盐或近似原料[18,19]。为了形成主要胶凝相的凝胶，反应组分允许的钙离子含量通常很低，以形成类沸石网络结构[20]而不是链状的水化硅酸钙。激发剂通常是碱金属氢氧化物或硅酸盐[21]。低钙粉煤灰和煅烧黏土在地质聚合物人工合成中是最普遍的原料[22]。注意"地质聚合物"一词也被一些工程人员（包括学术和商业领域）使用；广义讲，这通常出于营销（而不是科学）目的。

这些分类之间的区别如图 1-2 所示。这显然是混凝土成形体系化学一种高度简化的观点；任何试图精简化学体系例如 $CaO-Al_2O_3-SiO_2-M_2O-Fe_2O_3-SO_3-H_2O$（只列出最主要的组分），呈一个单一的三维视图，都不可避免地会造成重大遗漏。然而作为一种说明 AAM 分类的方法及其相对于 OPC 及硫铝酸钙水泥体系中的位置，它确实很有用。地质聚合物是 AAM 中 Al 浓度最高和 Ca 浓度最低的一类材料。我们还注意到，高炉矿渣（BFS）的一般分类自始至终在报告中也使用；这也包含通过各种可用方法冷却并造粒的矿渣。这里不详细讨论炉渣的产生和化学机理；有些信息将涉及特殊粒化的矿渣和粉磨粒化的矿渣，但它们都归类于一般意义上的 BFS。

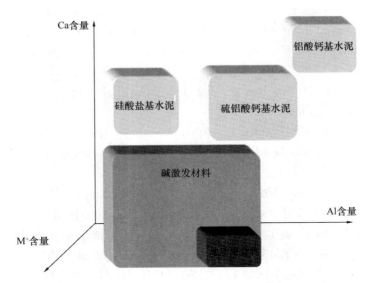

图 1-2　AAM 的分类与 OPC 和硫铝酸钙凝胶化学组成作比较
阴影厚度表示大致碱的含量；黑色阴影对应于较高浓度的 Na 或 K
（图由 I. Beleña 提供）

参考文献

[1] Kühl H. Slag. Cement and process of making the same. Compiler：US，900939. 1908.

[2] Purdon A. O. The action of alkalis on blast-furnace slag. J. Soc. Chem. Ind. Trans. Commun. 1940，59：191-202.

[3] Purdon A. O. Improvements in processes of manufacturing cement, mortars and concretes. Compiler：British，GB 427227. 1935.

[4] Wang S.-D., Pu X.-C., Scrivener K. L., Pratt P. L. Alkali-activated slag cement and concrete: a review of properties and problems. Adv. Cem. Res. 1995, 7(27): 93-102.

[5] Glukhovsky V. D. Gruntosilikaty (Soil Silicates). Gosstroyizdat, Kiev, 1959.

[6] Roy D. Alkali-activated cements-opportunities and challenges. Cem. Concr. Res. 1999, 29(2): 249-254.

[7] Davidovits J. Mineral polymers and methods of making them. Compiler: US, 4349386. 1982.

[8] Davidovits J. Geopolymer Chemistry and Applications. Institut Géopolymère, Saint-Quentin, 2008.

[9] Malone P. G., Randall C. J., Kirkpatrick T. Potential applications of alkali-activated aluminosilicate binders in military operations. Geotechnical Laboratory, Department of the Army, 1985, GL-85-15.

[10] Van Deventer J. S. J., Provis J. L., Duxson P., Brice D. G. Chemical research and climate change as drivers in the commercial adoption of alkali activated materials. Waste Biomass Valoriz, 2010, 1(1): 145-155.

[11] Buchwald A., Kaps C., Hohmann M. Alkali-activated binders and pozzolan cement binders-complete binder reaction or two sides of the same story? In: Proceedings of the 11th International Conference on the Chemistry of Cement, Durban, South Africa, 2003: 1238-1246.

[12] Shi C., Krivenko P. V., Roy D. M. Alkali-Activated Cements and Concretes. Taylor & Francis, Abingdon, 2006.

[13] Wagh A. S. Chemically Bonded Phosphate Ceramics. Elsevier, Oxford, 2004.

[14] Wagh A. S. Chemically bonded phosphate ceramics-a novel class of geopolymers. Ceram. Trans, 2005, 165: 107-118.

[15] Cross D., Stephens J., Vollmer J. Structural applications of 100 percent fly ash concrete. In: World of Coal Ash 2005, Lexington, KY. 2005, 131.

[16] Taylor R., Richardson I. G., Brydson R. M. D. Composition and microstructure of 20-yearold ordinary Portland cement-ground granulated blast-furnace slag blends containing 0 to 100 % slag. Cem. Concr. Res. 2010, 40(7): 971-983.

[17] Davidovits J. Geopolymers-inorganic polymeric new materials. J. Therm. Anal. 1991, 37(8): 1633-1656.

[18] Duxson P., Provis J. L., Lukey G. C., Separovic F., van Deventer J. S. J. 29Si NMR study of structural ordering in aluminosilicate geopolymer gels. Langmuir, 2005, 21(7): 3028-3036.

[19] Rahier H., Simons W., Van Mele B., Biesemans M. Low-temperature synthesized aluminosilicate glasses. 3. Influence of the composition of the silicate solution on production, structure and properties. J. Mater. Sci, 1997, 32(9): 2237-2247.

[20] Provis J. L., Lukey G. C., van Deventer J. S. J. Do geopolymers actually contain nanocrystalline zeolites? -A reexamination of existing results. Chem. Mater, 2005,

17(12): 3075-3085.

[21] Provis J. L. Activating solution chemistry for geopolymers. In: Provis, J. L., van Deventer, J. S. J. (eds.) Geopolymers: Structure, Processing, Properties and Industrial Applications, Woodhead, Cambridge. 2009: 50-71.

[22] Duxson P., Fernández-Jiménez A., Provis J. L., Lukey G. C., Palomo A., van Deventer J. S. J. Geopolymer technology: the current state of the art. J. Mater. Sci. 2007, 42(9): 2917-2933.

第 2 章 历史背景与概述

John L. Provis，Peter Duxson，Elena
Kavalerova，Pavel V. Krivenko，Zhihua
Pan，Francisca Puertas and Jannie
S. J. van Deventer

2.1 碱激发材料发展的必然性

2.1.1 全球视角：温室气体排放驱动

水泥和混凝土对世界经济体系至关重要，整个建筑行业在 2008 年为全球经济贡献了 3.3 万亿美元[1]。直接归因于材料成本的这一数字的比例在不同国家之间差异很大——特别是在发展中国家和发达国家之间。2008 年全球水泥产量约为 29 亿 t[2]，是世界上最大宗的商品之一。因此，混凝土是世界上仅次于水泥的第二大商品[3]。值得注意的是，除了混凝土生产之外，水泥类胶凝材料的应用还包括瓷砖灌浆、胶凝材料、密封剂、废弃物固化、陶瓷和其他相关领域，这些将在第 12、13 章中进行更详细地讨论，本章主要关注混凝土的大规模生产。

如此巨大的产量也伴随着非常显著的环境代价。在水泥生产过程中导致的全球二氧化碳排放量至少增加 5%~8%[4]，这主要是由于在水泥生产中需要高温分解石灰石（碳酸钙）来产生具有反应活性的硅酸盐相和铝酸盐相。随着世界经济的不断发展，全球对水泥的需求量正在迅速增长[5]，这意味着迫切需要可替代的胶凝材料以满足数十亿人的住房和基础设施需求，但不能进一步增加二氧化碳的排放。

普通硅酸盐水泥（OPC）通常是通过在回转窑中将原料混合物加热至约 1450℃，再冷却其半熔融材料形成固体熟料，然后与硫酸钙共同粉磨产生细粉制成的。使用的主要原料是石灰石（主要是 $CaCO_3$），其与诸如页岩或黏土等类似的材料共混以提供必要的氧化铝和二氧化硅含量。熟料主要为硅酸钙，其被快速冷却以稳定艾利特（$3CaO \cdot SiO_2$）和贝利特（$2CaO \cdot SiO_2$）混合相，具有少量（但很重要）的富含 CaO 的铝酸盐和铁铝酸四钙。然而，当石灰石在窑中加热时，根据式（2-1）的分解反应，每生产 1t 的 CaO 会释放 0.78t CO_2。

$$CaCO_3 \longrightarrow CaO + CO_2 \tag{2-1}$$

Gartner[6]计算了与每种水泥主要形成相相关的总原料 CO_2 排放量（即从碳酸盐前驱体释放的 CO_2 算起）。将典型 CEM I 水泥的相组成的这些值相加，根据精确的熟料组成，得出通过石灰石分解每吨熟料将释放约 0.5t 的 CO_2。Damtoft[7]等人计算得出，每吨熟料释放约 0.53t CO_2，与能源消耗相关的每吨水泥平均释放 0.34t CO_2。

据行业权威者估计，硅酸盐水泥生产目前最好的技术手段（包括使用可替代燃料，优化在混凝土中使用的水泥，回收利用以及与火山灰混合）可以减少来自水泥生产释放的二氧化碳约17%的排放量[7]。

尽管很少能得到充分证实，但碱激发材料（AAM）相对于传统硅酸盐水泥的一个经常声称的优点是，在与AAM生产相关的作业时CO_2的排放量会更低。AAM是目前正在讨论的几种可替代胶凝材料之一，以期待在建筑业中获得环保投资上的节约，其他还包括铝酸钙、硫铝酸盐、富硫酸化水泥和镁基胶凝材料[8]。在利用碱激发胶凝材料的情况下，CO_2的减排主要是由于避免了在灰分或炉渣合成AAM中的高温煅烧步骤。使用碱性溶液确实重新引入了一些温室成本，这是需要定量和优化的关键点。Duxson等人[9]提出了基于激发溶液的溶解固体（Na_2O+SiO_2）含量的函数，各种碱激发飞灰和偏高岭土掺合料的CO_2排放量的计算。CO_2排放量包含在氯碱法中生产Na_2O时产生的CO_2和作为硅酸钠水溶液的SiO_2用作主要投入物而产生的CO_2。经计算显示，与硅酸盐水泥相比，在碱激发胶凝材料/水泥胶凝材料的CO_2排放量减少了80%。在以上分析中，主要使用了基本单元模块的概念。当考虑胶凝材料性能时，"良好的"碱激发材料可以提供超过硅酸盐水泥标准性能之外的优点，但是相反，由于较低的强度（为了得到相同的结构性能必须使用更多的材料）和较差的耐久性（需要经常更换），通过使用新型胶凝材料获得的环境优势在全生命周期的考核下就不占优势了。

对于碱激发胶凝材料生命周期的分析，不同的配合比设计和胶凝材料类型会有不同的结果。本报告就不对作者们采用的不同生命周期方法的有效性进行评论了。文献中，从生命周期研究中引用的估算CO_2减排量（比较AAM与硅酸盐水泥）的范围从80%[10]到30%[11]，其他研究提供的值介于这两个边界值之间[12-16]。另一项近期的研究[17]提供了多个类别的详细排放分析，但在假设中使用一种相当低效的方法，即使用生产的硅酸钠作为激发剂，这导致基于这些AAM材料计算的全球变暖的优势将超过OPC混凝土，以及在可持续性的其他方面，特别是在环境有害性方面有着显著不利的结果。显然，在科学界和工程界之间，以及在生命周期分析专家之间达成共识之前，需要进一步的工作和一致性来理解生命周期分析（Life Cycle Analysis，LCA）方法。分析方法和数据需要有一致性和广泛性，再辅以科学坚实的（在当前碳税、经济条件下）论证，这些论证才能最终支持产业界更广泛地使用AAM。

2.1.2 全球视角：其他环境因素驱动

尽管在减少水泥制造所导致的环境影响方面取得了可观的和可持续的进展，但除了CO_2排放之外，这个行业对于世界工业部门的环境影响在多个领域还做出了非常显著的贡献。

首先，这与矿产储量枯竭的现状有关，特别是对于相对较小的欧洲和东亚国家来说，水泥消耗量相当高，但自然资源有限。受社会环境压力，各国政府越来越多地限制天然原料的开采和大宗固体废弃物的填埋。一方面是出于环境保护的原因，另一方面是出于生态美学原因，老百姓并不希望在附近有一个大型的采石场或粉煤灰堆场。虽然对AAM混凝土需要采集的骨料基本上与OPC混凝土相同，但是通过采用粉煤灰或矿渣替代石灰石原料在这两个领域里都有好处。在这方面，西欧在世界范围是领先的，他们将底灰和矿渣加入水泥混凝土中。欧洲的几个国家在过去十几年几乎100%利用了粉煤灰、炉底灰和矿渣[18,19]。目前在胶凝材料生产中，矿渣利用面临的主要问题是如何获得细粉状的玻璃体矿粉。对于缓慢冷却的矿渣逐渐晶化，反应活性较低，主要用作骨料。在大多数发达国家，粉磨的高炉矿渣由于其在硅酸盐水泥混凝土中的优异性质而得到充分利用[20]，矿粉可以与水泥混合球磨作混合材，也可在混凝土中加入作掺合料。[21]

水泥生产中的第二个主要问题是水泥窑会向大气中排放有害物质。在过去的几十年中，对水泥窑除尘和其他潜在排放物的捕获和回收利用取得了重大进展[22]，但即使如此，在排放的烟道气中还会存在不可忽略的有害成分，特别是在气体净化能力十分有限的旧窑炉中。在一些发展中国家，废弃物还被再次用作燃料。应该注意的是，用于碱激发的硅酸钠或氢氧化物激发剂的生产也会将非温室气体的排放（SO_x、NO_x、磷酸盐等）引入到生命周期的计算中[16,17]。由于初始计算表明这些问题对于碱激发胶凝材料比硅酸盐水泥的影响更大，所以关键是要开发出严密的计算方法，用于量化来自不同硅酸钠生产过程中产生的 CO_2 排放。

相反，由于节省了基于煅烧的制备工艺，AAM 的重金属（除了可能来自某些形式的氯碱法中汞的排放）排放量可能比硅酸盐水泥的更低。在某些情况下，可以得到论证，将来自填埋废料的碱激发材料通过在碱激发基体内固化这些金属，可以减少重金属的排放量。

碱激发技术还为不能在硅酸盐水泥中添加的固体废弃物提供了利用机会。例如，对镁铁渣[23]、镍铁矿[24]炉渣和钨矿固废[25]的研究表明，这些材料可以通过碱激发有效转化为有用材料，而这些固废不能用作水泥的矿物掺合料。这可能不是碱激发技术的主要应用领域，但确实在某种程度上，使这些固废在某些区域得到有效利用，推动环境治理。

欧洲环境政策的基本目标和原则在"罗马条约"第 130 号文件中规定，旨在预防、减少和消除环境损害，最好从源头对原材料储层做精细管理。最先进的建筑材料生产国家及公司主要位于欧洲，他们积极致力于技术前沿，包括开发未来的新型建筑材料[8,26,27]。欧洲环境政策就是试图引入诸如"最佳可用技术"的概念，以便限制排放，满足环境质量标准。出于环境和经济因素考量，水泥和混凝土行业的技术创新也是为了满足以上目标。在水泥行业，为进一步推进节能及环保政策，需要开发并商业化应用含更低水泥熟料的胶凝材料体系，从这一点来看，碱激发或复合碱激发水泥应该是一种较为合适的胶凝材料。

2.1.3 商业驱动

虽然目前环境因素可能不足以推动地质聚合物技术的广泛应用，但预计国家和国际碳交易计划以及"碳税"的推行将促使 CO_2 的排放成本增加。另外，饮用水成本的提高（这在非常干旱的地区可能很重要），也将成为商业化驱动的重要因素，这对建筑制品行业具有一定意义。

在建筑行业，一般不同水泥的市场竞争力主要是由硬化胶凝材料性能、新拌混凝土性质（工作性）、价格、标准化和经典工程案例所决定的。一代又一代的工程师们一直在尝试选择更低成本的胶凝材料及混凝土配合比来满足性能要求，而且由于硅酸盐水泥基胶凝材料易于使用、性能良好，绝大部分工程都选择水泥基材料体系。但是，如 2.1.2 节中讨论的环境问题，材料使用的优先权发生了变化，当前在发达国家选择建筑材料时优先考虑环保性能（包括能源消耗）。为了实现可持续发展，越来越多中等收入的发展中国家也关注于建筑材料的节能与环保。随着全球变暖议题更加深入大众，消费者对"绿色"产品的偏好以及相关的碳交易的关注，使得 AAM 在传统的水泥和混凝土行业中可以优先被大规模使用。

关注温室气体排放本身显然不足以推动 AAM 在世界所有地区的应用，特别是在许多发展中国家，水泥相对便宜，利润率低，成本效益至关重要。在这些地区，需要为 AAM 混凝土开发原材料供应链和市场,以便与水泥生产商竞争市场份额[5]。在这些地区使用本地的天然原料(如火山灰和黏土)作为 AAM 前躯体,配以合适的激发剂。在一些发达国家,原料的供应可能存在问题,其中 BFS 和粉煤灰广泛用于硅酸盐水泥的混合材,因此工业企业没有足够原料用于碱激发技术开发。也就是在这些发达国家,工业固废不能用于碱激发材料的开发,而天然原料可以用于碱激发材料的开发,且未来效益可观。

发达国家商业开发经验表明,将新材料引入建筑行业存在两大阻力:

(1) 在每个政府辖区推行新材料都需要有标准规范,而这些标准规范的制定和引入是一个渐近的过程。

(2) 关于混凝土耐久性方面的问题,考虑到结构混凝土至少需要几十年的评估,常规的耐久性数据根本不适用于新开发的材料。

RILEM TC 224-AAM 工作组专门致力于解决这方面的问题。标准、耐久性测试,特别是加速耐久性测试的方法,可以方便快捷地用于测试 AAM 混凝土和胶凝材料的性能,是本报告的核心内容。

在发达国家,胶凝材料性能测试都有具体的标准。这些标准都是约定俗成地使用基于 OPC 混凝土的性能标准,但标准中规定的"最低水泥用量"其实有些过度限制,即使是在 OPC 混凝土体系[28]。诸如 AAM 混凝土或其他替代胶凝材料,甚至一些高性能水泥基产品可能不单单是现有 OPC 技术的发展演变,还需要用不同的化学和工程观点来理解它们的性能,也不能完全执行现有的 OPC 标准,特别是新型材料的流变性能和化学成分与 OPC 混凝土有很大区别。水泥生产商和潜在的 AAM 市场开发者长期以来一直认为这是 AAM 技术在发达国家推行的主要阻力。然而,澳大利亚和美国的工程实践案例表明,通过与所有利益相关单位(包括监管机构、建筑师、保险公司、结构工程师、建筑公司和最终用户)沟通解释产品,只要满足监管标准,这些阻力可以克服。

如果能够建立一个实用有效的机制来满足 AAM 混凝土在市场上的准入标准,绝大部分开发单位和用户最关心的仍然是耐久性的问题。水泥、混凝土耐久性测试的大多数标准方法都是将小样品短期置于极端条件,近似等同于预测材料在正常环境下在几十年或更长时间内的性能[29]。这些耐久性的测试方法只能给出预期性能的数据,而不是实际长期的性能数据。因此,新材料应用是一个非常漫长的过程,可能需要等待二三十年的实际服役性能验证,对于大多数工程设计师来说,这些数据非常必要。

然而,发展中国家就没有这么多固定的标准,并且愿意接受一些创新的解决方案,例如在混凝土中使用替代胶凝材料,因为市场对胶凝材料的需求预计明显超过目前未来几十年可能的供应量。这也是苏联的碱激发胶凝材料开发的原始驱动力。因为水泥需求超过供应,而且近年来,在发展中国家出现大量"假冒"水泥产品,在当前经济条件下开发可替代的低成本水泥产品为其提供了额外推动力。发展中国家的市场,特别是在中国和印度,由于燃煤发电而产生大量的粉煤灰,与此同时伴随着的二氧化碳排放可能成为日益严重的政治问题,这可能就证明了碱激发技术在监管层面被接受的原因。

还值得注意的是,在某些通常相对独立的地区,利用当地的原材料开发 AAM 胶凝材料比进口硅酸盐水泥成本更具优势。一个特别的例子就是阿拉斯加,目前完全依赖进口水泥,但国内生产粉煤灰的比例很高,在这种特定市场中,使用两种技术制备低强度混凝土,则 AAM 比 OPC 更具有成本和性能优势[30]。在全世界使用燃煤发电的其他地区,由于硅酸盐水泥的产量有限，可能也存在这种情况。尽管如此，在每个区域都需要进行详细研究，确保当地产业界和观念上能接受AAMs。

2.1.4 技术驱动

接着2.1.3节的内容，如本报告后面的章节所述，碱激发水泥和碱激发混凝土不仅能够满足现有建筑应用中规定的性能要求，而且在耐酸侵蚀和防火方面具有优良的性能。传统的水泥混凝土标准大部分不包含碱激发技术，这是因为传统的工业企业和监管部门对该项技术缺乏商业利益的驱动。现在不同了，由于经济和环境因素的推动，碱激发技术将逐渐进入市场。随着时间的推移，AAM的应用价值逐渐凸显出来（特别在经济效益和环境效益方面）；与此同时，材料耐久性的验证数据也逐渐增多，并且有相关标准正在编制。相关内容在第8、9和10章中详细叙述，本节只作一般性讨论。

在苏联，碱激发胶凝材料开发主要使用地壳中存在的天然碱/碱土-铝硅酸盐矿物[31]。自20世纪50年代以来，研究人员通过不断研究，验证了混凝土生产条件下可以使用天然材料实现该类材料成相[32,33]。这项工作得出了以下结论：

（1）传统硅酸钙胶凝材料硬化的物理化学过程，在一定程度上类似于岩石的化学风化、沉积、无定形化变质过程；然而，反应过程因原料的碱度和理化特性不同而速率不同。

（2）与结构稳定的碱性岩石相比，钙基水泥的矿物组分碱度更高，稳定性较差，更容易快速地发生水化反应，生成憎水的水化硅酸盐和铝酸盐以及可溶的氢氧化钙。

（3）为了促进天然岩石或组成相似的人造碱和碱/碱土-铝硅酸盐反应生成水硬性胶凝材料，通常会加入碱性氧化物或氢氧化物，将它们从稳定的晶态转变为活性的亚稳态或玻璃态。从结果来看，高碱性物质的水化过程与长石和霞石岩石风化的自然过程类似。

（4）高碱性和低碱性物质的水化产物类似于天然矿物，特别是低溶解度的水化铝硅酸盐，包括层状双氢氧化物、沸石和低碱性钙硅酸盐，以及钠和钾的可溶性氢氧化物或硅酸盐。类似于以上矿物，纳米无序的晶粒会沉淀成硬化的凝胶，加入骨料可制备单块的碱激发混凝土。随着水化产物越来越稳定，混凝土中胶凝相的化学耐久性和热力学耐久性也越来越强。

许多概念最初是基于化学和地球化学理论提出的；反应机理和凝胶纳米结构的详细表征随着分析技术的进步而变得清晰明朗，并且大量文献已经论述，把碱激发过程的产物描述为类铝硅酸盐矿物化学风化[34]、纳米级层状双氢氧化物[35-37]、硅质水凝胶[38]，以及纳米级或晶粒尺寸范围的沸石[39,40]。

碱金属引入凝胶化合物的量远远超过了传统硅酸盐水泥标准的允许范围，这就必须考虑到，碱金属不仅作为激发剂加速硬化，也是铝硅酸盐胶凝材料体系中必要的独立组分。低钙体系的反应产物可能是类沸石的无定形态化合物[39,41,42]，而高钙体系的水化产物主要是被部分取代（特别是Al）的水化硅酸钙凝胶，并且与其他化合物的颗粒（主要是无序的化合物）共存，这些凝胶性能取决于铝硅酸盐前驱体和激发剂的化学特性[37,43-48]。碱在低钙体系的凝胶产物中尤为重要，体系反应会形成类沸石凝胶结构，且在孔溶液中的浓度较高。有些作者的研究表明，类沸石凝胶在碱激发BFS的高钙体系的固体反应中较少出现[49,50]，但在有些情况下也有含Na固相生成，特别是在低Mg含量下，会限制类水滑石产物的生成，从而消耗铝硅酸盐前驱体中的部分Al[51]。

图2-1给出了胶凝相化学、结构、稳定性和耐久性之间的关系图,这是由Glukhovsky和Krivenko[52]领导的研究小组对该领域的学术和工程领域50年来的研究总结。该图相关的临界值是反应产物溶解度与组分函数的关键。高碱水化产物比低碱水化产物更易受腐蚀影响;这种趋势在地球化学领域中是众所周知的[53,54],与矿物耐候性相关,并且能够为碱激发胶凝材料的稳定性提供热力学依据,其所述胶凝材料通常具有比硅酸盐水泥的水化产物更低的钙含量(并因此降低总碱度和更高的硅酸盐连接度)。近来,Gartner和Macphee[26]详细讨论了这些热力学过程在胶凝材料合成反应机理中的物理化学特性。不同的碱金属和碱土金属阳离子以及Al和Si的相对成键和断链过程会直接影响这些体系的反应机理和凝胶结构。对胶凝结构形成的热力学机理必须详细理解,但想让AAM胶凝材料被广泛接受还远远不够,世界各地对AAMs的研究还在持续进行。精确可用的热力学数据永远不能说服终端用户或工程师信赖一种新材料,但没有相关数据也无法使用户信服,因此该领域的科学进步对于在主要工程中采用AAM混凝土至关重要。

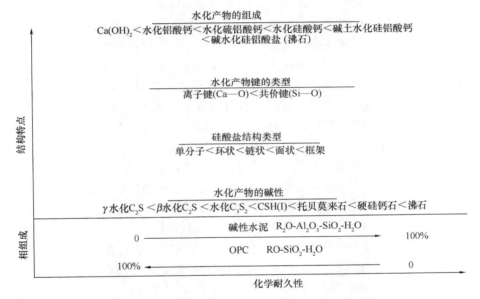

图2-1 矿物耐久性、水化产物的相组成及其结构之间的关系(引自Krivenko[52])

其他领域引入AAM技术的驱动还包括[9]:

(1) 强度(抗压、抗折和耐磨性),包括在特定养护条件下一些混合物快速、可控的强度增长;
(2) 耐高温和防火、低导热系数;
(3) 耐酸和化学侵蚀;
(4) 低碱硅反应;
(5) 硬化体积稳定性好;
(6) 与水泥/混凝土、陶瓷、玻璃和一些金属基材结合性能好;
(7) 新拌浆体pH值高,促进钢筋钝化;
(8) 低渗透性;

(9) 低成本。

以上每一个优点都将在第 8、9 和 10 章中详细讨论。以上优良特性可以通过使用正确设计的碱激发胶凝材料体系来实现,但不能通过一个设计的材料体系实现上述所有特性。碱激发铝硅酸盐胶凝材料已经应用于从航空航天领域到建筑领域,再到危险废弃物的管理中。用量最大的肯定还是建筑材料领域,这也是本报告的重点,但应该了解,在某些领域的高附加值应用可能比大宗混凝土产品具有更多的商业开发价值。

2.2 碱激发历史

2.2.1 碱激发在古代水泥中的应用

在过去的三十年里,从科学/工程和历史角度,对于碱在水泥中的作用做了大量讨论研究,例如古埃及和古罗马建筑,中东地区早期文明所属建筑中的一些材料,特别是在罗马帝国兴起之前的叙利亚和希腊建筑,碱激发材料都可能是这些建筑材料中的选择。Davidovits[55,56]提出了一系列详细理论,验证了"地质聚合物"材料与埃及金字塔的建筑材料有明显的联系,埃及人可能使用化学碱激发方式,现场浇筑制备大型砌块。相关文献对金字塔石块样品做了详细研究分析[57,58],结果表明金字塔石块中的碱金属或铝含量并不高,但确实存在无定形二氧化硅和其他成分,且可能是使用石灰石现场浇筑凝结形成的。迄今为止,科学和历史的研究既没有完全支持这一理论,也没有完全反驳。本报告提及此事是说明该领域在古代已发展到一定程度。

围绕古罗马混凝土[59]和现代碱激发胶凝材料[31]之间的潜在联系也有不少讨论。罗马混凝土在组成、性能和耐久性方面差别很大,在本报告中特别感兴趣的是基于火山灰材料[来自意大利南部的波佐利(原来的普托利奥)地区的火山灰]与含钙化合物混合,特别是石灰,其中由石灰反应产生的高 pH 值会引发火山灰材料的反应。在这些混凝土中使用的火山灰含有大量的碱[61],混凝土样品在 2000 年后检验,发现有方沸石存在。我们知道方沸石通常存在于火山灰中;检测掺有未反应的火山灰的罗马砂浆,其方沸石的含量高于未反应的火山灰中的方沸石[62],这表明混凝土内的碱性环境可能促进额外沸石的形成,这与火山灰暴露于碱性地质环境时的研究结果一致[63]。虽然分析过程不能把未反应的火山灰有效分离,但也提供了有力证据[64]。

与罗马混凝土相比,我们现在讨论的耐久性和强度保持就可以推断出来了。在 2000 年的时间里,这些混凝土在海水侵蚀环境下仍然屹立不倒,另外罗马万神殿的混凝土经过剧烈地震仍屹立在那。类似的水利工程混凝土基道路、多层楼房、地下建筑和穹顶建筑仍然屹立至今。尽管罗马混凝土结构经过几个世纪已经衰落,但这么多屹立不倒的建筑给后人从化学和设计角度提供很多经验。另外,在现代混凝土中观察到一些劣化现象主要是由于钢筋腐蚀导致的,而未增强的罗马建筑结构却没有受到这种劣化方式的影响。

早期的罗马砂浆似乎主要通过石灰碳化实现硬化和强度发展,后来发现含硅火山灰和/或煅烧黏土反应生产铝硅酸盐凝胶相可提高性能(通过实验或通过实施从周围文明获得的知识),即改变原料的矿物组成可以提高耐久性。古代水泥的矿物成分与现代硅酸盐水泥不同,古代胶凝材料的 Ca 含量更低并且富含碱、Si 和 Al[31,60,64]。在各种存在方沸石的古老水泥中都有佐证，即初始相经过长期水热反应转化而形成的更稳定的沸石相，这与图 2-1 中描述的一致。

还值得注意的是，尽管现代水泥经常被用于修复古建筑，但是现代水泥修复后暴露在相同环境下很少被证明与古代水泥一样耐久。也就是说，用低钙和高铝硅酸盐含量原料生成的胶凝材料具有耐久性更优的特点，或者如果修复材料选择不当，可能与现有材料和新材料之间（化学和机械）性能不兼容。虽然 C-S-H 凝胶是硅酸盐水泥浆体和许多古老水泥的主要成分，但不能完全证明 C-S-H 凝胶是提供结构耐久性的主要因素。材料在环境中的耐久性似乎与凝胶交联程度有关，如图 2-1 所示，钙含量降低和铝氧四面体含量升高（通常用碱来平衡电荷，沸石就是这种情况）具有更好的优势。因此，这是本报告讨论古代水泥和现代水泥的关键，可以从某些古老材料的分析中得出与现代水泥耐久性相关的有用信息，并且这些趋势对硅铝酸盐基胶凝材料（不同于硅酸钙基凝胶）现有的化学耐久性研究有利。

2.2.2 碱激发在现代水泥中的应用

大多数现代水泥标准都描述了水泥的最大允许碱含量，试图使碱-骨料反应的可能性最小化，并且有时提出关于这种限值（通常约 0.6wt%Na_2O 当量）是否充分限制的问题[65]。在硅酸盐水泥中加入碱，或者作为高碱熟料或在混合水中加入 NaOH，会明显影响混凝土的机械性能、收缩性能及损害混凝土的其他性能（但不一定普遍存在），并可以减少收缩[66]。当考虑孔溶液中碱度的发展与保持[67]、外加剂-水泥相容性[68]、碱在熟料的硅酸盐[69,70]和铝酸盐[71]相水化过程中早期反应机理所起的作用时，一些碱的含量就至关重要了。

众所周知，添加火山灰材料可以减少碱硅反应的可能性，额外加入与碱反应的 Al 和 Si 不会形成损害混凝土结构的非膨胀性凝胶产物[75]。而且高碱孔溶液环境有利于钢筋混凝土中钢筋的钝化。

结合以上观点和 2.1 节的讨论表明，很值得研发类火山灰材料与碱反应制备的碱-铝硅酸盐胶凝材料混凝土。体系中含有一定量的钙对于胶凝材料是有益的，上述讨论的所有水泥（硅酸盐和非硅酸盐）的化学机理都是基于碱性介质存在的。传统硅酸盐胶凝材料体系和非传统硅酸盐胶凝材料体系的反应过程在一定程度上是相似的，这使得研究人员能够利用钙基凝胶材料的水化硬化过程的现有技术数据建立碱激发水泥生产混凝土的理论基础。当然，一些研究的理论深度不够，也有一些研究阐释得不完全正确，会引起歧义（对于所有复杂的研究领域，人们必须保持谨慎的研究态度），但事实表明，可以从硅酸盐水泥及混凝土的大量文献中获得有价值的数据，帮助研究人员理解 AAM 胶凝材料和 AAM 混凝土。

2.3 碱激发溶液化学

这里必须简要描述一下铝硅酸盐前驱体制备新型胶凝材料的碱激发溶液化学。这里的论述并不详尽，但也提供了充分的背景介绍，为后边第 3、4 和 5 章的详细介绍作铺垫。后边三章都是基于对激发剂和固体前驱体的不同组合进行的讨论。对于硅酸盐和氢氧化物溶液化学的概述，读者还需参考相关文献[76]，涉及碳酸盐或硫酸盐激发剂溶液的反应过程，参见 Shi、Krivenko 和 Roy 的书籍[77]。

2.3.1 碱金属氢氧化物

最常用作碱激发剂的碱金属氢氧化物是钠和/或钾的碱金属氢氧化物，锂、铷和铯氢氧化物成本高、储量少，LiOH 在水中的溶解度相对较低（25℃的溶解度为 5.4mol/kg H_2O[78]），使这些氢氧化物不能大规模应用。通过比较，氢氧化钠和氢氧化钾 25℃时在水中的溶解度超过 20mol/kg H_2O[79,80]，通常浓度超过 5mol/kg H_2O 被广泛用于与碱激发有关的研究，特别是使用 F 类粉煤灰作前驱体时。

氢氧化钠主要通过氯-碱法与 Cl_2 平行生产。这对其在碱激发中的使用会产生较大的环境影响，无论是在温室气体排放（通过耗电）方面，还是在其他成分如汞的排放方面。使用上述方法的同时，有时也采用现代膜分离技术使得氢氧化钠的生产更加高效。类似地，通过电解 KCl 溶液生产氢氧化钾。

在使用过程中，除了需要注意它们的强腐蚀性外，还必须考虑浓碱溶液的黏度，以及固体碱溶解时的放热。即使是极浓碱溶液的黏度也很少超过水黏度的 10 倍[76]，这比碱硅酸盐溶液具有明显优势，相关内容将在 2.3.2 小节中作讨论。

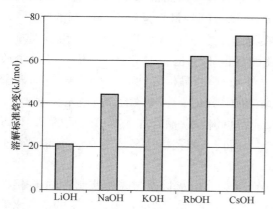

图 2-2 25℃下，MOH（M：碱金属）溶解到无限稀释溶液里的标准焓变（数据来源于 Gurvich 等人[81,82]的综述）

图 2-2 给出了碱金属氢氧化物在无限稀释溶液中溶解的标准焓变 $\Delta_{aq}H^0_{298.15K}$，焓变是阳离子尺寸的函数。更重要的是，在制备高浓度氢氧化物溶液时，温度明显升高，这在工业环境中可能存在问题。事实上，碱在浓溶液中的溶解放热值略低于无限稀释溶液的溶解放热值，这是因为碱溶液被无限稀释时会额外放热。然而，稀释作用占溶解焓的主要部分，约 90% 的总焓变绘制于图 2-2 中[76]。

图 2-2 中的数据可以用于计算当固体碱溶解成高浓度溶液时的近似温升值。如果假设将 10mol NaOH 溶解在 1L 水中释放的 90% 热量，移植到无限稀释的溶液中释放，大约有 400kJ。将 NaOH 溶液的平均热容大致等同于水的平均热容[83]，这个热量足以将水加热到 90℃以上。在大多数情况下，热量将损失到周围环境和/或通过蒸发一些溶液而消耗。然而，在工业生产装置中要充分考虑和监测在强腐蚀溶液中发生的温升现象。

泛碱（特别是由于白色碳酸盐或碳酸氢盐晶体的形成）也是用过高浓度的氢氧化物溶液激发胶凝材料中常见的问题，其中过量的碱与大气中的 CO_2 反应。泛碱使得材料看起来不美观，但并不总是对材料的结构完整性有害。在氢氧化物激发的胶凝材料中，通常 Na 比 K 的存在使得材料泛碱更为显著。

2.3.2 碱金属硅酸盐

跟氢氧化物类似，钠和钾的硅酸盐是在碱激发中使用最多的工业产品[76]，也将是本小节讨论的重点。硅酸锂在大多数碱激发胶凝材料体系中的溶解度不够；铷和铯硅酸盐成本高、产量少,限制了它们的大规模使用。碱硅酸盐通常由碳酸盐和二氧化硅通过煅烧,然后按设定比例溶解在水中而产生,这会消耗大量能量和排放CO_2。然而,由于在大多数碱激发胶凝材料中激发剂的含量(质量)相对低,因此碳酸盐煅烧所排放CO_2的量比硅酸盐水泥生产排放CO_2少得多[6],这是目前文献还在持续研究的领域。

图2-3显示了25℃下结晶等温线的Na_2O-SiO_2-H_2O相组成的一部分[76]。在Brown[84]的工作中,还有关于更低Si含量的化合物相图,以及在水热地质矿床中观察到的更多硅质相的潜在形成的讨论,但是其作用稳定范围仍然未知[84]。这个体系的全相图从来没有被确定过,主要是由亚稳态硅酸盐水溶液决定的,如果它们沉淀,那么沉淀速度将非常缓慢[84]。Weldes和Lange[85]给出的"典型的商业硅酸钠"表中显示,对于普通硅酸钠溶液给出的所有化合物是落在相图范围内的,其中$Na_2SiO_3·9H_2O$将被沉淀。

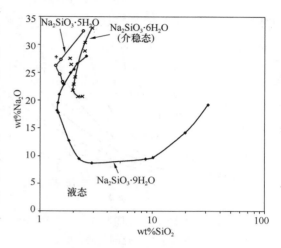

图2-3 25℃下Na_2O-SiO_2-H_2O水化化合物的一部分的结晶等温线,由Wills[86]绘制。未给出低硅高钠的$Na_3HSiO_4·5H_2O$相(改编自文献[76])

根据Vail[87]的分类,描述所有三元Na_2O-SiO_2-H_2O体系在不同区域中形成产物的性质,在碱激发胶凝材料合成中主要使用的硅酸盐为部分结晶混合物、高黏度溶液和/或易于结晶的水化偏硅酸钠。这需要仔细放置和储存这些溶液,在使用之前(在实验室实验或在大规模生产中试验)长时间储存硅酸盐溶液可能导致意想不到的和/或不利的结果。硅酸钾溶液通常不会遇到类似的问题,因为水化硅酸钾比硅酸钠不容易沉淀,并且均相水溶液的稳定范围更宽。

二氧化硅的化学性质(液态和固态)可能比除碳以外的其他元素的化学性质(液态和固态)更复杂,并且从基本观点来看,对它的理解相对较差。在高浓度碱溶液中,二氧化硅聚合成列状小尺寸化合物[88,89],其中几种的特性和结构现在仅能通过^{29}Si NMR[88,90-93]和质谱[94,95]来确定,红外光谱也有应用[96-98],但光谱特征的解释并不总是与NMR实验的结果一致。

在地质聚合反应过程中,现有的硅酸盐物相是可变的、不稳定的,不同相之间可能会转变(也是反应性)。在浓缩溶液中,与硅酸盐低聚物形成过程平行发生的是酸碱反应。单体二氧化硅,$Si(OH)_4$也称为"原硅酸",它在碱性条件下表现为弱酸性。

用于AAMs的激发剂溶液中主要的硅酸盐位点处出现一次或两次去质子化[99]。考虑到现有物相的分布和它们各自的pK_a值,大多数硅酸盐激发溶液通过硅酸盐去质子化平衡后,pH值被稀释至11~13.5,与相同pH值的碱溶液相比,可以提供更高的"可用碱度",这是因为现有碱性物相的稀释。显然,在与固体铝硅酸盐混合之前,碱激发溶液将具有比该固体铝硅酸盐更高的pH值,但是在反应过程中将适量的二氧化硅从固体前驱体溶解到这种溶液中将迅速使pH值升高。

图2-4和图2-5分别表示在室温下硅酸钠和硅酸钾溶液的黏度随组成的变化,数据来自文献[87]。在两个图中,黏度以对数坐标绘制,并且在较高二氧化硅含量下显著增加。重

要的是，硅酸钾溶液比硅酸钠的黏度低得多，硅酸钠也比硅酸钾黏稠得多[85]。因此，浇筑新拌硅酸钠激发混凝土可能会有问题，因为含钠的 AAM 胶凝材料会粘住混凝土模具，同时它也与砂和粗骨料颗粒粘附得很好。

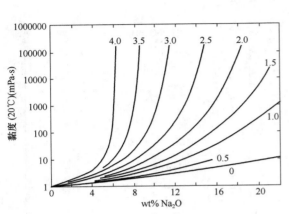

图 2-4　硅酸钠溶液的黏度随 SiO_2/Na_2O 质量比（标记于曲线上）变化的曲线，摩尔比比质量比大 3%

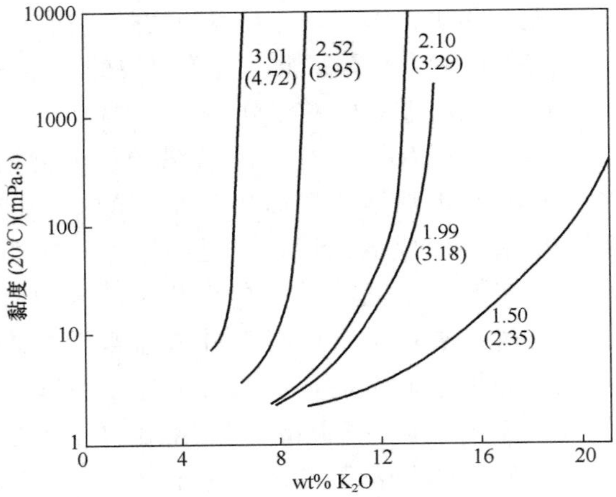

图 2-5　硅酸钾溶液的黏度随 SiO_2/K_2O 质量比（标记于曲线上）变化的曲线（源自 Vail[87]）

Yang 等人[102]研究发现，硅酸钠溶液的黏度随着温度的升高而明显降低。然而，一些偏硅酸钠相的溶解度在升温时会降低[87]，这意味着加热条件下不一定能保证提高固体偏硅酸盐激发溶液制备胶凝材料的性能。由于在极高浓度下溶解物质不能完全水化，固态偏硅酸钠溶解到高浓度溶液中的焓变也会随浓度的升高而降低[87]。

2.3.3　碱金属碳酸盐

利用碳酸盐溶液激发铝硅酸盐也有不少研究。碱碳酸盐通过 Solvay 方法或直接从碳酸盐矿床（除了世界各地大宗储量，美国怀俄明州的绿河流域中的单一沉积物中估计有 5×10^{10} t）开采得到[103]。这些方法产生的直接温室效应比生产氢氧化物或硅酸盐溶液的影响小，但碳酸盐溶液的碱度低于其他激发溶液的碱度，这意味着选择与碳酸酯激发剂一起使用的铝硅酸盐原料的使用面更窄，相应内容将在第 3、4、5 章中详细讨论。碳酸钾或其他碱金属的碳酸盐通常不用于制备碱激发胶凝材料。

图 2-6 给出了三元体系 Na_2CO_3-$NaHCO_3$-H_2O 的相图，其与碳酸盐激发剂的使用以及碱激发胶凝材料孔溶液的碳化有关。有必要考虑碳酸氢盐的潜在形成，甚至在固体 Na_2CO_3 与水结合以形成激发溶液的体系中，因为即使少量的碳酸盐水解或碳化都将导致碳酸氢盐形成。在该图中特别有意思的一点是形成含水碳酸钠，包括 $Na_2CO_3 \cdot 10H_2O$；如果允许形成，其将结合大量的水；如果它在混合期间形成，这可能引起配合比设计时无法确定需水量，而且如果其随后作为胶凝材料或其孔溶液的碳化产物生成，就会填充基体的孔隙。

在激发过程中，Na_2CO_3 和 $NaHCO_3$ 溶液的黏度不可能出现问题，因为饱和溶液的黏度都不会超过水的黏度的 5 倍[105]。可能问题其实是这些溶液会发生沉淀和/或碳化，如图 2-6 显示 Na_2CO_3 溶液中吸收 CO_2 会产生 $NaHCO_3$ 沉淀（该过程相对快速，这在碱碳酸盐作为

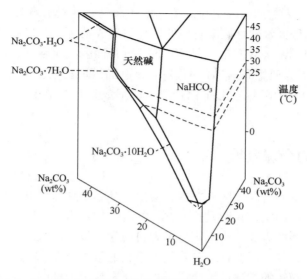

图 2-6 Na_2CO_3-$NaHCO_3$-H_2O 体系的三元相图（改编自 Hill 和 Bacon[104]）。三元共晶体在 $-3.3℃$，$Na_2CO_3 \cdot 10H_2O$-$NaHCO_3$-天然碱不变点在 $21.3℃$，$Na_2CO_3 \cdot 7H_2O$-天然碱共存线上的较低温度点和较高温度点分别为 $32.0℃$ 和 $35.2℃$，虚线是 $25℃$ 和 $30℃$ 的等温线

CO_2 的溶剂用于气体净化过程中被证明），并且 Na_2CO_3 的溶解度对温度很敏感。冷却浓碳酸钠溶液可能导致沉淀，并且高于 $35℃$ 的含水碳酸钠的溶解度也稍微逆行（随着温度升高而降低）。使用碳酸钠而不是硅酸盐作为激发剂也不能完全消除这些材料的黏性问题[106]，因为通过原料溶解释放的二氧化硅在反应的早期会出现类似问题。

2.3.4 碱金属硫酸盐

与碳酸盐溶液的情况类似，也可以通过加入硫酸钠来实现一些硅铝酸盐材料的碱激发。硫酸钠可以直接从采矿中获得（估计全世界存在超过 10^9 t 的自然资源[107]），也可来自许多其他工业化学品制造的副产物中；如果胶凝材料的性能能够达到令人满意的水平，硫酸钠激发确实可以为胶凝材料生产过程减排温室气体提供重要的选择方式。胶凝材料产品的性能主要围绕使用硫酸钠激发不含熟料的铝硅酸盐材料，通常需要添加熟料以产生足够的强度，在 3.6 节中会详细讲述。在所关注的温度和浓度范围内，硫酸钠水溶液的黏度比水的黏度高，不足 5 倍[108]。硫酸钾或其他碱金属的硫酸盐不常用于制备碱激发材料。

图 2-7 给出了 Na_2SO_4-H_2O 体系的相图，改

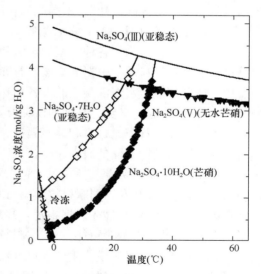

图 2-7 Na_2SO_4-H_2O 体系的相图（改编自文献[109]）；点是实验数据，并且由作者开发的热力学模型拟合成线，显示了亚稳态和稳定相平衡

编自 Steiger 和 Asmussen[109]的研究工作。该体系在水泥和混凝土耐久性的研究中也是重要的，因为在一些硫酸盐侵蚀水硬性胶凝材料的情况下形成钠盐。在与碱激发化学相关的相图中，存在各种有用的相。无水盐方钠石是 Na_2SO_4 的五种多晶化合物之一［另一种 Na_2SO_4（Ⅲ）在使用温度范围内为亚稳态］，另一个关键的稳定相是芒硝（格劳伯盐），其是十水化合物在 32.4℃发生热分解。亚稳态七水化合物相的形成也是可能的，特别是在快速冷却或蒸发的条件下[109]。

2.4 碱激发技术的全球发展

接下来的部分将从世界不同地区的技术、历史和社会角度探讨碱激发材料的开发。根据不同的经济、法规和气候因素，碱激发这一领域的科学技术由不同因素推动着发展，这在一定程度上反映了技术发展的方式。为了充分了解目前这一技术领域的发展现状，必须了解这种现有技术存在的研究背景，因此这些信息是本报告的重要方面。

还必须指出的是，简短的综述并不能详尽列出每个特定地区的科学家和工程师对碱激发水泥和碱激发混凝土的研究和开发所做出的贡献。这些描述将占用太多的篇幅，并且必然会带来一些误导。对不同时期研究团队的成果综述，有助于更全面地了解这些材料。虽然一些相关的名称可能被不经意地省略，但没有一个名称是不相关的。

2.5 碱激发材料在北欧、东欧和中欧地区的发展

20世纪 50 年代中期，在苏联对硅酸盐水泥替代品需求的背景下，V. D. Glukhovsky 开始研究在古罗马和埃及结构中使用的胶凝材料[31]，并发现了使用低碱性钙或无钙铝硅酸盐（黏土）和含碱金属的溶液制备胶凝材料的可能性[41]。他称胶凝材料为"土壤水泥"，称相应的混凝土为"土壤硅酸盐"。根据原材料的组成，这些胶凝材料可以分为两组：碱胶凝体系 Me_2O-Me_2O_3-SiO_2-H_2O 和碱土碱胶凝体系 Me_2O-MeO-Me_2O_3-SiO_2-H_2O。基于这些观察，他将铝硅酸盐废物如各种类型的矿渣和黏土与碱性工业废溶液结合以形成一类水泥替代胶凝材料。这些材料通常根据与天然沸石和低碱性钙铝硅酸盐类似的方法描述，并且能够使用已用于放置硅酸盐水泥混凝土的设备和专门技术来处理和放置。

从 1960 年代起，乌克兰基辅的研究所，现在以 Glukhovsky 教授命名，参与公共建筑、铁路枕木、路段、管道、排水和灌溉渠道、奶场的地板的建设，使用碱激发高炉矿渣[77,110]。随后对这些原始结构剖面的研究表明，这些材料具有高耐久性和致密的微观结构[111]。关于早期的矿渣混合物的大量专利和标准，现在在西方已经基本很难找到。基辅团队继续在 Kryvenko 的监督下开发不同原材料和应用的配合比设计[32,112,113]。使用其材料的几种应用细节将在第 11、12 章详细描述。据报道，1989 年以前在苏联浇筑了超过 $3\times10^6 m^3$ 的碱激发混凝土[114]。虽然这个数量可能占建筑材料市场总量份额很小，但肯定是一个大体积的材料生产，并展示了该技术规模化生产的可行性。

基辅国立土木工程和建筑大学于 1994 年和 1999 年在乌克兰基辅[115,116]、2007 年在捷克共和国布拉格组织了关于碱激发水泥和混凝土的国际会议（由欧盟、捷克共和国政府和布拉格市政府[117]共同组织）。这些会议的记录提供了特别有价值和难得的见解，展示了苏联进

行的大部分工作，这些研究工作没有以任何其他英语形式报道过。

在欧洲东部其他地方进行的大部分早期工作由 Talling 和 Brandstetr 详细描述[118]。从 1970 年代开始在波兰开发的专门技术[119-125]使得碱激发胶凝材料被发现用作坝中的灌浆料和砂浆[126]、废物固化[127]、铁路枕木和其他预应力制品[125]以及一系列其他应用。从 1974 年到 1979 年单独生产的混凝土超过 100000m^3[125]。在 1974 年，碱碳酸盐激发的 BFS 混凝土被浇筑，在使用 27 年之后分析了地板和工业仓库的外墙，并且显示出优异的耐久性和可用性[124]。

欧洲北部和东部的其他集团公司也在碱激发材料体系的开发初期取得了重大进展。在 20 世纪 80 年代和 90 年代初期，芬兰还写了大量的好文章[128-133]。特别值得关注的是瑞典的"F 混凝土"是超塑化的碱激发 BFS 体系[134]。然而，该区域发展似乎没有像早期那样稳健增长。在今天的捷克共和国，几位著名的研究人员一直在碱激发体系工作多年，这一地区的工作由 Brandstetr[118]启动，包括与加拿大同事合作发表的工作[135,136]。目前，最重要的团队之一是布拉格的团体，由 Škvára[137-141]领导，他们研究了 BFS 碱激发过程、粉煤灰和两种材料的混合物以及耐久性的关键技术。专攻 AAM 的公司中，ZPSVAS 公司还在从事商业规模的碱激发 BFS 混凝土的开发[142-144]和关于 AAM 耐火性[145,146]的研究。在捷克和斯洛伐克共和国，由 ALLDECO 公司开发的碱激发的铝硅酸盐胶凝材料已经发现可用于核废料的固化[147]。罗马尼亚自 20 世纪 80 年代[148]以来，一系列基于冶金渣的体系也被用于建筑领域。

在中欧地区，非传统水泥和混凝土[149-152]的四个国际会议组织已表现出对碱激发材料的浓厚兴趣，一个作为展示该地区最新科学进展的平台，专门针对碱激发材料[117]而设立。该地区的许多国家都希望利用冶金工业的副产品与廉价建筑材料的需求相结合，为碱激发技术的发展提供了强有力的驱动。

2.6 碱激发材料在中国的研究状况

近年来，碱激发矿渣制品——"碱矿渣水泥"[153]在中国已得到市场化推广。Wang[154]对早期碱激发材料技术发展进行了总结：致力于采用炼铁和有色冶金等冶金行业副产品，如矿渣和磷渣等，作为胶凝材料制备高强（高于 80MPa）低能耗的混凝土的研究。碱激发矿渣混凝土的凝结时间短和外加剂适用性差等问题已成为今后的研究方向。大量的研究结果[114]表明，碱激发技术已在中国、苏联和美国等地区取得重大的突破。潘教授和杨教授[155]总结了当前中国碱激发技术的发展进程并提出以下建议。

当前中国碱激发胶凝材料的发展主要受到两个非技术因素的影响：一方面，受到全球化科学材料技术历史发展的影响，另一方面，中国环境保护和混凝土胶凝材料的可持续发展对碱激发胶凝材料的发展也起着重要作用。新材料的可行性探索激发了大量学者的研究兴趣从而推动其发展。然而，近年来工业生产的快速发展所造成的环境问题得到了中国政府部门越来越多的重视和关注，已成为推动中国碱激发材料研究的最重要因素之一。耐久性优异的水泥和混凝土等方面的研究已成为当前重要的研究目标之一，包括推动碱激发胶凝材料技术在某些特定研究领域的发展，例如海工混凝土/涂层。

2.6.1 中国工业固体废弃物的产生

工业的迅速发展和制造业的飞速增长为中国碱激发技术的发展提供机遇。当前，固体废弃物因其潜在的火山灰活性而广泛地应用于碱激发材料制备的原材料。常见的固体废弃物有：高炉矿渣、钢渣、磷渣、粉煤灰、赤泥、煤矸石、电石渣、碱渣、磷石膏、烟气脱硫石膏等。同时其相应年产量见表2-1所示。

表2-1 可用于碱激发原料或硫酸盐激发的中国工业固废年产量

固体废弃物	年产量（百万 t/年）
高炉矿渣	230
钢渣	100
磷渣	10
粉煤灰	300
赤泥	3
煤矸石	100
电石渣	20
碱渣	15
磷石膏	20
烟气脱硫石膏	5

截至2009年，中国有13亿t粉煤灰和36亿t煤矸石等待进一步的工业和/或环境友好处理和/或再利用。与此相比，2006—2010年中国水泥以每年9%的增长率逐年增加，截至2009年已达16.5亿t。

如表2-1所示，高炉矿渣作为矿物掺合料广泛应用于混凝土中。因此，高炉矿渣在中国以及欧洲各国通常被认为是冶金副产品而非固体废弃物。目前虽然大约40%的粉煤灰、少量磷渣和石膏副产物已应用于建筑行业，但大部分固体废弃物仍未得到有效的再利用[155]。尽管大量的固体废弃物积累为碱激发材料的制备提供基础，但当前尚未形成合适的固体废弃物处置方法和工业化生产工艺，因此对于碱激发材料的工业化研究任重而道远。当前，中国除了对碱激发材料外加剂和基础化学领域的研究外，还对粉煤灰、高炉矿渣、磷渣、偏高岭土、赤泥和煤矸石等固体废弃物进行了系统的研究探索。

在中国，工业固体废物的再利用技术得到政府税收政策激励和鼓励[159,160]。工业固体废物（不含矿渣）使用量超过30%的产品可享受免税优惠[155]；国家也出台了相关的节能和降产的政策，旨在加强环境保护和推动可持续发展。因此，碱激发材料在化学和制造业等领域的相关研究符合国家政策要求，为其发展提供推动力，同时突出了其重要的研究价值[77]。

目前，中国约有50所从事碱激发材料的研究大学和研发机构[155]。自2004年以来，由中国硅酸盐学会建立的相应国家技术委员会组织已成功举办一系列以碱激发材料为主题的国家科技会议。

2.6.2 在中国碱激发材料和原材料供应的经济分析

与硅酸盐水泥价格相比，中国碱激发材料作为胶凝材料和混凝土应用的主要问题是生产成本较高。2009年，硅酸盐水泥的平均价格为300~350元/t（1EUR≈8CNY），高炉矿渣微粉的成本约为200元/t，商用液体水玻璃的成本约为4500元/t。因此，碱激发BFS胶凝材料的价格预计将在500元/t左右。较硅酸盐水泥而言，碱激发BFS价格较高[155]。高成本的问题也阻碍中国的碱激发技术的推广。虽然粉煤灰或偏高岭土作为碱激发材料成本较低，但其工业化和耐久性问题仍需克服。

2.7 西欧与北美碱激发水泥的发展

在西欧，冶金行业所产生的高炉矿渣长期以来被用作混凝土中的水泥成分。其发展至少可追溯到1860—1870年代的法国和德国[161]以及1905年的美国[162]。20世纪30年代，美国首次利用粉煤灰制备混凝土[163]。碱性材料、水硬性材料以及潜在火山灰材料作为胶凝材料的首次使用可追溯到20世纪初。当时德国科学家Kühl对研磨矿渣和碱性组分的混合物的凝固行为进行了研究。1937年，Chassevent[165]通过苛性钾和苏打溶液测定矿渣的反应活性，尽管此研究目的在于分析矿渣而非开发胶凝材料。1940年，Purdon[166]对有无石灰的两种炉渣和苛性钠制备的碱激发水泥进行了第一次广泛的实验室研究。随后的几十年中，碱激发材料的研究重点仅仅局限于其水泥与混凝土的生产和制备的优化。

20世纪70年代后期，法国化学工程师Davidovits以偏高岭土为原材料制备的碱激发材料重新点燃了学者们的兴趣，并首次以"地质聚合物"的概念来定义碱激发材料[167-171]。碱激发剂与添加剂的复合使用实现了地质聚合物早期抗压强度的快速增长。地质聚合物的高强特性引起了美国政府和工业组织的浓厚兴趣[172]。Davidovits[171]对地质聚合物合成的理想条件、生产反应产物的表征以及最终产物的特性、性质和应用前景的研究做出了突出贡献。20世纪80、90年代，法国地质聚合物研究所成功研发出一种名为Pyrament®的OPC/碱激发材料混合型混凝土[171,173]。同时由法国地质聚合物研究所[174-176]和澳大利亚[177]相关机构成功召开了以地质聚合物为主题的一系列会议。此类会议一直作为法国年度Geopolymer Camp活动。在此类会议中，Davidovits积极地宣传地质聚合物技术，推动其工业化发展进程。

20世纪80年代至90年代，Pyrament的研究表明，地质聚合物混凝土在标准养护4h后抗压强度可达20MPa，低至−2℃的温度4h抗压强度足以满足军事飞机场的使用要求[178]。1991年海湾战争中地质聚合物混凝土成功应用于快速安置跑道的建设。到1993年，Pyrament的研究成果已在美国50个工业设施、57个军事设施以及其他7个国家得到广泛使用[173,179]，最终因其优异的各项性能得到美国军方的强烈支持[180]。在1996年，Pyrament的研究成果因以下原因被其制造商的终止而失去了大部分运营厂商：一方面，Pyrament体系中并非纯碱激发硅铝酸盐体系，其中含有大量的水泥熟料导致其成本比硅酸盐水泥混凝土价格高约两倍[181]；另一方面，Pyrament体系的研究目的仅仅是满足技术性能要求，而非降低二氧化碳排放或成本，同时Pyrament体系结构的长期耐久性能并未研究。

在此基础上，英国[183]和荷兰[184]的学者提出了碱激发矿粉-粉煤灰共混物体系的地质聚合物：7%NaOH激发剂激发60%矿粉-40%粉煤灰共混物体系制备地质聚合物，结果表明其强度发展良好，但仍存在风化和碳化等问题。同时此实验也对固体废弃物生产水泥的Trief方法进行了深入的研究。此实验首次对粉煤灰-矿渣微粉碱激发材料体系进行深入研究得到的相关数据，奠定了粉煤灰-矿渣微粉碱激发体系的基础。

如上所述，20世纪80年代，Roy和其他美国学者着力于研究碱激发剂和古代水泥之间的联系，旨在研发一种可替代硅酸盐水泥的耐久性优异的材料[64,185-187]。路易斯安那州立大学对高钙粉煤灰的碱激发技术进行了深入的研究[188]，有助于理解碱激发的矿渣微粉的化学过程。与此同时，西欧学者致力于研究碱激发材料的科学基础。Palomo 和 Glasser 发表了一篇关于偏高岭土的碱激发的早期论文[190]。加拿大学者的研究成果[135,136,191-198]为这一领域的工程应用和科学研究提供了坚实的基础。然而随着时间的推移，大部分研究机构的研究重点已逐步转向高 SCM-OPC 混合物的制备。尽管最初高水平的重要成果仍在继续发布但此之后并未得到增长[199-200]。

1995 年以来，西班牙碱激发材料的研究主要以马德里的 Eduardo Torroja 建筑科学研究所（西班牙国家研究委员会机构）为代表，其主要致力于研究替代水泥的工业废弃物（高炉矿渣、粉煤灰、矿渣等）碱激发制备绿色环保建筑材料。与传统硅酸盐水泥相比，此研究目的在于利用工业废弃物及其副产品制备低能耗的耐久性和力学性能优异的建筑材料（水泥和混凝土）。1990 年代初期，Palomo 和 Puertas 研究团队对碱激发粉煤灰和高炉矿渣微粉制备地质聚合物的反应过程的化学微观结构[46,202-207]以及水化产物进行了深入的研究。结果表明，与 OPC 特征链型结构不同，矿粉的碱激发体系中的水化产物为层状 C-S-H 凝胶[208,209]，或称为"N-A-S-H 凝胶"[210,211]。其与硅酸盐水泥水化产物的 C-S-H 凝胶的碱铝硅酸盐类似。Palomo 和 Puertas 对碱激发材料的强度和耐久性进行了深入的研究，同时也研究其化学外加剂（超塑化剂、减缩剂）在地质聚合物体系中的适用性[212]，这将在第 6 章中详细讨论。

20 世纪 90 年代，英国学者研究了矿渣微粉在碱激发剂环境下的聚合反应。Wang 和 Scrivener 发表关于碱激发高炉矿渣相应文章[35,114,213,214]，其中他们涉及了当时活性炉渣体系承载强度的影响因素现状。

比利时 Wastiels 和 Rahier 教授领导的团队以偏高岭土和硅酸钠溶液之间相互作用的反应机理和产物为研究对象，发现其产物似于 Davidovits "地质聚合物"的玻璃状聚合物结构而非水泥状材料。这个术语似乎阻碍了其他学者对更广泛碱激发知识的吸收，但是此论文中呈现的信息对于理解碱激发材料产物结构是非常有价值的。在 20 世纪 80 年代至 90 年代，这个研究团队结合学术商业，研究出高强度粉煤灰基碱激发胶凝材料和纤维增强复合材料[222-224]及建筑（图 2-8）。这些产品在所需物理性能的开发方面在技术上是成功的，但由于粉煤灰供应限制，最终未达到大规模生产。

2000 年以来，欧洲研究机构对碱激发材料的研究兴趣再次高涨。在德国，多机构团队合作参与碱激发胶凝材料的研究[12,225-229]，主要包括生命周期分析和可持续发展相关领域的重要初始工作[12-14]。在荷兰，ASCEM 公司在过去 15 年里一直从事水玻璃碱激发的专用水泥。其在生产方法、强度发展和耐久性方面都取得了成功[230]。在瑞士，在固井系统的表征和热力学模型方面取得了关键进展，其中包括应用于水化矿渣微粉体系[50,231-236]。西班牙巴伦西亚的建筑技术研究所对黏土的碱激发研究作出了突

图 2-8　20 世纪 90 年代比利时开发的 AAM 建筑砌块（振捣压实，抗压强度达到 75MPa）

出的贡献[237]。在巴塞罗那,Querol 基于粉煤灰的碱激发胶凝材料和沸石形成等领域取得显著进展[238,239],意大利的研究人员揭示了火山岩的碱激发特性[240]。在葡萄牙,利用富含黏土的钨矿开采废弃物为原料,促使该领域的研究计划产生了一些有趣的结果和许多期刊论文[25,241,242]。目前欧盟生产商面临的主要障碍是明显限制性的标准制度。这一领域的一些概念的详细讨论将出现在第 7 章。

从 20 世纪 90 年代中期开始几年,美国碱激发材料的发展由 Balaguru 和 Rutgers 大学与政府机构合作专注于复合材料和涂料领域[243-246]。2000 年后,伊利诺伊大学的 Kriven 课题组在美国空军的部分资助下开展了一项关于煅烧黏土和合成铝硅酸盐前驱体的碱激发材料作为高温陶瓷型应用的低成本选择研究[42,247-251]。宾夕法尼亚州立大学的一个小组还开发了"水化陶瓷"(碱激发黏土水热固化)作为固化铯和含锶核废物流的潜在材料[252-255]。最近,粉煤灰在化学和工程领域的应用源于宾夕法尼亚州[256-258]、路易斯安那[259-261]、德克萨斯[262]和其他地方的工作,美国更加关注粉煤灰的碱激发而不是基于矿渣体系的碱激发。在美国基于水泥性能的标准的适用性似乎有利于碱激发材料在市场中的使用推广,但同时也受到相应的一些限制。

总而言之,欧洲和北美科学家对碱激发材料的研发取得了重大突破,发表了大量论文,在碱激发材料的发展中起到了至关重要的作用。世界各地技术的驱动因素,以及工业运行所面临的环境监管,促进了科学研究的投入,而东欧和中国则特别关注工程/应用领域的投入。

2.8 大洋洲碱激发水泥的发展

澳大利亚对碱激发材料的研究始于 20 世纪 90 年代中期,旨在开发处理和固化含有高浓度重金属污染物的采矿废物[263-265]。Provis 接替 Van Deventer 领导墨尔本大学的研究项目在该领域发表了超过 100 篇期刊论文。尽管对偏高岭土的碱激发材料已进行了系统研究,但澳大利亚未被充分利用的粉煤灰供应充足,因此地质聚合物的应用研究在很大程度上集中于粉煤灰的再利用[266-270]。此研究中首次引入了同步加速器和基于中子束线等先进的原位表征技术和各种计算方法对其纳米结构进行表征[34,42,89,99,273-279]。

Sanjayan 领导小组发表了许多关于碱激发矿渣/粉煤灰制备混凝土以及碱激发材料的高温性能的重要论文[280-286]。此项研究将结构工程和材料科学相结合,为大型结构和结构材料相关问题的分析提供了借鉴。

西澳的科廷大学分别对碱激发材料的结构[287-290]和理化性能[291-294]进行了研究。同时还通过可持续资源加工和开发组织中心与水泥和资源公司合作试点了一些规模化的混凝土,但尚未形成全面的商业化规模。澳大利亚国家研究机构 CSIRO[295-297] 和 ANSTO[298,299] 也在这一研究领域作出了重大贡献。澳大利亚是一个以资源为基础经济的发达国家,这意味着原材料的供应主要受限于供应链,而不是产率;如果这些问题可以解决,它似乎为碱激发技术的推广提供了良好的范围,标准体系也会得到长足发展。

在新西兰,MacKenzie 领导的研究项目促进了碱激发黏土的基本科学机理的发展[300-303];新西兰的粉煤灰供应相对有限,而可用的 C 类粉煤灰往往存在凝结时间和流变性能方面的问题[304]。

上述澳大利亚和新西兰进行研究的同时,墨尔本公司 Siloxo(1999—2005)和 Zeobond

(2005 年正在进行)[10,305-308]正在推动碱激发材料的商业开发。Siloxo 与当地水泥公司成功利用新西兰 C 类粉煤灰生产预拌混凝土。最近，Zeobond 技术项目应用于商业化活动（澳洲商标名称为 E-Crete，昆士兰州 Wagners 集团命名为"地球友好混凝土"）的现状使得碱激发材料在澳大利亚的接受度得到提高，同时领先于大多数发达国家对接受碱激发材料在民用基础设施中的应用。

2.9　拉丁美洲碱激发水泥的发展

拉丁美洲碱激发技术的发展主要建立在该地区丰富的矿渣微粉和偏高岭土资源的基础上。巴西[309-311]从 20 世纪 90 年代开始进行碱激发材料的研究，哥伦比亚[21,312-314]和墨西哥[315-317]从 21 世纪以来开始进行这方面的研究。拉丁美洲对建筑材料的需求高，原材料供应充足，独特环境和生物多样性方面的环保意识强烈，这为碱激发材料技术的广泛使用提供了一个良好的前景。

2.10　印度碱激发水泥的发展

印度作为粉煤灰生产大国，其 2011 年粉煤灰的产量已达 1.5 亿 t[318]。同时印度地区的煤质和燃烧条件导致其粉煤灰的反应活性相对较低[318,319]。这就促使了 AAMs 中粉煤灰机械化学活化方面的研究获得了重大进展[318,320-323]，进而进行碱激发粉煤灰混凝土混合料设计的经验性学术研究，粉煤灰碱激发材料里面砖的商业化已经在印度开始[323]。由于目前印度粉煤灰的利用水平低于世界其他大部分地区，但对建筑产品的需求不断增长，这一市场似乎为碱激发材料的发展提供了机会。

2.11　小结

在技术、经济和环境因素的驱动下，碱激发胶凝材料和碱激发混凝土的研发逐渐活跃。同时，碱激发胶凝材料在世界不同地区的市场条件和因素下的不同领域得到发展。碱激发技术的科学基础已基本建立，几十年的碱激发材料建筑与古法水泥建筑类似，具有很好的耐久性。历史已经证明，标准的建立和发展需要市场的推动。从历史的角度看，这些关键领域是本书其余章节的重点。

致谢：作者对 Jan Wastiels 教授就粉煤灰基 AAMs 的早期开发的讨论和提供的图 2-8 表示感谢。

参考文献

[1]　United Nations：UN national accounts main aggregates database. http：//un-stats. un. org/unsd/snaama/dnllist. asp (2009).

[2]　US Geological Survey：Mineral commodity summaries：cement. http：//minerals. usgs. gov/minerals/pubs/commodity/cement/mcs-2009-cemen. pdf (2009).

[3]　Aïtcin P. -C. Cements of yesterday and today；concrete of tomorrow. Cem. Concr. Res,

2000, 30: 1349-1359.

[4] Scrivener K. L. , Kirkpatrick R. J. Innovation in use and research on cementitious material. Cem. Concr. Res, 2008, 38(2): 128-136.

[5] Taylor M. , Tam C. , Gielen D. Energy efficiency and CO_2 emissions from the global cement industry. International Energy Agency, 2006.

[6] Gartner E. Industrially interesting approaches to "low-CO_2" cements. Cem. Concr. Res, 2004, 34(9): 1489-1498.

[7] Damtoft J. S. , Lukasik J. , Herfort D. , Sorrentino D. , Gartner E. Sustainable developmentand climate change initiatives. Cem. Concr. Res, 2008, 38(2): 115-127.

[8] Juenger M. C. G. , Winnefeld F. , Provis J. L. , Ideker J. Advances in alternative cementitious binders. Cem. Concr. Res, 2011, 41(12): 1232-1243.

[9] Duxson P. , Provis J. L. , Lukey G. C. , Van Deventer J. S. J. The role of inorganic polymer technology in the development of 'Green concrete'. Cem. Concr. Res, 2007, 37 (12): 1590-1597.

[10] Von Weizsäcker E. , Hargroves K. , Smith M. H. , Desha C. , Stasinopoulos P. Factor Five: Transforming the Global Economy Through 80% Improvements in Resource Productivity. Earthscan, London, 2009.

[11] Tempest B. , Sansui O. , Gergely J. , Ogunro V. , Weggel D. Compressive strength and embodied energy optimization of fly ash based geopolymer concrete. In: World of Coal Ash 2009, Lexington, KY. CD-ROM Proceedings, 2009.

[12] Buchwald A. , Dombrowski K. , Weil M. Evaluation of primary and secondary materials under technical, ecological and economic aspects for the use as raw materials in geopolymeric binders. In: Bilek, V. , Kersner, Z. (eds.) 2nd International Symposium on NonTraditional Cement and Concrete, Brno, Czech Republic, 2005: 32-40.

[13] Weil M. , Dombrowski K. , Buchwald A. . Life-cycle analysis of geopolymers. In: Provis J. L. , Van Deventer J. S. J. (eds.) Geopolymers: Structure, Processing, Properties and IndustrialApplications, Woodhead, Cambridge, 2009: 194-212.

[14] Weil M. , Jeske U. , Dombrowski K. , Buchwald A. . Sustainable design of geopolymers -evaluation of raw materials by the integration of economic and environmental aspects in the early phases of material development. In: Takata S. , Umeda Y. (eds.) Advances in Life Cycle Engineering for Sustainable Manufacturing Businesses, Tokyo, Japan, Springer, London, 2007: 279-283.

[15] McLellan B. C. , Williams R. P. , Lay J. , Van Riessen A. , Corder G. D. . Costs and carbon emissions for geopolymer pastes in comparison to ordinary Portland cement. J. Cleaner Prod, 2011, 19(9-10): 1080-1090.

[16] Stengel T. , Reger J. , Heinz D. . Life cycle assessment of geopolymer concrete-what is the environmental benefit? In: Concrete Solutions 09, Sydney, 2009.

[17] Habert G. , d'Espinose de Lacaillerie J. B. , Roussel N. . An environmental evaluation of geopolymer based concrete production: reviewing current research

trends. J. Cleaner Prod, 2011, 19(11): 1229-1238.

[18] Manz O. E.. Worldwide production of coal ash and utilization in concrete and other products, Fuel, 1997, 76(8): 691-696.

[19] Vom Berg W., Feuerborn H.-J.. CCPs in Europe. In: Proceedings of Clean Coal Day in Japan 2001, Tokyo, Japan. ECOBA(European Coal Combustion Products Association). http://www.energiaskor.se/rapporter/ECOBA_paper.pdf, 2001.

[20] Rai A., Rao D. B. N.. Utilisation potentials of industrial/mining rejects and tailings as building materials. Manag. Environ. Qual. Int. J, 2005, 16(6): 605-614.

[21] Neville A. M.. Properties of Concrete, 4th edn. Wiley, Harlow, 1996.

[22] Hewlett P. C.. Lea's Chemistry of Cement and Concrete, 4th edn. Elsevier, Oxford, 1998.

[23] Zosin A. P., Priimak T. I., Avsaragov K. B.. Geopolymer materials based on magnesia-ironslags for normalization and storage of radioactive wastes. Atom. Energy, 1998, 85(1): 510-514.

[24] Komnitsas K., Zaharaki D., Perdikatsis V.. Geopolymerisation of low calcium ferronickel slags. J. Mater. Sci, 2007, 42(9): 3073-3082.

[25] Pacheco-Torgal F., Castro-Gomes J., Jalali S.. Investigations about the effect of aggregates on strength and microstructure of geopolymeric mine waste mud binders. Cem. Concr. Res, 2007, 37(6): 933-941.

[26] Gartner E. M., Macphee D. E.. A physico-chemical basis for novel cementitious binders. Cem. Concr. Res, 2011, 41(7): 736-749.

[27] Shi C., Fernández-Jiménez A., Palomo A.. New cements for the 21st century: the pursuit of an alternative to Portland cement. Cem. Concr. Res, 2011, 41(7): 750-763.

[28] Hooton R. D.. Bridging the gap between research and standards. Cem. Concr. Res, 2008, 38(2): 247-258.

[29] Alexander M. J.. Durability indexes and their use in concrete engineering. In: Kovler, K., et al. (eds.) International RILEM Symposium on Concrete Science and Engineering: A Tribute to Arnon Bentur, Evanston, IL. 2004; 9-22. RILEM Publications, Bagneux, France.

[30] Sonafrank C.. Investigating 21st century cement production in interior Alaska using Alaskan resources. Cold Climate Housing Research Center, Report 012409, 2010.

[31] Glukhovsky V. D.. Ancient, modern and future concretes. In: Krivenko P. V. (ed.) Proceedings of the First International Conference on Alkaline Cements and Concretes, Kiev, Ukraine, 1994, 1: 1-9. VIPOL Stock Company.

[32] Krivenko P. V.. Alkaline cements. In: Krivenko, P. V. (ed.) Proceedings of the First International Conference on Alkaline Cements and Concretes, Kiev, Ukraine, 1994, 1: 11-129. VIPOL Stock Company.

[33] Krivenko P. V.. Alkaline cements: structure, properties, aspects of durability. In:

Krivenko P. V. (ed.) Proceedings of the Second International Conference on Alkaline Cements and Concretes, Kiev, Ukraine, 1999: 3-43. ORANTA.

[34] Provis J. L., Van Deventer J. S. J.. Geopolymerisation kinetics. 2. Reaction kinetic modelling. Chem. Eng. Sci, 2007, 62(9): 2318-2329.

[35] Wang S. D., Scrivener K. L.. Hydration products of alkali-activated slag cement. Cem. Concr. Res, 1995, 25(3): 561-571.

[36] Brough A. R., Atkinson A.. Sodium silicate-based, alkali-activated slag mortars: Part I. Strength, hydration and microstructure. Cem. Concr. Res, 2002, 32 (6): 865-879.

[37] Richardson I. G., Brough A. R., Groves G. W., Dobson C. M.. The characterization of hardened alkali-activated blast-furnace slag pastes and the nature of the calcium silicate hydrate(C-S-H) paste. Cem. Concr. Res, 1994, 24(5): 813-829.

[38] Fernández-Jiménez A., Vázquez T., Palomo A.. Effect of sodium silicate on calcium aluminate cement hydration in highly alkaline media: a microstructural characterization. J. Am. Ceram. Soc, 2011, 94(4): 1297-1303.

[39] Provis J. L., Lukey G. C., Van Deventer J. S. J.. Do geopolymers actually contain nanocrystalline zeolites? -a reexamination of existing results. Chem. Mater, 2005, 17 (12): 3075-3085.

[40] Fernández-Jiménez A., Monzó M., Vicent M., Barba A., Palomo A.. Alkaline activation of metakaolin-fly ash mixtures: obtain of zeoceramics and zeocements. Microporous Mesoporous Mater, 2008, 108(1-3): 41-49.

[41] Glukhovsky V. D.. Gruntosilikaty (Soil Silicates). Gosstroyizdat, Kiev, 1959.

[43] Bell J. L., Sarin P., Provis J. L., Haggerty R. P., Driemeyer P. E., Chupas P. J., Van Deventer J. S. J., Kriven W. M.. Atomic structure of a cesium aluminosilicate geopolymer: a pair distribution function study. Chem. Mater, 2008, 20 (14): 4768-4776.

[43] Richardson I. G.. Tobermorite/jennite-and tobermorite/calcium hydroxide-based models for the structure of C-S-H: applicability to hardened pastes of tricalcium silicate, β-dicalcium silicate, Portland cement, and blends of Portland cement with blast-furnace slag, metakaolin, or silica fume. Cem. Concr. Res, 2004, 34(9): 1733-1777.

[44] Chen W., Brouwers H.. The hydration of slag, Part 1: reaction models for alkali-activated slag. J. Mater. Sci, 2007, 42(2): 428-443.

[45] Puertas F., Palacios M., Manzano H., Dolado J. S., Rico A., Rodríguez J.. A model for the C-AS-H gel formed in alkali-activated slag cements. J. Eur. Ceram. Soc, 2011, 31(12): 2043-2056.

[46] Puertas F., Martínez-Ramírez S., Alonso S., Vázquez E.. Alkali-activated fly ash/slag cement. Strength behaviour and hydration products. Cem. Concr. Res, 2000, 30: 1625-1632.

[47] Puertas F.. Cementos de escoria activados alcalinamente: situación actualy perspec-

tivas defuturo. Mater. Constr, 1995, 45(239): 53-64.

[48] Myers R. J., Bernal S. A., San Nicolas R., Provis J. L.. Generalized structural description of calcium-sodium aluminosilicate hydrate gels: the crosslinked substituted tobermorite model. Langmuir, 2013, 29(17): 5294-5306.

[49] Wang S. D.. The role of sodium during the hydration of alkali-activated slag. Adv. Cem. Res, 2000, 12(2): 65-69.

[50] Lothenbach B., Gruskovnjak A.. Hydration of alkali-activated slag: thermodynamic modelling. Adv. Cem. Res, 2007, 19(2): 81-92.

[51] Bernal S. A., Provis J. L., Mejía de Gutierrez R., Rose V.. Evolution of binder structure in sodium silicate-activated slag-metakaolin blends. Cem. Concr. Compos, 2011, 33(1): 46-54.

[52] Krivenko P. V.. Alkaline cements: from research to application. In: Lukey, G. C. (ed.)Geopolymers 2002. Turn Potential into Profit, Melbourne, Australia. CD-ROM Proceedings. Siloxo Pty. Ltd. 2002.

[53] Goldich S. S.. A study in rock-weathering. J. Geol, 1938, 46(1): 17-58.

[54] Langmuir D.. Aqueous Environmental Geochemistry. Prentice Hall, Upper Saddle River, 2007.

[55] Davidovits J., Davidovits F.. The Pyramids: An Enigma Solved. 2nd Revised Ed. Éditions J. Davidovits, Saint-Quentin, France, 2001.

[56] Davidovits J.. Geopolymeric reactions in archaeological cements and in modern blended cements. In: Davidovits J., Orlinski J. (eds.) Proceedings of Geopolymer '88-First European Conference on Soft Mineralurgy, Compeigne, France, 1998, 1: 93-106. Universitede Technologie de Compeigne.

[57] Barsoum M. W., Ganguly A., Hug G.. Microstructural evidence of reconstituted limestone blocks in the Great Pyramids of Egypt. J. Am. Ceram. Soc, 2006, 89(12): 3788-3796.

[58] MacKenzie K. J. D., Smith M. E., Wong A., Hanna J. V., Barry B., Barsoum M. W.. Were the casing stones of Senefru's Bent Pyramid in Dahshour cast or carved?: multinuclear NMR evidence. Mater. Lett, 2011, 65(2): 350-352.

[59] Vitruvius: The Ten Books of Architecture. Dover, Trans M. H. Morgan. New York, 1960.

[60] Gotti E., Oleson J. P., Bottalico L., Brandon C., Cucitore R., Hohlfelder R. L.. A comparison of the chemical and engineering characteristics of ancient Roman hydraulic concrete with a modern reproduction of Vitruvian hydraulic concrete. Archaeometry, 2008, 50: 576-590.

[61] Brandon C., Hohlfelder R. L., Oleson J. P., Stern C.. The Roman Maritime Concrete Study(ROMACONS): the harbour of Chersonisos in Crete and its Italian connection. Rev. Geogr. Pays Méditerr, 2005, 104: 25-29.

[62] Sánchez-Moral S., Luque L., Cañaveras J.-C., Soler V., Garcia-Guinea J., Apari-

cio A.. Lime-pozzolana mortars in Roman catacombs: composition, structures and restoration. Cem. Concr. Res, 2005, 35(8): 1555-1565.

[63] Abe H., Aoki M., Konno H.. Synthesis of analcime from volcanic sediments in sodium silicate solution. Contrib. Mineral. Petrol, 1973, 42(2): 81-92.

[64] Roy D. M., Langton C. A.. Studies of ancient concrete as analogs of cementitious sealing materials for a repository in tuff, Report LA-11527-MS. Los Alamos National Laboratory, 1989.

[65] Nguyen B. Q., Leming M. L.. Limits on alkali content in cement-results from a field study. Cem. Concr. Aggr, 2000, 22(1). CCA10462J.

[66] Smaoui N., Bérubé M. A., Fournier B., Bissonnette B., Durand, B.. Effects of alkali addition on the mechanical properties and durability of concrete. Cem. Concr. Res, 2005, 35(2): 203-212.

[67] Chen W., Brouwers H. J. H.. Alkali binding in hydrated Portland cement paste. Cem. Concr. Res, 2010, 40(5): 716-722.

[68] Jiang S., Kim B.-G., Aïtcin P.-C.. Importance of adequate soluble alkali content to ensure cement/superplasticizer compatibility. Cem. Concr. Res, 1999, 29(1): 71-78.

[69] Way S. J., Shayan A.. Early hydration of a Portland cement in water and sodium hydroxide solutions: composition of solutions and nature of solid phases. Cem. Concr. Res, 1989, 19(5): 759-769.

[70] Martínez-Ramírez S., Palomo A.. OPC hydration with highly alkaline solutions. Adv. Cem. Res, 2001, 13(3): 123-129.

[71] Kirchheim A. P., Dal Molin D. C., Fischer P., Emwas A.-H., Provis J. L., Monteiro P. J. M.. Real-time high-resolution X-ray imaging and nuclear magnetic resonance study of the hydration of pure and Na-doped C3A in the presence of sulfates. Inorg. Chem, 2011, 50(4): 1203-1212.

[72] Thomas M.. The effect of supplementary cementing materials on alkali-silica reaction: a review. Cem. Concr. Res, 2011, 41(12): 1224-1231.

[73] Ramlochan T., Thomas M., Gruber K. A.. The effect of metakaolin on alkali-silica reaction in concrete. Cem. Concr. Res, 2000, 30(3): 339-344.

[74] Krivenko P. V., Petropavlovsky O., Gelevera A., Kavalerova E.. Alkali-aggregate reaction in the alkali-activated cement concretes. In: Bilek V., Keršner Z. (eds.) Proceedings of the 4th International Conference on Non-Traditional Cement & Concrete, Brno, Czech Republic. ZPSV, a. s. 2011.

[75] Chappex T., Scrivener K. L.. The influence of aluminium on the dissolution of amorphous silica and its relation to alkali silica reaction. Cem. Concr. Res, 2012, 42(12): 1645-1649.

[76] Provis J. L.. Activating solution chemistry for geopolymers. In: Provis J. L., Van Deventer J. S. J. (eds.) Geopolymers: Structure, Processing, Properties and Industrial Applications, 2009: 50-71. Woodhead, Cambridge.

[77] Shi C., Krivenko P. V., Roy D. M.. Alkali-Activated Cements and Concretes. Taylor & Francis, Abingdon, 2006.

[78] Monnin C., Dubois M.. Thermodynamics of the LiOH + H_2O system. J. Chem. Eng. Data, 2005, 50(4): 1109-1113.

[79] Pickering S. U.. The hydrates of sodium, potassium and lithium hydroxides. J. Chem. Soc. Trans, 1893, 63: 890-909.

[80] Kurt C., Bittner J.. Sodium hydroxide. In: Ullmann's Encyclopedia of Industrial Chemistry. Wiley-VCH Verlag, Weinheim, 2006.

[81] Gurvich L. V., Bergman G. A., Gorokhov L. N., Iorish V. S., Leonidov V. Y., Yungman V. S.: Thermodynamic properties of alkali metal hydroxides. Part 1. Lithium and sodium hydroxides. J. Phys. Chem, 1996, 25(4): 1211-1276.

[82] Gurvich L. V., Bergman G. A., Gorokhov L. N., Iorish V. S., Leonidov V. Y., Yungman V. S.. Thermodynamic properties of alkali metal hydroxides. Part 2. Potassium, rubidium, and cesium hydroxides. J. Phys. Chem, 1997, 26(4): 1031-1110.

[83] Simonson J. M., Mesmer R. E., Rogers P. S. Z.. The enthalpy of dilution and apparent molar heat capacity of NaOH(aq) to 523 K and 40 MPa. J. Chem. Thermodyn, 1989, 21: 561-584.

[84] Brown P. W.. The system Na_2O-CaO-SiO_2-H_2O. J. Am. Ceram. Soc, 1990, 73(11): 3457-3561.

[85] Weldes H. H., Lange K. R.. Properties of soluble silicates. Ind. Eng. Chem, 1969, 61(4): 29-44.

[86] Wills J. H.. A review of the system Na_2O-SiO_2-H_2O. J. Phys. Colloid Chem, 1950, 54(3): 304-310.

[87] Vail J. G.. Soluble Silicates: Their Properties and Uses. Reinhold, New York, 1952.

[88] Knight C. T. G., Balec R. J., Kinrade S. D.. The structure of silicate anions in aqueous alkaline solutions. Angew. Chem. Int. Ed, 2007, 46: 8148-8152.

[89] Provis J. L., Duxson P., Lukey G. C., Separovic F., Kriven W. M., Van Deventer J. S. J.. Modeling speciation in highly concentrated alkaline silicate solutions. Ind. Eng. Chem, 2005, 44(23): 8899-8908.

[90] Engelhardt G., Jancke H., Hoebbel D., Wieker W.. Strukturuntersuchungen an Silikatanionen in wäßriger Lösung mit Hilfe der 29Si-NMR-Spektroskopie. Z. Chem, 1974, 14(3): 109-110.

[91] Harris R. K., Knight C. T. G.. Silicon-29 nuclear magnetic resonance studies of aqueous silicate solutions. Part 5. First-order patterns in potassium silicate solutions enriched with silicon-29. J. Chem. Soc. Faraday Trans. II, 1983, 79(10): 1525-1538.

[92] Harris R. K., Knight C. T. G.. Silicon-29 nuclear magnetic resonance studies of aqueous silicate solutions. Part 6. Second-order patterns in potassium silicate solutions

enriched with silicon-29. J. Chem. Soc. Faraday Trans. II 1983, 79(10): 1539-1561.

[93] Cho H., Felmy A. R., Craciun R., Keenum J. P., Shah N., Dixon, D. A.. Solution state structure determination of silicate oligomers by 29Si NMR spectroscopy and molecular modeling. J. Am. Chem. Soc, 2006, 128(7): 2324-2335.

[94] Pelster S. A., Schrader W., Schüth F.. Monitoring temporal evolution of silicate species during hydrolysis and condensation of silicates using mass spectrometry. J. Am. Chem. Soc, 2006, 128(13): 4310-4317.

[95] Petry D. P., Haouas M., Wong S. C. C., Aerts A., Kirschhock C. E. A., Martens J. A., GaskellS. J., Anderson M. W., Taulelle F.. Connectivity analysis of the clear sol precursor of silicalite: are nanoparticles aggregated oligomers or silica particles? J. Phys. Chem. C, 2009, 113(49): 20827-20836.

[96] Halasz I., Agarwal M., Li R., Miller N.. Monitoring the structure of water soluble silicates. Catal. Today, 2007, 126: 196-202.

[97] Halasz I., Agarwal M., Li R., Miller N.. Vibrational spectra and dissociation of aqueous Na_2SiO_3 solutions. Catal. Lett, 2007, 117(1-2): 34-42.

[98] Halasz I., Agarwal M., Li R. B., Miller N.. What can vibrational spectroscopy tell about the structure of dissolved sodium silicates? Microporous Mesoporous Mater, 2010, 135(1-3): 74-81.

[99] White C. E., Provis J. L., Kearley G. J., Riley D. P., Van Deventer J. S. J.. Density functional modelling of silicate and aluminosilicate dimerisation solution chemistry. Dalton Trans, 2011, 40(6): 1348-1355.

[100] Nordström J., Nilsson E., Jarvol P., Nayeri M., Palmqvist A., Bergenholtz J., Matic, A.. Concentration and pH-dependence of highly alkaline sodium silicate solutions. J. Colloid Interface Sci, 2011, 356(1): 37-45.

[101] Phair J. W., Van Deventer J. S. J.. Effect of the silicate activator pH on the microstructural characteristics of waste-based geopolymers. Int. J. Miner. Proc, 2002, 66(1-4): 121-143.

[102] Yang X., Zhu W., Yang Q.. The viscosity properties of sodium silicate solutions. J. Solut. Chem, 2008, 37(1): 73-83.

[103] Kostick D. S.. Mineral Commodity Summaries-Soda Ash. U. S. Geological Survey, 2011.

[104] Hill A. E., Bacon L. R.. Ternary systems. VI. Sodium carbonate, sodium bicarbonate, and water. J. Am. Chem. Soc, 1927, 49(10): 2487-2495.

[105] Ozdemir O., Celik M. S., Nickolov Z. S., Miller J. D.. Water structure and its influence on the flotation of carbonate and bicarbonate salts. J. Colloid Interface Sci, 2007 314(2): 545-551.

[106] Byfors K., Klingstedt G., Lehtonen H. P., Romben L.. Durability of concrete made with alkali-activated slag. In: Malhotra V. M. (ed.) 3rd International Conference on Fly Ash, SilicaFume, Slag and Natural Pozzolans in Concrete, ACI SP114,

Trondheim, Norway, 1989: 1429-1444. American Concrete Institute.

[107] Kostick D. S.. Mineral Commodity Summaries-Sodium Sulfate. U. S. Geological Survey, 2011.

[108] Abdulagatov I. M., Zeinalova A., Azizov N. D.. Viscosity of aqueous Na_2SO_4 solutions at temperatures from 298 to 573 K and at pressures up to 40 MPa. Fluid Phase Equilib, 2005, 227(1): 57-70.

[109] Steiger M., Asmussen S.. Crystallization of sodium sulfate phases in porous materials: the phase diagram Na_2SO_4-H_2O and the generation of stress. Geochim. Cosmochim. Acta, 2008, 72(17): 4291-4306.

[110] Rostovskaya G., Ilyin V.. Blazhis A.. The service properties of the slag alkaline concretes. In: Ertl Z. (ed.) Alkali Activated Materials-Research, Production and Utilization, Prague, Czech Republic, 2007: 593-610. Česká Rozvojová Agentura.

[111] Xu H., Provis J. L., Van Deventer J. S. J., Krivenko P. V.. Characterization of aged slag concretes. ACI Mater. J, 2008, 105(2): 131-139.

[112] Talling B., Krivenko P. V.. Blast furnace slag-the ultimate binder. In: Chandra S. (ed.) Waste Materials Used in Concrete Manufacturing, 1997: 235-289. Noyes, Park Ridge.

[113] Krivenko P. V.. Alkali-activated aluminosilicates: past, present and future. Chem. List, 2008 102: s273-s277.

[114] Wang S.-D., Pu X.-C., Scrivener K. L., Pratt P. L.. Alkali-activated slag cement and concrete: a review of properties and problems. Adv. Cem. Res, 1995, 7(27): 93-102.

[115] Krivenko P. V. (ed.). Proceedings of the First International Conference on Alkaline Cements and Concretes. VIPOL Stock Company, Kiev, 1994.

[116] Krivenko P. V. (ed.). Proceedings of the Second International Conference on Alkaline Cements and Concretes. Oranta, Kiev, 1999.

[117] Ertl Z. (ed.). Proceedings of the International Conference on Alkali Activated Materials-Research, Production and Utilization. Česká rozvojová agentura, Prague, 2007.

[118] Talling B., Brandstetr J.. Present state and future of alkali-activated slag concretes. In: Malhotra V. M. (ed.) 3rd International Conference on Fly Ash, Silica Fume, Slag and Natural Pozzolans in Concrete, ACI SP114, Trondheim, Norway, 1989, 2: 1519-1546. American Concrete Institute.

[119] Slota R. J.. Utilization of water glass as an activator in the manufacturing of cementitious materials from waste by-products. Cem. Concr. Res, 1987, 17(5): 703-708.

[120] Deja J., Małolepszy J.. Long-term resistance of alkali-activated slag mortars to chloride solution. In: 3rd International Conference on Durability of Concrete, Nice, France, 1994: 657-671.

[121] Małolepszy J., Deja J.. The influence of curing conditions on the mechanical prop-

erties of alkali-activated slag binders. Silic. Ind, 1988, 53(11-12): 179-186.

[122] Mozgawa W., Deja J.. Spectroscopic studies of alkaline activated slag geopolymers. J. Mol. Struct, 2009, (924-926): 434-441.

[123] Brylicki W., Małolepszy J., Stryczek S.. Alkali activated slag cementitious material for drilling operation. In: 9th International Congress on the Chemistry of Cement, New Delhi, India, 1992, 3: 312-316.

[124] Deja J.. Carbonation aspects of alkali activated slag mortars and concretes. Silic. Ind, 2002, 67(1): 37-42.

[125] Małolepszy J., Deja J., Brylicki W.. Industrial application of slag alkaline concretes. In: Krivenko P. V. (ed.) Proceedings of the First International Conference on Alkaline Cements and Concretes, Kiev, Ukraine, 1994, 2: 989-1001. VIPOL Stock Company.

[126] Dziewański J., Brylicki W., Pawlikowski M.. Utilization of slag-alkaline cement as a grouting medium in hydrotechnical construction. Bull. Eng. Geol. Environ, 1980, 22(1): 65-70.

[127] Deja J.. Immobilization of Cr^{6+}, Cd^{2+}, Zn^{2+} and Pb^{2+} in alkali-activated slag binders. Cem. Concr. Res, 2002, 32(12): 1971-1979.

[128] Kukko H., Mannonen R.. Chemical and mechanical properties of alkali-activated blast furnace slag (F-concrete). Nord. Concr. Res, 1982, 1: 16.1-16.16.

[129] Metso J.. The alkali reaction of alkali-activated Finnish blast furnace slag. Silic. Ind, 1982, 47(3-4): 123-127.

[130] Forss B.. Experiences from the use of F-cement-a binder based on alkali-activated blastfurnace slag. In: Idorn G. M., Rostam S. (eds.) Alkalis in Concrete, Copenhagen, Denmark, 1983: 101-104. Danish Concrete Association.

[131] Häkkinen T.. The influence of slag content on the microstructure, permeability and mechanical properties of concrete: Part 1. Microstructural studies and basic mechanical properties. Cem. Concr. Res, 1993, 23(2): 407-421.

[132] Häkkinen T.. The influence of slag content on the microstructure, permeability and mechanical properties of concrete: Part 2. Technical properties and theoretical examinations. Cem. Concr. Res, 1993, 23(3): 518-530.

[133] Häkkinen T.. Durability of alkali-activated slag concrete. Nord. Concr. Res, 1987, 6(1): 81-94.

[134] Kutti T.. Hydration products of alkali activated slag. In: 9th International Congress on the Chemistry of Cement, New Delhi, India, 1922, 2: 468-478.

[135] Douglas E., Bilodeau A., Brandstetr J., Malhotra V. M.. Alkali activated ground granulated blast-furnace slag concrete: preliminary investigation. Cem. Concr. Res, 1991, 21(1): 101-108.

[136] Douglas E., Brandstetr J.. A preliminary study on the alkali activation of ground granulated blast-furnace slag. Cem. Concr. Res, 1990, 20(5): 746-756.

[137] Škvára F., Bohuněk J.. Chemical activation of substances with latent hydraulic properties. Ceram. -Silik, 1999, 43(3): 111-116.

[138] Škvára F., Bohuněk J., Marková A.. Alkali-activated fly-ash. In: Proceedings of 14th IBAUSIL, Weimar, Germany, 2000, 1: 523-533.

[139] Mináříková M., Škvára F.. Fixation of heavy metals in geopolymeric materials based on brown coal fly ash. Ceram. -Silik, 2006, 50(4): 200-207.

[140] Allahverdi A., Škvára F.. Nitric acid attack on hardened paste of geopolymeric cements -Part 1. Ceram. -Silik, 2001, 45(3): 81-88.

[141] Allahverdi A., Škvára F.. Nitric acid attack on hardened paste of geopolymeric cements -Part 2. Ceram. -Silik, 2001, 45(4): 143-149.

[142] Bilek V., Szklorzova H.. Freezing and thawing resistance of alkali-activated concretes for the production of building elements. In: Malhotra V. M. (ed.) Proceedings of 10th CANMET/ACI Conference on Recent Advances in Concrete Technology, Supplementary Papers, Seville, Spain, 2009: 661-670.

[143] Bilek V.. Alkali-activated slag concrete for the production of building elements. In: Ertl Z. (ed.) Proceedings of the International Conference on Alkali Activated Materials-Research, Production and Utilization, Prague, Czech Republic, 2007: 71-82. Česká rozvojová agentura.

[144] Bilek V., Urbanova M., Brus J., Kolousek D.. Alkali-activated slag development and their practical use. In: Beaudoin J. J. (ed.) 12th International Congress on the Chemistry of Cement, Montreal, Canada. CD-ROM Proceedings, 2007.

[145] Rovnaník P., Bayer P., Rovnaníková P.. Properties of alkali-activated aluminosilicate composite after thermal treatment. In: Bílek V., Keršner Z. (eds.) Proceedings of the 2nd International Conference on Non-Traditional Cement and Concrete, Brno, Czech Republic, 2005: 48-54. Brno University of Technology & ZPSV Uhersky Ostroh, a. s.

[146] Rovnaník P., Bayer P., Rovnaníková P.. Role of fiber reinforcement in alkali-activated aluminosilicate composites subjected to elevated temperature. In: Bílek V., Keršner Z. (eds.)Proceedings of the 2nd International Conference on Non-Traditional Cement and Concrete, Brno, Czech Republic, 2005: 55-60. Brno University of Technology & ZPSV Uhersky Ostroh, a. s.

[147] Majersky D.. Removal and solidification of the high contaminated sludges into the aluminosilicate matrix SIAL during decommissioning activities. In: CEG Workshop on Methods and Techniques for Radioactive Waste Management Applicable for Remediation of Isolated Nuclear Sites, Petten. IAEA, 2004.

[148] Teoreanu I., Volceanov A., Stoleriu S.. Non Portland cements and derived materials. Cem. Concr. Compos, 2005, 27: 650-660.

[149] Bílek V., Keršner Z. (eds.). Proceedings of the 1st International Conference on NonTraditional Cement and Concrete, Brno, Czech Republic, 2002.

[150] Bílek V., Keršner Z. (eds.). Proceedings of the 2nd International Conference on NonTraditional Cement and Concrete. Brno University of Technology & ZPSV Uhersky Ostroh, a. s., Brno, 2005.

[151] Bílek V., Keršner Z. (eds.). Proceedings of the 3rd International Conference on NonTraditional Cement and Concrete. ZPSV a. s., Brno, 2008.

[152] Bílek V., Keršner Z. (eds.). Proceedings of the 4th International Conference on NonTraditional Cement and Concrete, Brno, Czech Republic, 2011.

[153] Dong J.. A review of research and application of alkaline slag cement and concrete in China. In: Krivenko P. V. (ed.) Proceedings of the Second International Conference on Alkaline Cements and Concretes, Kiev, Ukraine, 1999: 705-711. ORANTA.

[154] Bílek V., Keršner Z. (eds.). Proceedings of the 2nd International Conference on Non- Traditional Cement and Concrete. Brno University of Technology & ZPSV Uhersky Ostroh, a. s., Brno, 2005.

[155] Bílek V., Keršner Z. (eds.). Proceedings of the 3rd International Conference on Non- Traditional Cement and Concrete. ZPSV a. s., Brno, 2008.

[156] Bílek V., Keršner Z. (eds.). Proceedings of the 4th International Conference on Non- Traditional Cement and Concrete, Brno, Czech Republic, 2011.

[157] Dong J.. A review of research and application of alkaline slag cement and concrete in China. In: Krivenko P. V. (ed.) Proceedings of the Second International Conference on Alkaline Cements and Concretes, Kiev, Ukraine, 1999: 705-711. ORANTA,

[158] Wang S. D.. Review of recent research on alkali-activated concrete in China. Mag. Concr. Res, 1991, 43(154): 29-35.

[159] Pan Z., Yang N.. Updated review on AAM research in China. In: Shi C., Shen X. (eds.) First International Conference on Advances in Chemically-Activated Materials, Jinan, China, 2010: 45-55. RILEM.

[160] Zhang Z., Yao X., Zhu H.. Potential application of geopolymers as protection coatings for marine concrete: I. Basic properties. Appl. Clay Sci, 2010, 49(1-2): 1-6.

[161] Zhang Z., Yao X., Zhu H.. Potential application of geopolymers as protection coatings for marine concrete: II. Microstructure and anticorrosion mechanism. Appl. Clay Sci, 2010, 49(1-2): 7-12.

[162] APP China Cement Task Force: Status report of China cement industry. In: 8th CTF Meeting, Vancouver, 2010.

[163] Rogers A.. Chapter 8: Waste. In: Taking Action: An Environmental Guide for You and Your Community. United Nations, New York, 1996.

[164] Barnes I., Moedinger F.. Novel products-from concept to market. In: Cox M., Nugteren H., Janssen-Jurkovičová M. (eds.) Combustion Residues: Current, Novel

and Renewable Applications, 2008: 379-418. Wiley, Chichester.

[165] Sprung S.. Cement. Ullmann's Encyclopedia of Industrial Chemistry. Wiley-VCH Verlag GmbH & Co. KGaA, Weinheim, 2000.

[166] ACI Committee 233: Ground Granulated Blast-Furnace Slag as a Cementitious Constituent in Concrete. American Concrete Institute, 2000.

[167] Davis R. E., Carlson R. W., Kelly J. W., Davis H. E.. Properties of cements and concretes containing fly ash. J. Am. Concr. Inst, 1937, 33: 577-612.

[168] Kühl H.. Slag cement and process of making the same. Compiler: U. S., 900, 939 [P]. 1908.

[169] Chassevent L.. Hydraulicity of slags. Compt. Rend, 1937, 205: 670-672.

[170] Purdon A. O.. The action of alkalis on blast-furnace slag. J. Soc. Chem. Ind. Trans. Commun, 1940, 59: 191-202.

[171] Davidovits J.. Mineral polymers and methods of making them. Gmpiler: U. S., 4349386[P]. 1982.

[172] Davidovits J.. The need to create a new technical language for the transfer of basic scientific information. In: Gibb J. M., Nicolay D. (eds.) Transfer and Exploitation of Scientific and Technical Information, EUR 7716, 1982: 316-320. Commission of the European Communities, Luxembourg.

[173] Davidovits J.. Synthetic mineral polymer compound of the silicoaluminates family and prep- aration process. Compiler: U. S, 4472199[P]. 1984.

[174] Davidovits J., Sawyer J. L.. Early high-strength mineral polymer. Compiler: U. S, 4509985[P]. 1985.

[175] Davidovits J.. Geopolymers-inorganic polymeric new materials. J. Therm. Anal, 1991, 37(8): 1633-1656.

[176] Malone P. G., Randall C. J., Kirkpatrick T.. Potential applications of alkali-activated alumi- nosilicate binders in military operations. Geotechnical Laboratory, Department of the Army, 1985, GL-85-15.

[177] Davidovits J.. Geopolymer Chemistry and Applications. Institut Géopolymère, Saint- Quentin, 2008.

[178] Davidovits J. (ed.). Proceedings of the World Congress Geopolymer 2005-Geopolymer, Green Chemistry and Sustainable Development Solutions. Institut Géopolymère, 2005.

[179] Davidovits J., Davidovits R., James C. (eds.). Proceedings of Second International Conference Geopolymer '99. Institut Géopolymère, 1999.

[180] Davidovits J., Orlinski J. (eds.). Proceedings of Geopolymer '88-First European Conference on Soft Mineralurgy. Universite de Technologie de Compeigne, 1988.

[181] Lukey G. C. (ed.): Geopolymers 2002. Turn Potential into Profit, Melbourne, Australia. CD-ROM Proceedings. Siloxo Pty. Ltd., 2002.

[182] Bennett D. F. H.. Innovations in Concrete. Thomas Telford, London, 2002.

[183] Wheat H. G.. Corrosion behavior of steel in concrete made with Pyrament® blended cement. Cem. Concr. Res, 1992, 22: 103-111.

[184] Husbands T. B., Malone P. G., Wakeley L. D.. Performance of concretes proportioned with Pyrament blended cement, U. S. Army Corps of Engineers Construction Productivity Advancement Research Program, Report, 1994, CPAR-SL-94-2.

[185] MaGrath A. J.. Ten timeless truths about pricing. J. Bus. Ind. Mark. 1891, 6(3-4): 15-23.

[186] Geopolymer Institute: PYRAMENT cement good for heavy traffic after 25 years. http://www.geopolymer.org/news/pyrament-cement-good-for-heavy-traffic-after-25-years, 2011.

[187] Smith M. A., Osborne G. J.. Slag/fly ash cements. World Cem. Technol, 1977, 1(6): 223-233.

[188] Bijen J., Waltje H.. Alkali activated slag-fly ash cements. In: Malhotra V. M. (ed.) 3rd International Conference on Fly Ash, Silica Fume, Slag and Natural Pozzolans in Concrete, ACI SP114, Trondheim, Norway, 1989, 2: 1565-1578. American Concrete Institute.

[189] Langton C. A., Roy D. M.. Longevity of borehole and shaft sealing materials: characteriza- tion of ancient cement-based building materials. In: McVay G. (ed.) Materials Research Society Symposium Proceedings, vol. 26, Scientific Basis for Nuclear Waste Management, 1986: 543-549. North Holland, New York.

[190] Roy D.. Alkali-activated cements-opportunities and challenges. Cem. Concr. Res, 1999, 29(2): 249-254.

[191] Roy D. M.. New strong cement materials: chemically bonded ceramics. Science 235(4789), 1987: 651-658.

[192] Roy A., Schilling P. J., Eaton H. C.. Alkali activated class C fly ash cement. Compiler: U. S, 5565028[P]. 1996.

[193] Roy A., Schilling P. J., Eaton H. C., Malone P. G., Brabston W. N., Wakeley L. D.. Activation of ground blast-furnace slag by alkali-metal and alkaline-earth hydroxides. J. Am. Ceram. Soc, 1992, 75(12): 3233-3240.

[194] Palomo A., Glasser F. P.. Chemically-bonded cementitious materials based on metakaolin. Br. Ceram. Trans. J. 1992, 91(4): 107-112.

[195] Douglas E., Bilodeau A., Malhotra V. M.. Properties and durability of alkali-activated slag concrete. ACI Mater. J. 1992, 89(5): 509-516.

[196] Cheng Q.-H., Tagnit-Hamou A., Sarkar S. L.. Strength and microstructural properties of water glass activated slag. Mater. Res. Soc. Symp. Proc. 1991, 245: 49-54.

[197] Gifford P. M., Gillott J. E.. Alkali-silica reaction (ASR) and alkali-carbonate reaction (ACR) in activated blast furnace slag cement (ABFSC) concrete. Cem. Concr. Res, 1996, 26(1): 21-26.

[198] Gifford P. M., Gillott J. E.. Freeze-thaw durability of activated blast furnace slag cement concrete. ACI Mater. J, 1996, 93(3): 242-245.

[199] Gifford P. M., Gillott J. E.. Behaviour of mortar and concrete made with activated blast furnace slag cement. Can. J. Civil Eng, 1997, 24(2): 237-249.

[200] Shi C., Day R. L.. Acceleration of the reactivity of fly ash by chemical activation. Cem. Concr. Res, 1995, 25(1): 15-21.

[201] Shi C., Day R. L.. Selectivity of alkaline activators for the activation of slags. Cem. Concr. Aggr, 1996, 18(1): 8-14.

[202] Shi C.. Strength pore structure and permeability of alkali-activated slag mortars. Cem. Concr. Res, 1996, 26(12): 1789-1799.

[203] Shi C.. Corrosion resistance of alkali-activated slag cement. Adv. Cem. Res, 2003, 15(2): 77-81.

[204] Shi C., Stegemann J. A.. Acid corrosion resistance of different cementing materials. Cem. Concr. Res, 2000, 30(5): 803-808.

[205] Day R. L., Moore L. M., Nazir M. N.. Applications of chemically activated blended cements with very high proportions of fly ash. In: Beaudoin J. J. (ed.) 12th International Congress on the Chemistry of Cement Montreal Canada. CD-ROM Proceedings, 2007.

[206] Fernández-Jiménez A., Puertas F.. Alkali-activated slag cements: kinetic studies. Cem. Concr. Res, 1997, 27(3): 359-368.

[207] Fernández-Jiménez A., Puertas F.. Influence of the activator concentration on the kinetics of the alkaline activation process of a blast furnace slag. Mater. Constr, 1997, 47(246): 31-42.

[208] Palomo A., Grutzeck M. W., Blanco M. T.. Alkali-activated fly ashes-a cement for the future. Cem. Concr. Res, 1999, 29(8): 1323-1329.

[209] Fernández-Jiménez A., Palomo J. G., Puertas F.. Alkali-activated slag mortars. Mechanical strength behaviour. Cem. Concr. Res, 1999. 29: 1313-1321.

[210] Fernández-Jiménez A., Palomo A., Sobrados I., Sanz J.. The role played by the reactive alumina content in the alkaline activation of fly ashes. Microporous Mesoporous Mater, 2006, 91(1-3): 111-119.

[211] Fernández-Jiménez A., Palomo A., Criado M.. Microstructure development of alkali-activated fly ash cement: a descriptive model. Cem. Concr. Res, 2005, 35(6): 1204-1209.

[212] Fernández-Jiménez A., Puertas F., Sobrados I., Sanz J.. Structure of calcium silicate hydrates formed in alkaline-activated slag: influence of the type of alkaline activator. J. Am. Ceram. Soc, 2003, 86(8): 1389-1394.

[213] Puertas F., Fernández-Jiménez A., Blanco-Varela M. T.. Pore solution in alkali-activated slag cement pastes. Relation to the composition and structure of calcium silicate hydrate. Cem. Concr. Res, 2004, 34(1): 139-148.

[214] Fernández-Jiménez A., Vallepu R., Terai T., Palomo A., Ikeda K.. Synthesis and thermal behavior of different aluminosilicate gels. J. Non-Cryst. Solids, 2006, 352: 2061-2066.

[215] Palomo A., Alonso S., Fernández-Jiménez A., Sobrados I., Sanz J.. Alkaline activation of fly ashes: NMR study of the reaction products. J. Am. Ceram. Soc, 2004, 87(6): 1141-1145.

[216] Palacios M., Puertas F.. Effect of superplasticizer and shrinkage-reducing admixtures on alkali-activated slag pastes and mortars. Cem. Concr. Res, 2005, 35(7): 1358-1367.

[217] Wang S.-D., Scrivener K. L., Pratt P. L.. Factors affecting the strength of alkali-activated slag. Cem. Concr. Res, 1994, 24(6): 1033-1043.

[218] Wang S.-D., Scrivener K.L.. 29Si and 27Al NMR study of alkali-activated slag. Cem. Concr. Res, 2003, 33(5): 769-774.

[219] Richardson I. G., Groves G. W.. Microstructure and microanalysis of hardened cement pastes involving ground granulated blast-furnace slag. J. Mater. Sci, 1992, 27: 6204-6212.

[220] Richardson I. G., Brough A. R., Brydson R., Groves G. W., Dobson C. M.. Location of aluminum in substituted calcium silicate hydrate (C-S-H) gels as determined by 29Si and 27Al NMR and EELS. J. Am. Ceram. Soc, 1993, 76(9): 2285-2288.

[221] Rahier H., Van Mele B., Biesemans M., Wastiels J., Wu X.. Low-temperature synthesized aluminosilicate glasses. 1. Low-temperature reaction stoichiometry and structure of a model compound. J. Mater. Sci, 1996, 31(1): 71-79.

[222] Rahier H., Van Mele B., Wastiels J.. Low-temperature synthesized aluminosilicate glasses. 2. Rheological transformations during low-temperature cure and high-temperature properties of a model compound. J. Mater. Sci, 1996, 31(1): 80-85.

[223] Rahier H., Simons W., Van Mele B., Biesemans M.. Low-temperature synthesized alumi- nosilicate glasses. 3. Influence of the composition of the silicate solution on production structure and properties. J. Mater. Sci, 1997, 32(9): 2237-2247.

[224] Rahier H., Denayer J. F., Van Mele B.. Low-temperature synthesized aluminosilicate glasses. Part IV. Modulated DSC study on the effect of particle size of metakaolinite on the production of inorganic polymer glasses. J. Mater. Sci, 2003, 38(14): 3131-3136.

[225] Rahier H., Wastiels J., Biesemans M., Willem R., Van Assche G., Van Mele B.. Reaction mechanism kinetics and high temperature transformations of geopolymers. J. Mater. Sci, 2007, 42(9): 2982-2996.

[226] Faignet S., Bauweraerts P., Wastiels J., Wu X.. Mineral polymer system for making proto- type fibre reinforced composite parts. J. Mater. Proc. Technol, 1995, 48: 757-764.

[227] Patfoort G., Wastiels J., Bruggeman P., Stuyck L.. Mineral polymer matrix composites. In: Brandt A. M., Marshall I. H. (eds.) Proceedings of Brittle Matrix Composites 2 (BMC 2), Cedzyna Poland, 1989: 587-592. Elsevier.

[228] Wastiels J., Wu X., Faignet S., Patfoort G.. Mineral polymer based on fly ash. J. Resour. Manag. Technol, 1994, 22(3): 135-141.

[229] Buchwald A., Hilbig H., Kaps C.. Alkali-activated metakaolin-slag blendsperformance and structure in dependence on their composition. J. Mater. Sci, 2007, 42(9): 3024-3032.

[230] Buchwald A., Schulz M.. Alkali-activated binders by use of industrial by-products. Cem. Concr. Res, 2005, 35(5): 968-973.

[231] Kaps C., Buchwald A.. Property controlling influences on the generation of geopolymeric binders based on clay. In: Lukey G. C. (ed.) Geopolymers 2002. Turn Potential into Profit Melbourne. CD-ROM Proceedings. Siloxo Pty. Ltd. 2002.

[232] Buchwald A., Kaps C., Hohmann M.. Alkali-activated binders and pozzolan cement binders-complete binder reaction or two sides of the same story? In: Grieve G., Owens G. (eds.) Proceedings of the 11th International Conference on the Chemistry of Cement Durban South Africa, 2003: 1238-1246.

[233] Buchwald A.. What are geopolymers? Current state of research and technology, the opportunities they offer, and their significance for the precast industry. Betonw. Fert. Technol, 2006, 72(7): 42-49.

[234] Buchwald A., Wierckx J.. ASCEM cement technology-alkali-activated cement based on synthetic slag made from fly ash. In: Shi C., Shen X. (eds.) First International Conference on Advances in Chemically-Activated Materials, Jinan, China, 2010: 15-21. RILEM.

[235] Gruskovnjak A., Lothenbach B., Holzer L., Figi R., Winnefeld F.. Hydration of alkali- activated slag: comparison with ordinary Portland cement. Adv. Cem. Res, 2006, 18(3): 119-128.

[236] Winnefeld F., Leemann A., Lucuk M., Svoboda P., Neuroth M.. Assessment of phase formation in alkali activated low and high calcium fly ashes in building materials. Constr. Build. Mater, 2010, 24(6): 1086-1093.

[237] Ben Haha M., Le Saout G., Winnefeld F., Lothenbach B.. Influence of activator type on hydration kinetics, hydrate assemblage and microstructural development of alkali activated blast-furnace slags. Cem. Concr. Res, 2011, 41(3): 301-310.

[238] Ben Haha M., Lothenbach B., Le Saout G., Winnefeld F.. Influence of slag chemistry on the hydration of alkali-activated blast-furnace slag-Part I: effect of MgO. Cem. Concr. Res, 2011, 41(9): 955-963.

[239] Le Saoût G., Ben Haha M., Winnefeld F., Lothenbach B.. Hydration degree of alkali- activated slags: a 29Si NMR study. J. Am. Ceram. Soc, 2011, 94(12): 4541-4547.

[240] Ben Haha M., Lothenbach B., Le Saout G., Winnefeld F.. Influence of slag chemistry on the hydration of alkali-activated blast-furnace slag-Part II: effect of Al_2O_3. Cem. Concr. Res, 2012, 42(1): 74-83.

[241] Beleña I., Tendero M. J. L., Tamayo E. M., Vie D.. Study and optimizing of the reaction parameters for geopolymeric material manufacture. Bol. Soc. Esp. Cerám. Vidr, 2004, 43(2): 569-572.

[242] Querol X., Moreno N., Alastuey A., Juan R., Andrés, J. M., López-Soler A., Ayora C., Medinaceli A., Valero A.. Synthesis of high ion exchange zeolites from coal fly ash. Geol. Acta, 2005, 5(1): 49-57.

[243] Izquierdo M., Querol X., Davidovits J., Antenucci D., Nugteren H., Fernández-Pereira C.. Coal fly ash-slag-based geopolymers: microstructure and metal leaching. J. Hazard. Mater, 2009, 166(1): 561-566.

[244] Kamseu E., Leonelli C., Perera D. S., Melo, U. C., Lemougna P. N.. Investigation of volcanicash based geopolymers as potential building materials. Interceram 2009, 58(2-3): 136-140.

[245] Pacheco-Torgal F., Castro-Gomes J., Jalali S.. Properties of tungsten mine waste geopoly- meric binder. Constr. Build. Mater, 2008, 22(6): 1201-1211.

[246] Pacheco-Torgal F., Castro-Gomes J., Jalali S.. Tungsten mine waste geopolymeric binder: preliminary hydration products investigations. Constr. Build. Mater, 2009, 23(1): 200-209.

[247] Giancaspro J., Balaguru P. N., Lyon R. E.. Use of inorganic polymer to improve the fire response of balsa sandwich structures. J. Mater. Civil Eng, 2006, 18(3): 390-397.

[248] Lyon R. E., Balaguru P. N., Foden A., Sorathia U., Davidovits J., Davidovics M.. Fire- resistant aluminosilicate composites. Fire Mater, 1997, 21(2): 67-73.

[249] Papakonstantinou C. G., Balaguru P., Lyon R. E.. Comparative study of high temperature composites. Compos. B, 2001, 32(8): 637-649.

[250] Papakonstantinou C. G., Balaguru P.. Fatigue behavior of high temperature inorganic matrix composites. J. Mater. Civil Eng, 2007, 19(4): 321-328.

[251] Comrie D. C., Kriven W. M.. Composite cold ceramic geopolymer in a refractory application. Ceram. Trans, 2003, 153: 211-225.

[252] Kriven W. M., Bell J. L., Gordon M.. Microstructure and microchemistry of fully-reacted geopolymers and geopolymer matrix composites. Ceram. Trans, 2003, 153: 227-250.

[253] Bell J. L., Gordon M., Kriven W. M.. Use of geopolymeric cements as a refractory adhesive for metal and ceramic joins. Ceram. Eng. Sci. Proc, 2005, 26(3): 407-413.

[254] Gordon M., Bell J. L., Kriven W. M.. Comparison of naturally and synthetically derived, potassium-based geopolymers. Ceram. Trans, 2005, 165: 95-106.

[255] Kriven W. M., Kelly C. A., Comrie D. C.. Geopolymers for structural ceramic ap-

[256] Bao Y., Kwan S., Siemer D. D., Grutzeck M. W.. Binders for radioactive waste forms made from pretreated calcined sodium bearing waste. J. Mater. Sci, 2003, 39(2): 481-488.

[257] Bao Y., Grutzeck M. W., Jantzen C. M.. Preparation and properties of hydroceramic waste forms made with simulated Hanford low-activity waste. J. Am. Ceram. Soc, 2005, 88(12): 3287-3302.

[258] Brenner P., Bao Y., DiCola M., Grutzeck M. W.. Evaluation of new tank fill materials for radioactive waste management at Hanford and Savannah River. The Pennsylvania State University. http://www.personal.psu.edu/gur/Second%20tank%20fill%20report.pdf, 2006.

[259] Siemer D. D.. Hydroceramics, a "new" cementitious waste form material for US defense-type reprocessing waste. Mater. Res. Innov, 2002, 6(3): 96-104.

[260] Rostami H., Brendley W.. Alkali ash material: a novel fly ash-based cement. Environ. Sci. Technol, 2003, 37(15): 3454-3457.

[261] Miller S. A., Sakulich A. R., Barsoum M. W., Jud Sierra E.. Diatomaceous earth as a pozzolan in the fabrication of an alkali-activated fine-aggregate limestone concrete. J. Am. Ceram. Soc, 2010, 93(9): 2828-2836.

[262] Sakulich A. R., Miller S., Barsoum M. W.. Chemical and microstructural characterization of 20-month-old alkali-activated slag cements. J. Am. Ceram. Soc, 2010, 93(6): 1741-1748.

[263] Diaz E. I., Allouche E. N.. Recycling of fly ash into geopolymer concrete: creation of a data-base. In: Green Technologies Conference 2010, IEEE, Grapevine, TX, USA. CD-ROM Proceedings, 2010.

[264] Diaz E. I., Allouche E. N., Eklund S.. Factors affecting the suitability of fly ash as source material for geopolymers. Fuel, 2010, 89: 992-996.

[265] Diaz-Loya E. I., Allouche E. N., Vaidya S.. Mechanical properties of fly-ash-based geopolymer concrete. ACI Mater. J, 2011, 108(3): 300-306.

[266] Chancey R. T., Stutzman P., Juenger M. C. G., Fowler D. W.. Comprehensive phase characterization of crystalline and amorphous phases of a Class F fly ash. Cem. Concr. Res, 2010, 40(1): 146-156.

[267] Van Jaarsveld J. G. S., Van Deventer J. S. J.. The effect of metal contaminants on the formation and properties of waste-based geopolymers. Cem. Concr. Res, 1999, 29(8): 1189-1200.

[268] Van Jaarsveld J. G. S., Van Deventer J. S. J., Lorenzen L.. The potential use of geopolymeric materials to immobilise toxic metals. 1. Theory and applications. Miner. Eng, 1997, 10(7): 659-669.

[269] Van Jaarsveld J. G. S., Van Deventer J. S. J., Schwartzman A.. The potential use of geopoly-meric materials to immobilise toxic metals: Part II. Material and leaching

characteristics. Miner. Eng, 1999, 12(1): 75-91.

[270] Lee W. K. W., Van Deventer J. S. J.. Structural reorganisation of class F fly ash in alkaline silicate solutions. Colloids Surf. A, 2002, 211(1): 49-66.

[271] Rees C. A., Provis J. L., Lukey G. C., Van Deventer J. S. J.. Attenuated total reflectance Fourier transform infrared analysis of fly ash geopolymer gel aging. Langmuir, 2007, 23(15): 8170-8179.

[272] Lloyd R. R., Provis J. L., Van Deventer J. S. J.. Microscopy and microanalysis of inorganic polymer cements. 1: remnant fly ash particles. J. Mater. Sci, 2009, 44(2): 608-619.

[273] Provis J. L., Yong C. Z., Duxson P., Van Deventer J. S. J.. Correlating mechanical and thermal properties of sodium silicate-fly ash geopolymers. Colloids Surf, 2009, A 336(1-3): 57-63.

[274] Sofi M., Van Deventer J. S. J., Mendis P. A., Lukey G. C.. Engineering properties of inorganic polymer concretes (IPCs). Cem. Concr. Res, 2007, 37(2): 251-257.

[275] Duxson P., Lukey G. C., Separovic F., Van Deventer J. S. J.. The effect of alkali cations on aluminum incorporation in geopolymeric gels. Ind. Eng. Chem. Res, 2005, 44(4): 832-839.

[276] Duxson P., Provis J. L., Lukey G. C., Mallicoat S. W., Kriven W. M., Van Deventer J. S. J.. Understanding the relationship between geopolymer composition, microstructure and mechanical properties. Colloids Surf, 2005, A 269(1-3): 47-58.

[277] White C. E., Provis J. L., Proffen T., Riley D. P., Van Deventer J. S. J.. Combining density functional theory (DFT) and pair distribution function (PDF) analysis to solve the structure of metastable materials: the case of metakaolin. Phys. Chem. Chem. Phys, 2010, 12(13): 3239-3245.

[278] Provis J. L., Van Deventer J. S. J.. Geopolymerisation kinetics. 1. In situ energy dispersive X-ray diffractometry. Chem. Eng. Sci, 2007, 62(9): 2309-2317.

[279] Provis J. L., Rose V., Bernal S. A., Van Deventer J. S. J.. High resolution nanoprobe X-ray fluorescence characterization of heterogeneous calcium and heavy metal distributions in alkali activated fly ash. Langmuir, 2009, 25(19): 11897-11904.

[280] Hajimohammadi A., Provis J. L., Van Deventer J. S. J.. Time-resolved and spatially-resolved infrared spectroscopic observation of seeded nucleation controlling geopolymer gel forma- tion. J. Colloid Interface Sci, 2011, 357(2): 384-392.

[281] Provis J. L., Rose V., Winarski R. P., Van Deventer J. S. J.. Hard X-ray nanotomography of amorphous aluminosilicate cements. Scr. Mater, 2011, 65(4): 316-319.

[282] Rees C. A., Provis J. L., Lukey G. C., Van Deventer J. S. J.. In situ ATR-FTIR study of the early stages of fly ash geopolymer gel formation. Langmuir, 2007, 23(17): 9076-9082.

[283] White C. E., Provis J. L., Proffen T., Van Deventer J. S. J.. The effects of temperature on the local structure of metakaolin-based geopolymer binder: a neutron pair distribution function investigation. J. Am. Ceram. Soc, 2010, 93 (10), 3486-3492.

[284] Collins F., Sanjayan J. G.. Early age strength and workability of slag pastes activated by NaOH and Na_2CO_3. Cem. Concr. Res, 1998, 28(5): 655-664.

[285] Bakharev T., Sanjayan J. G., Cheng Y. B.. Effect of elevated temperature curing on properties of alkali-activated slag concrete. Cem. Concr. Res, 1999, 29 (10): 1619-1625.

[286] Collins F. G., Sanjayan J. G.. Workability and mechanical properties of alkali activated slag concrete. Cem. Concr. Res, 1999, 29(3): 455-458.

[287] Bakharev T., Sanjayan J. G., Cheng Y. B.. Effect of admixtures on properties of alkali- activated slag concrete. Cem. Concr. Res. 2000, 30(9): 1367-1374.

[288] Collins F., Sanjayan J.. Prediction of capillary transport of alkali activated slag cementitious binders under unsaturated conditions by elliptical pore shape modeling. J. Porous. Mater, 2010, 17(4): 435-442.

[289] Bakharev T.. Geopolymer materials prepared using Class F fly ash and elevated temperature curing. Cem. Concr. Res, 2005, 35(6): 1224-1232.

[290] Guerrieri M., Sanjayan J., Collins F.. Residual compressive behavior of alkali-activated concrete exposed to elevated temperatures. Fire Mater, 2009, 33(1): 51-62.

[291] Hardjito D., Rangan B. V.. Development and properties of low-calcium fly ash-based geo- polymer concrete. Curtin University of Technology, Research Report GC1, 2005.

[292] Wallah S. E., Rangan B. V.. Low-calcium fly ash-based geopolymer concrete: Long-term properties. Curtin University of Technology, Research Report GC2, 2006.

[293] Sumajouw D. M. J., Rangan B. V.. Low-calcium fly ash-based geopolymer concrete: Reinforced beams and columns, Research Report GC3. Curtin University of Technology, 2006.

[294] Rangan B. V.. Engineering properties of geopolymer concrete. In: Provis J. L., Van Deventer J. S. J. (eds.) Geopolymers: Structure, Processing, Properties and Industrial Applications, 2009: 213-228. Woodhead, Cambridge.

[295] Chen-Tan N. W., Van Riessen A., Ly C. V., Southam D. C.. Determining the reactivity of a fly ash for production of geopolymer. J. Am. Ceram. Soc, 2009, 92(4): 881-887.

[296] Temuujin J., Van Riessen A.. Effect of fly ash preliminary calcination on the properties of geopolymer. J. Hazard. Mater, 2009, 164(2-3): 634-639.

[297] Rickard W. D. A., Williams R., Temuujin J., Van Riessen A.. Assessing the suitability of three Australian fly ashes as an aluminosilicate source for geopolymers in

high temperature applications. Mater. Sci. Eng, 2011, A 528(9): 3390-3397.

[298] Williams R. P., Hart R. D., Van Riessen A.. Quantification of the extent of reaction of metakaolin-based geopolymers using X-ray diffraction, scanning electron microscopy, and energy-dispersive spectroscopy. J. Am. Ceram. Soc, 2011, 94(8): 2663-2670.

[299] Singh P. S., Trigg M., Burgar I., Bastow T.. Geopolymer formation processes at room temperature studied by 29Si and 27Al MAS-NMR. Mater. Sci. Eng, 2005, A 396(1-2): 392-402.

[300] De Silva P., Sagoe-Crentsil K., Sirivivatnanon V.. Kinetics of geopolymerization: role of Al_2O_3 and SiO_2. Cem. Concr. Res, 2007, 37: 512-518.

[301] De Silva P., Sagoe-Crentsil K.. Medium-term phase stability of Na_2O-Al_2O_3-SiO_2-H_2O geopolymer systems. Cem. Concr. Res, 2008, 38(6): 870-876.

[302] Blackford M. G., Hanna J. V., Pike K. J., Vance E. R., Perera D. S.. Transmission electron microscopy and nuclear magnetic resonance studies of geopolymers for radioactive waste immobilization. J. Am. Ceram. Soc, 2007, 90(4): 1193-1199.

[303] Perera D. S., Cashion J. D., Blackford M. G., Zhang Z., Vance E. R.. Fe speciation in geo-polymers with Si/Al molar ratio of ~2. J. Eur. Ceram. Soc, 2007, 27(7): 2697-2703.

[304] Barbosa V. F. F., MacKenzie K. J. D., Thaumaturgo C.. Synthesis and characterisation of materials based on inorganic polymers of alumina and silica: sodium polysialate polymers. Int. J. Inorg. Mater, 2000, 2(4): 309-317.

[305] Barbosa V. F. F., MacKenzie K. J. D.. Thermal behaviour of inorganic geopolymers and composites derived from sodium polysialate. Mater. Res. Bull, 2003, 38(2): 319-331.

[306] Fletcher R. A., MacKenzie K. J. D., Nicholson C. L., Shimada S.. The composition range of aluminosilicate geopolymers. J. Eur. Ceram. Soc, 2005, 25(9): 1471-1477.

[307] MacKenzie K. J. D., Brew D. R. M., Fletcher R. A., Vagana R.. Formation of aluminosilicate geopolymers from 1∶1 layer-lattice minerals pre-treated by various methods: a comparative study. J. Mater. Sci, 2007, 42(12): 4667-4674.

[308] Nicholson C. L., Murray B. J., Fletcher R. A., Brew D. R. M., MacKenzie K. J. D., Schmücker M.. Novel geopolymer materials containing borate structural units. In: Davidovits J. (ed.) World Congress Geopolymer 2005, Saint-Quentin France, 2005: 31-33. Geopolymer Institute.

[309] Van Deventer J. S. J., Provis J. L., Duxson P., Brice D. G.. Chemical research and climate change as drivers in the commercial adoption of alkali activated materials. Waste Biomass Valoriz, 2010, 1(1): 145-155.

[310] Lukey G. C., Mendis P. A., Van Deventer J. S. J., Sofi M.. Advances in inorganic polymer concrete technology. In: Day K. W. (ed.) Concrete Mix Design, Quality

Control and Specification, 3rd edn. Routledge, London, 2006. Appendix A.

[311] Duxson P., Provis J. L.. Designing precursors for geopolymer cements. J. Am. Ceram. Soc, 2008, 91(12): 3864-3869.

[312] Provis J. L., Duxson P., Van Deventer J. S. J.. The role of particle technology in developing sustainable construction materials. Adv. Powder Technol, 2010, 21(1): 2-7.

[313] Oliveira C. T. A., John V. M., Agopyan V.. Pore water composition of clinker free granulated blast furnace slag cements pastes. In: Krivenko P. V. (ed.) Proceedings of the Second International Conference on Alkaline Cements and Concretes, Kiev, Ukraine, 1999: 109-119. ORANTA.

[314] Silva F. J., Thaumaturgo C.. Fibre reinforcement and fracture response in geopolymeric mortars. Fatigue Fract. Eng. Mater. Struct, 2003, 26(2): 167-172.

[315] Penteado Dias D., Thaumaturgo C.. Fracture toughness of geopolymeric concretes reinforced with basalt fibers. Cem. Concr. Compos, 2005, 27(1): 49-54.

[316] Puertas F., Mejía de Gutierrez R., Fernández-Jiménez, A., Delvasto S., Maldonado J.. Alkaline cement mortars. Chemical resistance to sulfate and seawater attack. Mater. Constr, 2002, 52: 55-71.

[317] Rodríguez E., Bernal S., Mejía de Gutierrez R., Puertas F.. Alternative concrete based on alkali-activated slag. Mater. Constr, 2008, 58(291): 53-67.

[318] Bernal S. A., Mejía de Gutierrez R., Pedraza A. L., Provis J. L., Rodríguez E. D., Delvasto S.. Effect of binder content on the performance of alkali-activated slag concretes. Cem. Concr. Res, 2011, 41(1): 1-8.

[319] Escalante-Garcia J. I., Gorokhovsky A. V., Mendoza G., Fuentes A. F.. Effect of geothermal waste on strength and microstructure of alkali-activated slag cement mortars. Cem. Concr. Res, 2003, 33(10): 1567-1574.

[320] Escalante García, J. I., Campos-Venegas, K., Gorokhovsky, A., Fernández A.. Cementitious composites of pulverised fuel ash and blast furnace slag activated by sodium silicate: effect of Na_2O concentration and modulus. Adv. Appl. Ceram, 2006, 105(4): 201-208.

[321] Marín-López C., Reyes Araiza J., Manzano-Ramírez A., Rubio Avalos J., Perez-Bueno J., Muñiz-Villareal M., Ventura-Ramos E., Vorobiev Y.. Synthesis and characterization of a concrete based on metakaolin geopolymer. Inorg. Mater, 2009, 45(12): 1429-1432.

[322] Chatterjee A. K.. Indian fly ashes: their characteristics and potential for mechano-chemical activation for enhanced usability. J. Mater. Civil Eng, 2011, 23(6): 783-788.

[323] Sharma R. C., Jain N. K., Ghosh S. N.. Semi-theoretical method for the assessment of reactivity of fly ashes. Cem. Concr. Res, 1993, 23(1): 41-45.

[324] Kumar S., Kumar R., Alex T. C., Bandopadhyay A., Mehrotra S. P.. Influence

of reactivity of fly ash on geopolymerisation. Adv. Appl. Ceram, 2007, 106(3): 120-127.

[325] Kumar R., Kumar S., Mehrotra S. P.. Towards sustainable solutions for fly ash through mechanical activation. Resour. Conserv. Recycl, 2007, 52(2): 157-179.

[326] Kumar S., Kumar R.. Mechanical activation of fly ash: effect on reaction, structure and properties of resulting geopolymer. Ceram. Int, 2011, 37(2): 533-541.

[327] Kumar S., Sahoo D. P., Nath S. K., Alex T. C., Kumar R.. From grey waste to green geopolymer. Sci. Cult, 2012, 78(11-12): 511-516.

第 3 章 胶凝材料化学——高钙碱激发胶凝材料

Susan A. Bernal，John L. Provis，Ana Fernández-Jiménez，
Pavel V. Krivenko，Elena Kavalerova，Marta Palacios
and Caijun ShiJohn L. Provis

3.1 介绍

正如前章所述，作为碱激发材料的高炉矿渣或其他富钙的工业副产物的研究已经进行了一个多世纪[1-3]。然而，近年来随着科学手段优化以及激发条件需求的驱使，碱激发胶凝材料的微观结构得到广泛研究，进而制备高强、稳定的胶凝材料，并最终得到高性能的碱激发混凝土材料（AAM），同时实现适宜的工作性和较小的环境污染。对于碱激发材料结构的详细、科学的理解是制定标准的技术基础，同时有利于其商业化应用[4-5]。

矿渣基碱激发胶凝材料的化学和基础工程方面的研究已有大量的科学文献报道[6-18]。但对其他富钙的胶凝材料的研究文献相对较少，这也将在本章的后面部分进行总结。

碱激发矿渣形成的凝胶结构很大程度上取决于各种化学因素。这些化学因素控制着材料反应机理、力学性能的发展和耐久性能。这些因素可以大致分为两类：激发剂及原料的物化特性。这两种类型的因素将在以下章节中详细讨论。

3.2 矿渣基碱激发胶凝材料的结构与化学特性

矿渣基碱激发高性能材料的结构发展是一个复杂的多相反应过程，其主要由四步反应控制：玻璃体颗粒的溶解、初始固相的成核和生长、新相在界面处的机械结合和相互作用以及在固化初期反应产物的扩散和化学平衡[19-21]。

众所周知，高炉矿渣碱激发过程所形成的反应产物主要以铝代 C-A-S-H 型凝胶[8,22-27]，具有类似的托勃莫来石 C-S-H（I）无定形结构为主。伴随着副反应产物如 AFm 相的形成（主要来自以 NaOH 为激发剂的情况）[23,28,29]、含硅的 AFm（来自以硅酸盐为激发剂的情况）[30,31]、水滑石（来自有较高 MgO 含量的碱矿渣水泥的情况）[25,32,33]和类沸石如斜方钙沸石和十字沸石［来自高 Al_2O_3 含量和低（<5%）氧化镁的碱矿渣水泥的情况］[21,34,35]。

高炉矿渣碱激发产物 C-A-S-H 的结构和组成极大程度上取决于所用的激发剂性质。与硅酸盐激发相比，NaOH 激发具有更高的 Ca/Si 比和更有序的结构[23,24]。这就导致硅酸盐激发体系的孔溶液中硅酸盐物相的含量增加。Puertas 等[36]人认为碱金属硅酸盐激发的高炉矿渣胶凝材料中的 C-A-S-H 型凝胶可能与 11Å 和 14Å 共存类硅镁石相具有相似结构。Myers 等[27]人基于在不同硅镁石单元的交联和非交联结构中固有的约束（图 3-1）提出一种结构模型来描述这类凝胶。该模型能够计算这些更复杂结构的链长、铝硅比和交联度，但这对于非交联的托勃莫来石的 C-S-H 凝胶则不能进行较为全面的描述。

最近的研究[37]表明，NaOH 激发（图 3-2）或硅酸盐激发（图 3-3）的碱矿渣胶凝材料中，

图 3-1　交联和非交联的托勃莫来石结构，代表了 C-(N)-A-S-H 型凝胶的普遍结构（图片由谢菲尔德大学的 R. J. Myers 提供）

C-A-S-H 的一些化学结合的 Ca^{2+} 可能被 Na^+ 替代，导致形成 C-(N)-A-S-H 内部凝胶。C-(N)-A-S-H 型凝胶也会出现在低钙硅比的硅酸盐激发矿渣混凝土中，主要集中在硅质骨料和胶凝材料之间[38,39]的界面过渡区中，在胶凝材料中较少。

碱矿渣水泥结构的研究有助于理解无碱激发剂矿渣与水反应形成的胶凝材料的化学性质。在无碱激发剂的条件下，经过 20 年的充足水化后，22% 的粗磨高炉矿渣反应形成的凝胶提供了足够的活性成分，并且形成 Al 取代的 C-S-H 相、Mg-Al 层状双氢氧化物和无序层状氢氧化铝相[41]。C-S-H 反应 Al 的能力有限，受硅酸盐链的几何形状和离子取代的热力学限制，最大铝硅比略小于 $0.20^{[27,42\text{-}45]}$。

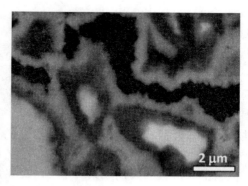

图 3-2　在经过 180d 的固化后，激发 BFS 的背散射电子显微图。黑暗的区域围绕着浅灰色的没有反应的钠含量较高的 BFS 粒子，明亮区域的物相 Na 含量低（改编自文献[37]）

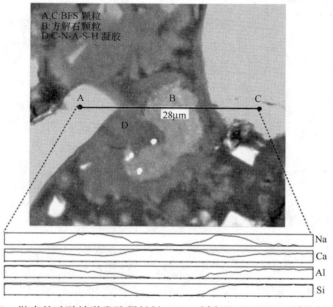

图 3-3　抛光的硅酸钠激发胶凝材料（80wt%高炉矿渣/20wt%偏高岭土）28d 后两个未反应的 BFS 颗粒的扫描电子显微图和元素分布（改编自文献[40]）

然而，除了在不同组成的胶凝材料中存在水滑石、AFm和类沸石相之外，对于碱激发矿渣体系中二次相形成的确切化学机理（动力学和平衡关系）还没有清晰的解释。已发表的许多研究，通过 XRD 和/或 NMR 等技术手段对特定碱激发矿渣体系的第二相的分析和鉴定已取得初步进展[32,46,47]。但由于炉渣地域性、铁矿石矿物学以及高炉操作的差异性，造成对合成渣的碱激发产品进行分析较为困难，需跨越更广泛的组合范围[46,48]，从而为系统分析矿渣化学对胶凝材料的形成影响提供更大的机会。

然而，激发剂的选择对胶凝材料化学性能的影响在文献中已有相应的研究，这将在下文中予以介绍。

3.3 矿渣体系的激发剂

正如在3.2节提到的，矿渣与水需要经过较长时间反应才能形成硬化体。因此，碱激发剂在 AAM 中的最关键作用就是加快反应进程，使其成为合理施工时间范围内可以使用的工程材料，而且在高 pH 值下更容易实现。第2章概述了 AAM 中最常使用的碱激发剂的化学性质。碱硅酸盐和氢氧化物在这些常见的激发剂中会形成高的 pH 值，而碳酸盐和硫酸盐则产生中等碱性环境，并且通过来自矿渣的钙反应生成用于激发过程的游离氢氧化物。

在反应过程的早期（第一个 24～48h），碱硅酸盐、碳酸盐或氢氧化物激发的矿渣胶凝材料的反应动力学采用等温量热法评估，确定其结构发展发生过程为五个阶段（诱导期、预诱导期、加速期、减速期和终止期），形成的放热曲线与常规硅酸盐水泥的预期大致相似[19,22]。然而，碱激发矿渣样品凝结的第一个小时内，每个放热阶段的持续时间和强度取决于使用的激发剂的类型（图3-4）。

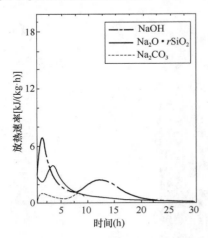

图3-4 碱激发矿渣胶凝材料的等温量热曲线，作为碱激发剂性质的函数（源自文献[49]）

在更广泛的研究中，针对激发剂类型的影响，Shi 和 Day[50-52]提出了 AAM 的三种反应模型：第一个是在反应第一分钟内观察到的唯一的峰，通常在用水或用 Na_2HPO_4 水化矿渣时出现。第二个模型首先出现一个诱导前的峰，随后是诱导期和第二个峰，与加速和相应的反应产物沉淀有关。这种现象通常可在 NaOH 激发矿渣胶凝材料中观察到，与硅酸盐水泥的水化过程比较，其中初始凝固时间和最终凝固时间直接受到放热加速期的影响，从而形成反应产物[33]。第三个模型为初始双峰，随后是诱导期，然后是反应加速期的第三个峰。当矿渣由包括硅酸盐、碳酸盐、磷酸盐和氟化物在内的弱盐激发时，预期为此模型[50]。与硅酸盐激发的胶凝材料相比，用 NaOH 或 Na_2CO_3 激发矿渣通常导致第一个 24h 的反应总热释量降低，这与反应过程较慢相关[19]。

碱激发矿渣胶凝材料的反应产物，以及由此而产生的性能强烈依赖于所使用的激发剂的性质和浓度[53]。可能的激发溶液包括碱氢氧化物[ROH、$Ca(OH)_2$]、弱盐（R_2CO_3、R_2S、RF）、强酸盐（Na_2SO_4、$CaSO_4 \cdot 2H_2O$）和碱金属硅酸盐 $R_2O \cdot rSiO_2$，其中 R 是 Na^+，或较不常用的 Li^+、Cs^+ 或 Rb^+。在这些物质中，用于制备碱激发矿渣胶凝材料的常用激发剂是氢氧化钠(NaOH)、硅酸钠($Na_2O \cdot rSiO_2$)、碳酸钠(Na_2CO_3)和硫酸钠(Na_2SO_4),正如在第 2 章所讨论的[18,6,53,54]一样。Shi 和 Day[55]确定,所有腐蚀性碱和其阴离子可与 Ca^{2+} 反应,产生比 $Ca(OH)_2$ 更难溶解的富 Ca 碱性化合物,可作为矿渣的激发剂。在初始反应期间,碱激发剂的阴离子组分与从矿渣溶解的 Ca^{2+} 反应,形成富含 Ca 的产物。

Royet 等人[56]用不同的氢氧化物溶液[LiOH、NaOH、KOH、$Ca(OH)_2$、$Sr(OH)_2$ 和 $Ba(OH)_2$]激发矿渣,证实了在相同 pH 值条件下、不同体系中产生了相似的反应产物,其与所添加阳离子的离子半径或价态无关。碱激发矿渣胶凝材料作为该类型的激发剂所产生的抗压强度如图 3-5 所示。

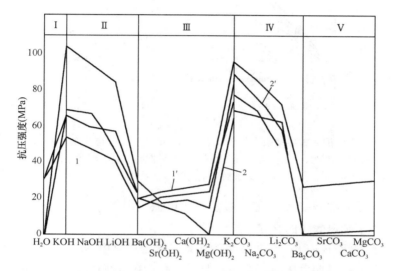

图 3-5 碱激发剂类型和矿渣前驱体组成的水泥组合物的抗压强度
1 和 1′的样品组基于 Mb=1.13（Al_2O_3 质量分数 6.75%）的矿渣;
2 和 2′的样品组基于 Mb=0.85（Al_2O_3 质量分数 15.85%）的矿渣;
样品组 1 和 2 蒸汽养护（T=90℃±5℃,3h+7h+2h）,并且 1′和 2′在高压釜中养护
（T=173℃,3h+7h+2h）（数据由 P. V. Krivenko 提供）

pH 值对激发剂的效率有很大的影响,因为其控制前驱体的初始溶解和随后的缩合反应[33,50,55-57]。pH 值对激发矿渣的效率也强烈依赖于激发剂的类型,因为在高 pH 值下,钙的溶解度降低,而二氧化硅和氧化铝的溶解度会增加。与不同类型激发剂激发的矿渣相比较,虽然 NaOH 激发溶液具有比碱浓度相似的硅酸钠溶液高得多的 pH 值,但是硅酸盐激发的胶凝材料通常比 NaOH 激发的矿渣发展出更高的力学强度[37,53,58,59]。这是因为体系中额外供应的硅酸盐物质与矿渣溶解的 Ca^{2+} 阳离子反应,形成了致密的 C-A-S-H 反应产物[24],并从溶液相中除去多余溶解的 Ca。因此,选择最合适的碱激发剂需要考虑钙物质在新拌浆体孔溶液 pH 值下的溶解度以及与由激发剂提供的阳离子的相互作用,其可以促进特定反应产物的形成。

Fernández-Jiménez 等人[58]通过设计统计实验,确定激发剂的性质是控制碱激发矿渣材料力学性能的主要因素。与其他研究（例如文献 [53]）获得的结果一致,具有中等比表面积（$450m^2/kg$）的激发矿渣导致使用硅酸钠激发剂时产生更高的力学强度。当使用 NaOH 和 Na_2CO_3 作为超细矿渣的激发剂时可改善力学强度[55,58]。

硅酸盐激发的矿渣材料力学性能的发展取决于激发剂溶液的模数（$Ms=SiO_2/Na_2O$ 摩尔比，见第 2 章的详细讨论）和所使用的矿渣的性质，如图 3-6 所示[53]。用 NaOH 激发和 Na_2CO_3 激发矿渣胶凝材料，水胶比对硅酸盐激发矿渣胶凝材料的反应动力学影响较小[52]。Palacios 和 Puertas[60] 认为混合时间对硅酸盐激发矿渣胶凝材料的力学强度发展也有显著的影响，因为较长的混合时间（最多 30min）可以使强度提高到 11%，而且，由于小孔体积减小会使体系的渗透性降低。

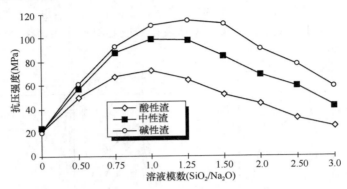

图 3-6 在 20℃下养护的硅酸盐激发的矿渣砂浆的 28d 抗压强度作为激发溶液的模数的函数
矿渣的细度为(450±30)m^2/kg，碱溶液/矿渣比为 0.41，砂/矿渣比为 2.0
（数据来自文献[53]）

已经确定，使用硅酸钠-碳酸锂复合激发剂比单独使用 $Na_2O \cdot rSiO_2$ 基激发剂反应程度更高，这部分归因于延迟效应的存在[61,62]。

相比 NaOH 激发和 $Na_2O \cdot rSiO_2$ 激发的体系，学者对使用 Na_2SO_4 激发矿渣的关注度更低，主要是由于在养护早期力学强度发展较低。然而，在这些体系中获得的孔溶液的 pH 值相对较低，有更多的钙矾石形成，给出了诸如核废料固化的应用所需的性能。在这种应用中，力学强度要求远低于常规民用基础设施应用，而钙矾石化学结合大量游离水，因此降低了金属腐蚀的风险[63,64]。使用矿粉，加入少量的石灰或硅酸盐熟料，可以在硫酸盐激发的矿渣胶凝材料中提供高得多的力学强度[53,65,66]。这些复合胶凝物将在第 5 章中详细讨论。

3.3.1 胶凝体系——氢氧化物激发

氢氧化物激发的矿渣体系中，胶凝材料的性质主要由 C-A-S-H 凝胶控制。例如层状双氢氧化物（通常类似于水滑石基团）和/或钙（硅）铝酸盐水化物[28,56,57,67-69]。此类紧密混合的富 Al 次生相受矿渣源的 Al 和 Mg 含量的影响。与大多数硅酸盐水泥水化产品相比，C-A-S-H 凝胶的结构有序度往往相对较高[67]，这种凝胶相显示出类似于 14Å 雪硅钙石的结构特征[70]。凝胶的内部产物和外部产物区域（即分别由矿渣颗粒或溶液最初占据的区域中形成的凝胶）倾向于具有相似的组成[70]。如在硅酸盐水泥体系中，C-A-S-H 凝胶包括具有 dreierketten 结构的四面体配位的硅酸盐链层（图 3-1），每个链包含 $3n-1$ 个四面体，整数值为 n。中间层区域富含 Ca，并且还含有结合水。C-A-S-H 凝胶中的铝主要存在于链内的四面体桥位中，在复合硅酸盐水泥体系中已经确定[45,71,72]，并且当 Al^{3+} 代替 Si^{4+} 时，Na^+ 可以平衡产生的负电荷。这种四面体配位的铝是硅酸盐链之间发生交联的主要位点[73-75]。Al 仅被替换成这些桥接位点，导致 C-A-S-H 结构中 Al 的取代程度被限制（尽管精确上限取决于链长度和交联度[27]），从而导致上述的富 Al 次生相的形成。

在 NaOH 激发矿渣中形成的 C-A-S-H 型凝胶倾向于具有相对低的交联度[24,76]，尽管 Palacios 和 Puertas[77]也在凝胶中鉴定了少量的 Q^3 单元。所使用的激发剂的类型也影响凝胶的化学组成；因为没有由氢氧化物激发剂提供的额外的 Si，体系（因此生成的 C-A-S-H 凝胶本身）的 Ca/(Si + Al) 比率将高于硅酸盐激发的胶凝材料。然而，独立于所用的激发剂，通过矿渣激发形成的 C-A-S-H 型凝胶具有比水化硅酸盐水泥体系更低的 Ca 含量，其 Ca/Si 比通常在 1.5 和 2.0 之间[9]。

3.3.2 胶凝体系——硅酸盐激发

自从 Purdon 的初始工作[3]以来，硅酸盐激发成为生产基于矿渣的 AAM 的最广泛的方法，这是由于该方法的通用性和所生产的胶凝材料具有较高性能。市售硅酸钠溶液或喷雾干燥粉末是最常见的用于生产基于矿渣的胶凝材料的硅酸盐激发剂。激发剂通常作为溶液包含在混合物中，然而，它也可以以固体状态掺入，与矿渣混合或球磨[53]。固体激发剂的添加可以形成更低、更可变的早期强度，这是由于在反应进程期间碱的缓慢释放[53,78-80]。此外，激发剂具有吸湿性，可能会在加入混合水之前引起部分反应[53]。

然而在某些情况下，有可能在价格和/或环境方面提供具有优势的激发剂的替代来源。Živicaand 等人[81-83]使用化学改性硅粉结合 NaOH 作为碱激发剂生产高性能碱激发矿渣胶凝材料，与使用商业硅酸钠溶液生产的胶凝材料相比，其具有高度致密的结构和更高的力学强度。Bernal 等人[84]通过化学改性的硅灰激发的矿渣/偏高岭土胶凝材料的研究表明，改性硅灰有利于其力学强度及结构特征的改善。硅的替代来源如稻壳灰[84]和纳米二氧化硅[85]也已被评估为这些替代激发剂中的二氧化硅替代源，表明碱与高度无定形含 Si 前驱体的组合可成功地用作 AAM 生产中的激发剂。

众所周知，碱-硅酸盐激发的矿渣胶凝材料中的主要反应产物是低结晶度的 C-A-S-H 型凝胶[36,68,86,87]，其结构受矿渣的化学性质和激发剂的成分影响[32,47]。外部产物快速形成，其通过在最初的流体填充区域中存在高浓度的二氧化硅（因此具有强烈的过饱和趋势）而增强[25,88]。孔结构为封闭型或相当开放型，这取决于配合比设计和样品制备及测量的方法，因为外压或不适当的干燥方式将在分析之前（或期间）导致样品损坏[89,90]。增加激发剂的量通常会形成更精细的孔隙网络[91]。

Si 和 Al 核磁共振的结果证明，使用硅酸盐型激发剂可诱发生成具有相对高含量的 Q^2 和 Q^3 位点的 C-A-S-H 凝胶，这是凝胶的高度交联和高致密化的表现[24,25,77,87,92]。Fernández-Jiménez[73]观察到，随着硅酸钠激发剂浓度的降低，桥接位置中 Al 的浓度降低，以及 Si 桥连单元增加。与在 25℃下养护的胶凝材料相比，在 45℃下养护会形成具有较低交联度［未观察到 Q^3 和 Q^3（1Al）位点］的高度均匀的 C-S-H 产物。硅酸盐激发矿渣在热养护条件下可以加速凝结，但并不总是有利于长期性能，特别是如果受热不均匀会导致产生微裂纹，或者介质反应所需的水从材料中损失，正在进行的反应就会受到限制[93]。

当使用的矿渣含有足够的 Mg 时，可观察到反应产物为类水滑石 Mg-Al 层状双氢氧化物[24,25,32,67,68,87]。沸石产物有时也可在具有较低 Mg 含量的样品中观察到，通常在具有不同程度的 Na/Ca 取代的托勃莫来石结构中[21,34]，而在老化的激发矿渣浆体中也观察到了镁沸石[94]。在硅酸盐激发的矿渣中观察到的其他产物包括硅质水凝胶，例如钙钛矿[87]，有时还包括结晶的 AFm 相以及石英岩[30,31]。

然而，AFm 相并不总是可以通过 XRD 检测，结合凝胶内的一些铝作为 AFm 存在，紧密混合到 C-S-H 结构中，因此不能通过 XRD 区分[29,68,87]。文献对碱激发矿渣胶凝材料中 AFm 相形成机理的详细分析非常少，尽管从化学计量的角度来看，这些相可能与高浓度的 Ca 和 Al 相关[30,86]。图 3-7 显示了从低模数硅酸钠溶液碱激发矿渣的热力学模型计算的复合相的实例，其中预测形成的相通常与上述实验观察的十分一致[30]。

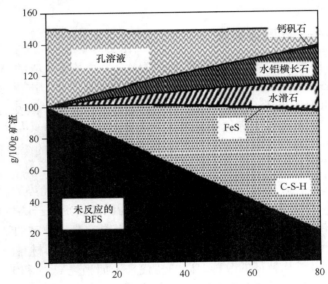

图 3-7　通过用硅酸钠激发矿渣形成的相组成的热力学模型，作为矿渣水化产物成分的函数（源自文献［30］）

图 3-8 显示了矿渣和碱硅酸盐激发的矿渣胶凝材料的 Al 核磁共振光谱。未反应的矿渣的光谱表现为 $50 \sim 80 \times 10^{-6}$ 之间的宽共振，以 68×10^{-6} 为中心，分配给玻璃状矿渣中的四面体 Al。激发后，养护 14d 和 3 年的样品显示出比未反应的矿渣更尖锐的 Al（IV）带，并以 74×10^{-6} 为中心形成相对较窄的共振。68×10^{-6} 以下的峰与残留的未反应的矿渣相关，也可能与 C-A-S-H 凝胶中的交联桥接四面体相关[45]。74×10^{-6} 的尖峰归属于键合到 Q^2（1Al）位点的非交联桥接四面体中的 Al（IV）[44,45,95,96]。

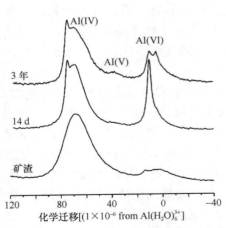

图 3-8　矿渣及与硅酸钠碱激发产物的 ^{27}Al MAS NMR 光谱（模数为 1.0，经 14d 和 3 年的养护）（数据来自文献［87］）

在八面体 Al 区域（从 $-10 \sim 20 \times 10^{-6}$），养护 14d 的碱激发矿渣样品以 10×10^{-6} 为中心的高强度窄峰，以及以 4.5×10^{-6} 为中心的小肩，其可以归属于 AFm 和水滑石型相[28,86,87]，因为这些类型的相的峰位置重叠，并且取决于相内的取代离子的性质和程度。

在图 3-8 所示的 3 年样品中，3×10^{-6} 峰的强度的增加与无序相的发展一致，例如 Andersen 等人的"第三铝酸盐水化物"[96,97]，其是含水无定形氧化铝结构并且可以与层状 Mg-Al 相紧密混合[41]。

图 3-9 显示了相同的 14d 养护的碱硅酸盐激发的矿渣样品的 ^{29}Si MAS NMR 光谱[87]。解卷积是指组分亚峰和识别未反应的矿渣的贡献能够直接计算矿渣的反应程度[87,98]，在这里所示的样品中约为 75%[87]。图 3-9 所示的反应产物亚峰的位置和相对强度与雪硅钙石样 C-A-S-H 凝胶的化学性质一致，位点连接性范围为 Q^1 至 Q^3，以及不同程度的 Al 取代和不同的电荷平衡种类导致每个连接类型的多个峰的存在[87]。

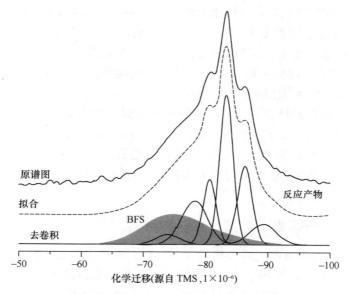

图 3-9　解卷积的养护 14d 碱激发矿渣的 ^{29}Si MAS NMR 光谱
暗灰色带代表残余无水矿渣的贡献（数据来自文献 [87]）

碱硅酸盐激发的矿渣和细磨的石灰石的组合能提供可接受的强度发展和耐久性能，并且石灰石通常以低于矿渣或硅酸钠的成本获得，这意味着为 AAM 提供了潜在的成本效益路线[99]。

3.3.3　胶凝体系——碳酸盐激发

矿渣的碳酸钠激发在东欧和中欧已经应用了五十多年[16,46,100]，比氢氧化物或硅酸盐激发剂成本更低、更环保。与许多碱激发的胶凝材料体系相比，该胶凝材料的使用促进了较低 pH 值的发展，在职业健康和安全考虑方面是有益的。在处理核废料中重要的活性金属固化等应用也可以受益于对腐蚀过程的敏感性降低[64]。

然而，对碳酸盐激发矿渣结构发展的认识是非常有限的，而且与其他激发的矿渣体系相比，碳酸盐激发的胶凝材料未能吸引到更多学术界和工业界的注意，因为当与 NaOH 或硅酸钠激发剂相比时，其通常延迟硬化与强度发展[59,101,102]。将细研磨的石灰石掺入 Na_2CO_3 激发的胶凝材料中还可以再提高性能并降低成本[103,104]，与硅酸盐水泥相比，可将温室排放减少97%的胶凝材料已经开发，3d强度超过40MPa。

研究表明，碳酸钠激发[24,33,54,58,59,101]矿渣比NaOH激发矿渣具有更高的力学强度，但低于硅酸盐激发矿渣，并且会延长凝结时间。Małolepszy[105]确定Na_2CO_3激发对含有镁黄长石的矿渣更有效，而当使用含有钙黄长石的矿渣时，用NaOH激发更合适。Wang等人[53]报道当激发剂与矿渣一起研磨或加入溶液中时，Na_2CO_3激发的矿渣混凝土中的力学强度相似，Collins和Sanjayan[106,107]也观察到$NaOH/Na_2CO_3$激发的矿渣胶凝材料与具有相近胶凝材料含量的普通硅酸盐水泥以相似的速率发展早期力学性能。在该研究中观察到，在胶凝材料配比中使用更高的Na_2CO_3、NaOH比率会提高和易性。

在反应的早期阶段，Na_2CO_3激发矿渣形成碳酸钙和碳酸钠/碳酸钙复盐，这是来自激发剂的CO_3^{2-}与来自溶解矿渣的Ca^{2+}相互作用的结果。在更长的养护时间内形成Al取代的C-S-H型凝胶，其能促进浆体的硬化[102]。在Na_2CO_3激发矿渣中形成的C-A-S-H具有包括Q^3位点的高度交联的结构，并且与碳酸盐平行形成，包括单斜钠钙石、$Na_2Ca(CO_3)_2 \cdot 5H_2O$[24]。Xu等人[100]评估用$Na_2CO_3$和$Na_2CO_3/NaOH$共混物激发不同龄期的矿渣混凝土，其在数年至数十年的期间内持续获得强度。研究表明，此反应的主要反应产物为具有相对低的Ca/Si比、高度交联的C-S-H型相的外部产物和包含碳酸根阴离子的内部产物。在Na_2CO_3激发的矿渣胶凝材料中，$CaCO_3$在Na_2CO_3提供缓冲的碱性环境下逐渐溶解维持体系中CO_3^{2-}的水平，同时释放Ca与来自矿渣溶解的硅酸盐反应，从而形成了循环水化的反应过程[100]。然而，还没有详细的证据表明这种机制在Na_2CO_3激发胶凝材料第一个月的反应中是如何进行的。

3.3.4 胶凝体系——硫酸盐激发

石灰-矿渣水泥是由矿渣制成的最早的水泥材料，在德国首先生产并随后传播到许多其他国家，但与现代硅酸盐水泥相比，由于其凝结时间长、早期强度低和储存快速劣化而失去了普及性[108]。然而，这种类型的水泥具有优异的耐硫酸盐地下水，并且与其强度发展相关的一些缺点已经通过使用碱激发剂例如硫酸钠来克服，如图3-10所示[46]，通过使用复合Na_2SO_4-硅酸钠-$Ca(OH)_2$激发剂[109]，56d后的抗压强度高达60MPa。

这些胶凝材料的结构表征表明，片状C-A-S-H以及C_4AH_{13}作为无硫酸钠的石灰-矿渣水泥中的主要反应产物，而在胶凝材料中包含的Na_2SO_4促进了钙矾石的形成，牺牲了C_4AH_{13}和AFm型相[46]。在不含Na_2SO_4的石灰-矿渣体系中，在养护的早期有利于水化钙（铝）硅酸盐的形成，并且这些水化产物覆盖矿渣颗粒的表面。向石灰-矿渣胶凝材料中加入Na_2SO_4加速了矿渣的溶解，但似乎延迟了$Ca(OH)_2$的溶解[46]。Na_2SO_4的存在加速了早期水化，由矿渣的初始溶解控制，在矿渣颗粒周围形成更广泛的C-S-H产物以及AFt针。这些因素有助于Na_2SO_4-石灰-矿渣浆体的早期强度。

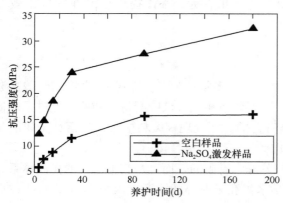

图3-10 与空白样品对照，在50℃养护的硫酸钠激发石灰-矿渣水泥的强度发展（源自文献[46]）

然而，在 Na_2SO_4 中的矿渣颗粒周围形成更多的 C-S-H 产物浆体抑制后来的水化，其通过水由水化层扩散到未反应的核心矿渣颗粒控制[46]。

以煅烧黏土或天然火山灰为铝硅酸盐原料，掺加少量的石灰或硅酸盐水泥熟料，这类使用硫酸盐激发制备的胶凝材料具有很好的性能[66,110-113]，这样的材料将在讨论共混胶凝材料体系的第5章中得到更详细的讨论。

3.4 孔溶液化学

对于任何用于钢筋混凝土生产的胶凝材料，保持孔溶液 pH 值在足够高的水平下使嵌入钢筋钝化，相当重要。几乎所有的碱激发胶凝体系都要考虑最初的高 pH 值，表面上看起来，这应该不用考虑太多。然而，缺乏 pH 值缓冲相，如硅酸盐水泥硬化胶凝材料中的氢氧化钙，意味着高碱度的保留，也可以防止外部酸性组分侵入或孔溶液碱浸出[114]。也就意味着碱激发胶凝材料的孔结构是决定其耐久性的关键因素。耐久性测试将在本书的第 8、9 和 10 章作详细论述。

碱激发矿渣胶凝材料的孔溶液化学是由 pH 值一般在 13 至 14 的碱金属氢氧化物控制的，成熟的胶凝材料中的溶解态硅酸盐的浓度降至 <10mm 水平[88,114-117]。Puertas 等人[117] 在养护 7d 过程中，于高压下提取孔溶液，并观察到溶解的钠离子和二氧化硅的浓度都降低了。Lloyd 等[114]研究用更高活性的激发剂养护 28d 之后的样品，发现二氧化硅浓度低于 Puertas 等人[117]的。可观察到文献[114]中铝浓度<1mm，而文献[117]中大多数的样品为 5mm 以下，相比可知更高浓度激发剂的使用提高了 Al 的嵌入量。类似的趋势在 Faucon 等人[118]的纯 C-S-H 样品中也可观察到。铝浓度在碱硅酸盐激发和氢氧化钠激发矿渣孔溶液中随着时间的推移而减少[88,116]。Gruskovnjak 等人[88]观察到 1d 浓度高达 7mm，但 180d 后减少到 3mm。

矿渣基胶凝材料孔溶液化学主要与材料内部的氧化还原环境有关。其中矿渣中的硫化物含量会使胶凝体系处于强还原态，氧化还原电势值达到 $-400mV$（图 3-11）[119]。硫化物和硫酸盐之间的平衡取决于矿渣的风化程度和氧化程度，这发生于材料作用期间。Gruskovnjak 等人[88]发现在密封条件下，在第一个 180d 钠硅酸激发矿渣加工处理中硫酸的浓度大致保持 40mm，而硫化物浓度 7d 后几乎可达到 10 倍高（350mm），在孔溶液中显现出强还原环境。X 光吸收光谱也表明[120]，较弱的碱性环境使矿渣-碱体系中硫的还原形式更稳定，这对于确定孔溶液和碱激发剂中任何嵌入钢筋元素之间的关系很重要，但无疑是一个需要进一步详细研究的领域。

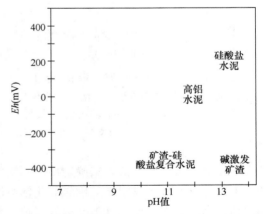

图 3-11 各种类型用作水泥的胶凝材料的孔溶液 pH-Eh 关系图（源自文献[119]）

3.5 BFS 性能的影响

3.5.1 细度

基于颗粒-流体的反应过程分析,矿渣细度是影响碱激发剂的反应[121]、凝结、强度发展、最终的微观结构的关键因素。Wang 等人[53]和 Puertas[18]表明,对于 AAM 的生产,矿渣的最佳细度范围为 $400\sim550m^2/kg$ 之间。Brough 和 Atkinson[25]发现,增加矿渣的细度可以促进抗压强度的提高,这与更小粒度会增强材料的反应相一致。

然而,凝结速率会受到矿渣细度的强烈影响。Talling 和 Brandstetr[122]报道,磨细矿渣的细度超出 $450m^2/kg$,凝结时间在 $1\sim3min$,这意味着这种材料的工作性很差。相反,Collins 和 Sanjayan[123]认定,部分(10%)常规矿渣由超细矿渣($1500m^2/kg$)替代只是略微减少 AAS 混凝土的和易性,但显著增加力学强度,所以一天养护混凝土的抗压强度达 20MPa。Lim 等人[124]使用 $1500m^2/kg$ "纳米级矿渣"作为氢氧化钠激发混凝土的唯一硅酸盐前驱体,与对比硅酸盐水泥混凝土相比,获得类似的凝结速度、强度发展和半绝热温度曲线,并显著降低矿渣基材料的二氧化碳排放量,尽管由于超细矿粉的能源成本显然较高。与Taylor 等人的报道测量矿渣反应程度为 22% 的粗矿渣经过水中 20 年的水化相比较[41],Kumar 等人[125]的报告表明矿渣的反应性可以通过使用破碎机机械活化来提高。这样机械处理矿渣可以在缺乏碱激发剂的条件下几天后便可与水完全反应。超细矿渣在水中形成的主要水化反应产物是水化硅酸钙型产物,即使起始矿渣包含质量分数为 8.8% 的 MgO,但并没有观察到铝碳酸镁的形成。水化 BFS 胶凝相的孔隙形成是由于材料的比表面积以及球磨过程中的 zeta 电势。矿渣水化随着时间的推移 pH 值会显著增加,如图 3-12 所示,有利于原材料的溶解和反应产物的形成。

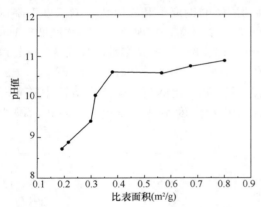

图 3-12 在水化磨细矿渣的溶液中 pH 值的变化(100g 机械处理的矿渣浸入带有水的玻璃管中),经过 14d 的反应之后,作为比表面积的函数(源自文献[125])

Shi 和 Day[55]评估了标准测试方法 ASTMC 1073[126]矿渣+氢氧化钠,在 50℃ 养护 24h 作为质量控制方法,来确定使用不同类型的矿渣为前驱体生产碱激发胶凝材料的适用性。考虑到这种方法仅限于使用氢氧化钠溶液,不是普遍适用于作为碱激发剂前驱体的矿渣的评估,当材料专门使用其他碱激发可提高性能。因此,建议改进质量控制测试方法的适用性,碱的选择应基于最优激发剂测试而不是指定氢氧化钠作为唯一的激发剂。

3.5.2 BFS 化学

虽然可以通过不同的冶金过程激发矿渣(基于此类矿渣的一些结果将在下一部分讨论),球磨粒状高炉矿渣(BFS)在碱激发材料生产中的使用最广泛[18,46,53,122],因为它的化学成分和非晶型性质有利于反应产物的形成，以在适度的养护周期及相对较低的需水量下形成高强度[50]。矿渣的化学成分通常为四元体系 CaO-MgO-Al_2O_3-SiO_2，其他组分包括锰、硫和钛，这取决于使用的铁矿石的化学成分；缓慢冷却矿渣往往是晶体，没有反应活性，但许多快速冷却（粒状）矿渣也可以包含晶态杂质。矿渣中夹杂物的性质和含量被炉渣成分（液相线温度）和加工条件所影响，其主要在低含量 Al_2O_3（10%～15% Al_2O_3）体系中受 CaO 和 MgO 含量所影响[127]。

矿渣的水硬性通过碱度系数（CaO＋MgO）/（SiO_2＋Al_2O_3）和质量系数（CaO＋MgO＋Al_2O_3）/（SiO_2＋TiO_2）[53,122] 等参数进行测量，另外也有在特定的情况下新开发和应用各种替代的关系式[6,46]。然而，碱激发剂的力学性能与这些参数之间的相关性并不总是很好。一般来说，CaO/SiO_2 比值在 0.50～2.0 之间和 Al_2O_3/SiO_2 比在 0.1～0.6 之间的玻璃矿渣，被认为最适合用于碱激发[122]。不同来源的矿渣材料之间的化学和矿物学差异限制了对 AAM 结构发展机制的完全理解。

Douglas 等人[109] 报道了使用 MgO 含量（9wt%、12wt%、18wt%）不同的硅酸盐激发矿渣的 28d 抗压强度，MgO 质量分数为 18% 的矿渣的抗压强度比 MgO 含量为 9% 的激发矿渣高 3 倍。这符合最近由 Ben Haha 等人研究的结果[32]，他们使用 MgO 含量在 8%～13% 的激发矿渣，被认定的是，增加 MgO 的含量使得反应更快、抗压强度更高，因为有更多量的铝碳酸镁类产物形成。这降低了当硅酸钠用作激发剂时形成的 C-S-H 型凝胶中的 Al 掺入程度。然而，这并非 NaOH 激发矿渣胶凝材料的情况，MgO 仅观察到轻微的强度变化。

在高炉矿渣有不同的氧化铝含量（7wt%～17wt%），观察[47] 到增加矿渣中氧化铝的含量会降低反应的早期养护程度，从而降低了激发态矿渣胶凝材料的抗压强度。这与在具有较高含量的 Al_2O_3 的激发矿渣胶粘剂中形成的水滑石中的 Mg/Al 比率的减少以及在 C（A）-S-H 型产物中增加 Al 的掺入导致形成晶体相关。然而，在 28d 的养护后，具有不同 Al_2O_3 含量的炉渣的碱激发矿渣胶凝材料之间没有显著的结构或力学差异[47]。

3.6 C-S-H 中的碱（包括碱激发硅酸盐水泥）

碱在矿渣和铝硅酸盐前驱体的碱激发期间形成的产物的性质和结构中起关键作用[46]。许多作者在合成凝胶和"真实"凝胶体系中研究了碱性氧化物（特别是 Na_2O）对硅酸钙水化物凝胶的组成和微观结构的影响。本节将简要讨论这一领域的现有结果，其中一篇综述涵盖了至 1990 年的文献[128]。在 20 世纪 40 年代，Kalousek[129] 确定了可以加入到 C-S-H 型凝胶中的 Na 的最大含量是 Na_2O/SiO_2 比为 0.25，形成 $Na_2O \cdot CaO \cdot SiO_2 \cdot xH_2O$（NCSH）型结构。已观察到少量 Na（甚至在总凝胶质量的 0.6% Na_2O 的量级）的存在会影响硅酸盐和水化硅酸盐的稳定性以及它们的水化速率[130]。

通常一致认为，碱进入 C-S-H 凝胶的程度随着凝胶中 Ca/Si 比的降低而增加。Taylor[131] 报道了 Na/Ca 比例约为 0.01，在水化硅酸盐水泥浆中，其 Ca/Si 比率在 1.3～2.3 之间，而 Hong 和 Glasser[132] 确定了 C-S-H 凝胶中 Na/Si 和 Ca/Si 比之间的负线性关系，在 0.85＜Ca/Si＜1.8 的范围内。

如 Stade[133] 所述，将碱结合到 C-S-H 型凝胶中的过程由三个主要步骤控制：
（1）中和 Si—OH 酸基团（在具有低 Ca/Si 比的 C-S-H 型凝胶中容易发生）。

(2) 交换 M^+ 和 Ca^{2+} 阳离子（特别是在具有高 Ca/Si 比的凝胶中）。

(3) 由于碱的存在（取决于凝胶的比表面积），Si—O—Si 键的最终溶解。

由于体系的复杂性，可能在任何时间的多种机制影响着碱离子的摄取。

Atkins 等人[134]通过 ^{29}Si MAS NMR 光谱测定，C-S-H 型凝胶中的 Q^2/Q^1 比在体系中 Na 含量增加时下降，表明碱的存在降低了所形成的 C-S-H 凝胶中的链长度，与上述第三步骤一致。Chen 和 Brouwers[135]评估了大量关于 C-S-H 中碱结合能力的文献数据，发现 Na^+ 的结合与溶液中的碱浓度呈线性比例。发现 K^+ 的数据更加分散，并且建议通过 Freundlich 型（幂律）等温线近似拟合，尽管具有低相关系数[135]。

已观察到，与仅基于纯 C-S-H 型产物的胶凝材料相比，将 Al 掺入到 C-S-H 型凝胶中，导致 C-A-S-H 型产物的形成增强了体系中碱的吸收，这与 C-A-S-H 凝胶中的 Ca/(Si + Al) 比降低一致[136]。碱结合通过价态补偿机制发生，其中通过将三价四面体 Al 置换为四面体 Si 位点而产生的电荷不平衡，通过加入碱来平衡。还提出了由于与通过包含四面体 Al 改性骨架增强的连接而增强的硅烷醇结合能力（如对于 C-S-H 型产品所述）的可能性[136]。Skibsted 和 Andersen[137]确定了在 C-S-H 中引入的 Al 的量也部分与体系中碱离子的可用性相关，因为碱作为电荷平衡器（吸附/键合）在桥接 AlO_4 位点附近的 C-S-H 相的中间层被取代成硅酸盐链。

最近使用合成凝胶体系进行的研究表明，由于解聚的 C-N-S-H 型凝胶的形成，高浓度的碱在短反应时间（72h）引起新鲜 C-S-H 凝胶的劣化。碱的包含也受所形成的凝胶的结晶度的影响；在高度结晶的凝胶中，可能的碱位点的数量有限，而在无序体系中，碱的结合将主要受电中性的约束。

该温度还对 C-S-H 型凝胶结构中的碱掺入具有重要的影响。对于给定的 Ca/Si 比，在高温下形成的凝胶更易结晶并具有更小的比表面积。因此，它们可以比在较低温度下形成的凝胶捕获更少的碱。在水热条件下合成的凝胶中，观察到了毛状体的形成[140,141]。该相是稳定的硅酸钙，其可与在硅酸盐水泥水化中形成的其他硅酸钙如托勃莫来石和（在较高温度下）硬硅钙石共存。一项比较研究评估了不同碱金属阳离子的影响，报告称在水化水泥浆中钠的结合能力明显比钾更强[131]。注意到这是与碱金属铝硅酸盐地质聚合物凝胶中所述的效果相反的，其中 K 优先于 Na 与凝胶结合[142]。

因此，显然在富含 Ca 的 AAM 中，碱可以以下几种形式存在[50]：

(1) 并入 C-S-H；

(2) 物理吸附在水化产物表面；

(3) 在孔溶液中游离。

在这些体系中，预计在反应的初始阶段，当可用于反应的液相中的有效 Ca/Si 比非常低，且反应区中的 Na/Ca 比非常高时，可在 C-S-H 型凝胶中引入碱。然而，考虑到 Ca/Si 比率和凝胶的结晶度将随时间和体系组成而变化，这种机理将取决于所使用的激发剂的类型。虽然 NaOH 激发矿渣可能形成具有比使用硅酸盐基激发剂时更高的 Ca/Si 比的 C-S-H 型凝胶[37]，但是预期在那些凝胶中碱的吸收仍将远高于在常规硅酸盐水泥体系中形成的 C-S-H 凝胶。

3.7 非高炉矿渣前驱体

3.7.1 钢渣

除了在从其矿石中提取铁期间产生的矿渣之外,将铁转化为钢的过程中还会产生各种类型的矿渣。由于组成取决于具体的工艺路线以及再循环材料掺入炼钢工艺的程度,这些钢渣的化学成分变化更大,不同工厂组成会有变化,甚至在同一工厂不同批次组成也会有变化。钢渣类型主要包括电弧炉(EAF)钢渣、碱氧气炉(BOF)钢渣、钢包钢渣和转炉钢渣。由合金或不锈钢的生产产生的钢渣在组成上非常不同,与大多数其他钢渣相比,呈现较低的 FeO 和较高的 Cr 含量。这是这些钢渣在某些管辖区被归类为危险废物的主要原因。

钢渣中的主要矿物相包括橄榄石、钼铁矿、C_3S、β-C_2S、γ-C_2S、C_4AF、C_2F、"RO"相(CaO-FeO-MnO-MgO 固溶体)、游离 CaO 和游离 MgO[143,144]。这些矿渣中特别令人感兴趣的相是 γ-C_2S,其在冷却期间从 β-C_2S 中获得。β-C_2S 至 γ-C_2S 的转化伴随着体积增加接近 10%,并且导致晶体粉碎成粉尘,使得材料有效地自粉碎,因此显著降低了在研磨过程方面的成本。然而,γ-C_2S 不是水硬性的,因此对水泥状胶凝材料形成的反应过程没有显著贡献。随着矿渣的碱性增加,C_3S、C_2S、C_4AF 和 C_2F 的存在确实为这些矿渣提供了一些弱的水泥性质[46]。

基于钢铁渣混合物的水泥,主要由钢渣、矿渣、水泥熟料和石膏组成,另外添加或不添加碱激发剂,这种水泥已在中国市场销售了二十多年[145]。

这些碱激发胶凝材料的一些应用将在第 4 章描述。钢渣基水泥的强度取决于钢渣的碱度,这些材料在各种中国规定性标准下已被批准用于一般建筑应用[46]。

钢渣本身在适当的碱激发剂的作用下显示出非常好的胶凝性能。几个研究已经证实,使用碱激发剂可以增加钢渣水泥的早期强度和其他性能[146-148]。通常,一些其他材料如矿渣或粉煤灰与钢渣一起使用,以消除与尺寸稳定性相关的问题。碱激发钢渣-矿渣水泥可以表现出非常高的强度和腐蚀阻力[46,146,149-152],已经发现矿渣和高碱性钢渣的组合赋予尺寸稳定性和受控热量的良好平衡,其受到钢渣中存在的一些含钙结晶相的影响[153]。

对于这种胶凝材料,蒸汽养护后强度高达 120MPa,标准室温条件下 3d 强度为 62MPa,28d 强度为 96MPa。偏高岭土和钢渣的碱激发胶凝材料也表现出良好的性能,作为粘结砂浆夹芯材料[144];基于钢渣和偏高岭土的 AAM 也显示出很好的强度发展、微结构性能和耐高温性[154,155]。

3.7.2 磷渣

从磷矿石生产元素磷也产生富含硅酸钙的炉渣称为磷渣,其在世界范围内的产生量已超过 500 万 t。在北美,过去几十年中产生的大部分磷渣(其中一些已在建筑中作为骨料重复使用)含有较高含量的天然放射性物质,因此产生的辐射高于监管限制水平,这限制了对其的再利用。然而,中国和俄罗斯的磷渣具有较低的辐射水平,并且被广泛用作水泥掺合物,这促进了使用粒状磷渣作为辅助水泥材料的标准开发并用于生产硅酸盐-粒状磷矿渣水泥[14]。

Shi[156,157]开始了最早的关于碱激发磷渣胶凝材料的学术研究。据报道,硅酸钠激发剂溶液的模数或剂量的增加导致凝固时间缩短,并且包含在熔渣中的可溶性磷对胶凝材料的凝固没有任何显著影响。硅酸钠激发剂的最佳溶液模数在 1.2~1.5 的范围内,形成具有良好工作性的更高的强度发展。在此条件下,固化过程中温度敏感性较高的砂浆样品[156,157]的 28d 抗压强度仍可达到 120MPa。在激发磷渣体系中,使用 NaOH 作为激发剂能促进早期强度发展,而基于硅酸盐的激发剂在较长的养护时间产生较高的强度发展[14]。Fang 等人[158]还开发了基于使用硅酸盐激发的磷渣粉煤灰复合 AAM 胶凝材料,在砂浆中的 28d 抗压强度高达 98MPa,并且与硅酸盐水泥相比具有更高的耐腐蚀性和抗冻融性,但干燥收缩较高。

3.7.3 其他冶金渣

在世界范围内有许多其他的火法冶金工艺生产有色金属(特别是 Pb、Zn、Ni 和 Cu),这导致产生各种类型的硅酸盐基矿渣。约有一半的有色金属矿渣是镍渣,约三分之一是铜渣[46]。这些矿渣通常富含 SiO_2 和 Fe 氧化物,通常具有 5%~15% 的 Al_2O_3 和 CaO,加上少量硫、MgO、Cr_2O_3 和其他氧化物,这取决于矿石的组成和加工工艺[46,159,160]。这些矿渣通常也主要是玻璃质的,其中少量结晶组分有时富含 Fe,并倾向于显示火山灰性质。这意味着它们至少在一定程度上潜在适用于碱激发。已经公开了对有色金属矿渣的碱激发的各种研究,但是大部分工作的焦点在于,使用在特定区域中可用的特定矿渣的配合比设计。此类工作已经出版,其中包括:

(1) 铜渣[14,161];
(2) 镍渣[162];
(3) 氧化镁-铁渣[163];
(4) 磷渣和铬铁合金渣[164];
(5) 含钛矿渣;
(6) Cu-Ni 熔渣[166]。

由 Komnitsas 领导的研究项目已经开发了一系列基于低 Ca 铁镍渣可行的高强度 AAM 配方[167-170],将在本报告的 4.6 节中进行更为详细的描述。Pontikes 等人[171]最近还公布了来自中试规模等离子体反应器的富含 Fe、Al 合成渣体系的研究,以期开发出将这些材料用作 AAM 前驱体所需的科学基础。该研究的关键发现是,对于给定的矿渣组成,水冷淬火提供了一种产品,其在与空气冷却或组合空气—水冷却相比,在碱激发环境中更具反应性[171]。

还提出了粒状有色金属的水化活性可以使用质量系数(式 3-1)来评估矿渣中铁的氧化态的贡献[46],并且 K 值越高碱激发的适应性越好。

$$K = \frac{CaO + MgO + Al_2O_3 + Fe_2O_3 + 1/2FeO}{SiO_2 + 1/2FeO} \tag{3-1}$$

这个概念已被用于根据它们反应性的可能开发有色金属渣的分级体系,并因此在乌克兰和俄罗斯制定基于这些胶凝材料的 AAM 混凝土的一套标准[46]。

3.7.4 底灰和市政固废焚烧灰

从燃煤锅炉获得的底灰也能够用于碱激发,这些比飞灰更粗糙,因此在碱激发之前需要研磨[172,173]。在建筑应用中使用这种材料的潜在问题是底灰中积累的重金属[172]。来自流化床煤燃烧的底灰在化学和矿物学性质上是类似的,并且还需要在用于碱激发之前从最初相当大的粒度（高达几个毫米）进行研磨。其热历程不同于大多数煤灰,因为在标准锅炉中燃烧温度仅为 800~900℃,而不是大于 1300℃,这意味着不会普遍形成莫来石。然而,一旦研磨,其高的 Si 和 Al 含量非常适用于作为 AAMs 的原材料[174-176],尽管在建筑应用中需要考虑其潜在的高重金属含量[175]。

城市固体废物焚烧（MSWI）灰略有火山灰活性,能够在碱激发期间反应,但也需要仔细考虑其有高含量的毒金属和氯化物。因此其在 AAM 中更通常是用于废物处理和防止浸出的稳定化/固化过程,而不是用作真正的铝硅酸盐 AAM 前驱体[177-179]。

3.8 小结

本章总结了富 Ca 碱激发胶凝材料的基本化学体系,在激发剂选择、前驱体化学和多种范围的 Ca-铝硅酸盐前驱体的可用性方面,超过应用最广泛的基于矿渣的胶凝材料体系。这是一个需要进一步开发和优化的领域,因为用于 AAM 生产的天然及工业副产物材料的范围的扩大,肯定需要鉴定和优化每种前驱体的最佳的激发剂化学（或前驱体的共混物）。到目前为止,该领域已经发表的大部分工作集中在每项研究的单一前驱体,适用范围有限。资源之间的矿渣化学变异性的影响尚未被特别好地理解,为这一领域的研究提供了强大的动力。富钙的 AAM 体系是世界上大多数地区最成熟的商业应用 AAM 材料体系,因为钙在增强 AAM 混凝土的耐久性等相关性质中的作用开始被很好地理解。然而,商业开发和先进研究之间的持续相互作用,对于使这些材料达到国家和国际标准化所需的真正技术成熟来说似乎至关重要。

参考文献

[1] Kühl H.. Slag cement and process of making the same. Compiler：U. S, 900939[P]. 1908.

[2] Purdon A. O.. Improvements in processes of manufacturing cement, mortars and concretes. Compiler：British. GB427227[P]. 1935.

[3] Purdon A. O.. The action of alkalis on blast-furnace slag. J. Soc. Chem. Ind. Trans. Commun. 1940, 59：191-202.

[4] Van Deventer J. S. J., Provis J. L., Duxson P., Brice D. G.. Chemical research and climate change as drivers in the commercial adoption of alkali activated materials. Waste Biomass Valoriz, 2010, 1(1)：145-155.

[5] Van Deventer J. S. J., Provis J. L., Duxson, P.. Technical and commercial progress in the adoption of geopolymer cement. Miner. Eng, 2012, 29：89-104.

[6] Talling B., Krivenko P. V.. Blast furnace slag-the ultimate binder. In：Chandra, S. (ed.) Waste Materials Used in Concrete Manufacturing, 1997：235-289. Noyes Publications, Park Ridge.

[7] Krivenko P. V.. Alkaline cements-from research to application. In：Lukey, G. C. (ed.) Geopolymers 2002. Turn Potential into Profit, Melbourne. CD-ROM Proceed-

ings. Siloxo Pty. Ltd. 2002.

[8] Wang S. D., Pu X. C., Scrivener K. L., Pratt P. L.. Alkali-activated slag cement and concrete: a review of properties and problems. Adv. Cem. Res, 1995, 7(27): 93-102.

[9] Richardson I. G.. The nature of C-S-H in hardened cements. Cem. Concr. Res, 1999, 29: 1131-1147.

[10] Juenger M. C. G., Winnefeld F., Provis J. L.. Ideker J.. Advances in alternative cementitious binders. Cem. Concr. Res, 2011, 41(12): 1232-1243.

[11] Yang N.. Physical chemistry basis for the formation of alkali activated materials I- J. Chin. Ceram. Soc. 1996, 24(2): 209-215.

[12] Yang N.. Physical chemistry basis for the formation of alkali activated materials-II. J. Chin. Ceram. Soc. 1996, 24(4): 459-465

[13] Li C., Sun H., Li L.. A review: The comparison between alkali-activated slag(Si + Ca) and metakaolin(Si + Al) cements. Cem. Concr. Res. 2010, 40(9): 1341-1349.

[14] Shi C., Qian J.. High performance cementing materials from industrial slags-a review. Resour. Conserv. Recycl, 2000, 29: 195-207.

[15] Wang S. D.. Review of recent research on alkali-activated concrete in China. Mag. Concr. Res. 1991, 43(154): 29-35.

[16] Krivenko P. V.. Alkaline cements. In: Krivenko P. V. (ed.) Proceedings of the First International Conference on Alkaline Cements and Concretes, Kiev, Ukraine. 1994, 1: 1. 11-129. VIPOL Stock Company.

[17] Krivenko P. V.. Alkaline cements: structure, properties, aspects of durability. In: Krivenko P. V. (ed.) Proceedings of the Second International Conference on Alkaline Cements and Concretes, Kiev, Ukraine, 1999: 3-43. ORANTA.

[18] Puertas F.. Cementos de escoria activados alcalinamente: situación actualy perspectivas defuturo. Mater. Constr, 1995, 45(239): 53-64.

[19] Fernández-Jiménez A., Puertas F.. Alkali-activated slag cements: kinetic studies. Cem. Concr. Res, 1997, 27(3): 359-368.

[20] Fernández-Jiménez A., Puertas F., Arteaga A.. Determination of kinetic equations of alkaline activation of blast furnace slag by means of calorimetric data. J. Therm. Anal. Calorim, 1998, 52(3): 945-955.

[21] Bernal S. A., Provis J. L., De Mejía Gutierrez R., Rose, V.. Evolution of binder structure in sodium silicate-activated slagmetakaolin blends. Cem. Concr. Compos. 2011, 33(1): 46-54.

[22] Zhou H., Wu X., Xu Z., Tang, M.. Kinetic study on hydration of alkali-activated slag. Cem. Concr. Res, 1993, 23(6): 1253-1258.

[23] Escalante-Garcia J., Fuentes A. F., Gorokhovsky A., Fraire-Luna P. E., Mendoza-Suarez G.. Hydration products and reactivity of blast-furnace slag activated by various alkalis. J. Am. Ceram. Soc, 2003, 86(12): 2148-2153.

[24] Fernández-Jiménez A., Puertas F., Sobrados I., Sanz, J.. Structure of calcium sili-

cate hydrates formed in alkaline-activated slag: influence of the type of alkaline activator. J. Am. Ceram. Soc. 2003, 86(8): 1389-1394.

[25] Brough A. R. Atkinson A.. Sodium silicate-based, alkali-activated slag mortars: Part I. Strength, hydration and microstructure. Cem. Concr. Res, 2002, 32(6): 865-879.

[26] Richardson I. G., Groves G. W.. Microstructure and microanalysis of hardened cement pastes involving ground granulated blast-furnace slag. J. Mater. Sci. 1992, 27(22): 6204-6212.

[27] Myers R. J., Bernal S. A., San Nicolas R., Provis J. L.. Generalized structural description of calcium-sodium aluminosilicate hydrate gels: the cross linked substituted tobermorite model. Langmuir, 2013, 29(17): 5294-5306.

[28] Schilling P. J., Butler L. G., Roy A., Eaton H. C.. 29Si and 27 Al MAS-NMR of NaOH-activated blast-furnace slag. J. Am. Ceram. Soc, 1994, 77(9): 2363-2368.

[29] Bonk F., Schneider J., Cincotto M. A., Panepucci H.. Characterization by multinuclear high-resolution NMR of hydration products in activated blast-furnace slag pastes. J. Am. Ceram. Soc, 2003, 86(10), 1712-1719.

[30] Lothenbach B., Gruskovnjak, A.. Hydration of alkali-activated slag: thermodynamic modelling. Adv. Cem. Res, 2007, 19(2): 81-92.

[31] Chen W., Brouwers H.. The hydration of slag, part 1: reaction models for alkali-activated slag. J. Mater. Sci. 2007, 42(2): 428-443.

[32] Ben Haha M., Lothenbach B., Le Saout G., Winnefeld F.. Influence of slag chemistry on the hydration of alkali-activated blast-furnace slag-Part I: effect of MgO. Cem. Concr. Res, 2011, 41(9): 955-963.

[33] Fernández-Jiménez A., Puertas F.. Effect of activator mix on the hydration and strength behaviour of alkali-activated slag cements. Adv. Cem. Res. 2003, 15(3): 129-136.

[34] Bernal S. A., De Mejía Gutierrez R., Rose V., Provis J. L.. Effect of silicate modulus and metakaolin incorporation on the carbonation of alkali silicate-activated slags. Cem. Concr. Res, 2010, 40(6): 898-907.

[35] Zhang Y. J., Zhao Y. L., Li H. H., Xu D. L.. Structure characterization of hydration products generated by alkaline activation of granulated blast furnace slag. J. Mater. Sci, 2008, 43: 7141-7147.

[36] Puertas F., Palacios M., Manzano H., Dolado J. S., Rico A., Rodríguez J.. A model for the C-A-S-H gel formed in alkali-activated slag cements. J. Eur. Ceram. Soc. 2011, 31(12): 2043-2056.

[37] Ben Haha M., Le Saout G., Winnefeld F., Lothenbach B.. Influence of activator type on hydration kinetics, hydrate assemblage and microstructural development of alkali activated blast-furnace slags. Cem. Concr. Res, 2011, 41(3): 301-310.

[38] San Nicolas R., Provis J. L.. Interfacial transition zone in alkali-activated slag concrete. In: Twelfth International Conference on Recent Advances in Concrete Technology and Sustainability Issues, Prague, Czech Republic. Supplementary Papers CD-

ROM. American Concrete Institute, Detroit, 2012.

[39] Bernal S. A., San Nicolas R., Provis J. L., Mejía de Gutiérrez R., Van Deventer J. S. J.. Natural carbonation of aged alkali-activated slag concretes. Mater. Struct. (2013, in press). doi: 10.1617/s11527-013-0089-2.

[40] Bernal S. A., Provis J. L., Rose V., De Mejía Gutiérrez R.. High-resolution X-ray diffraction and fluorescence microscopy characterization of alkali-activated slag-metakaolin binders. J. Am. Ceram. Soc. 2013, 96(6): 1951-1957.

[41] Taylor R., Richardson I. G., Brydson R. M. D.. Composition and microstructure of 20-year-old ordinary Portland cement-ground granulated blast-furnace slag blends containing 0 to 100% slag. Cem. Concr. Res, 2010, 40(7): 971-983.

[42] Chen X., Pochard I., Nonat A.. Thermodynamic and structural study of the substitution of Si by Al in C-S-H. In: Beaudoin J. J. (ed.) 12th International Congress on the Chemistry of Cement, Montreal. CD-ROM proceedings, 2007.

[43] Pardal X., Pochard I., Nonat, A.. Experimental study of Si-Al substitution in calcium-silicate-hydrate(C-S-H) prepared under equilibrium conditions. Cem. Concr. Res, 2009, 39: 637-643.

[44] Richardson I. G., Brough A. R., Brydson R., Groves G. W., Dobson C. M.. Location of aluminum in substituted calcium silicate hydrate(C-S-H) gels as determined by 29 Si and 27 Al NMR and EELS. J. Am. Ceram. Soc, 1993, 76(9): 2285-2288.

[45] Sun G. K., Young J. F., Kirkpatrick R. J.. The role of Al in C-S-H: NMR, XRD, and compositional results for precipitated samples. Cem. Concr. Res, 2006, 36(1): 18-29.

[46] Shi C., Krivenko P. V., Roy D. M.. Alkali-Activated Cements and Concretes. Taylor & Francis, Abingdon, 2006.

[47] Ben Haha M., Lothenbach B., Le Saout G., Winnefeld F.. Influence of slag chemistry on the hydration of alkali-activated blast-furnace slag-Part II: effect of Al_2O_3. Cem. Concr. Res, 2012, 42(1): 74-83.

[48] Nocu -Wczelik W.. Heat evolution in alkali activated synthetic slag metakaolin mixtures. J. Therm. Anal. Calorim. 2006, 86(3): 739-743.

[49] Fernandez-Jimenez A., Puertas F., Arteaga A.. Determination of kinetic equations of alkaline activation of blast furnace slag by means of calorimetric data. J. Therm. Anal. Calorim. 1998, 52(3): 945-955.

[50] Shi C.. On the state and role of alkalis during the activation of alkali-activated slag cement. In: Grieve G., Owens G. (eds.) Proceedings of the 11th International Congress on the Chemistry of Cement, Durban, South Africa. Tech Books International, New Delhi, India, 2003: 2097-2105.

[51] Shi C., Day R. L.. A calorimetric study of early hydration of alkalislag cements. Cem. Concr. Res, 1995, 25(6): 1333-1346.

[52] Shi C., Day R. L.. Some factors affecting early hydration of alkalislag cemen-

ts. Cem. Concr. Res, 1996, 26(3): 439-447.

[53] Wang S. D., Scrivener K. L., Pratt, P. L.. Factors affecting the strength of alkali-activated slag. Cem. Concr. Res, 1994, 24(6): 1033-1043.

[54] Živica V.. Effects of type and dosage of alkaline activator and temperature on the properties of alkali-activated slag mixtures. Constr. Build. Mater. 2007, 21(7): 1463-1469.

[55] Shi C., Day R. L.. Selectivity of alkaline activators for the activation of slags. Cem. Concr. Aggress, 1996, 18(1): 8-14.

[56] Roy A., Schilling P. J., Eaton H. C., Malone P. G., Brabston W. N., Wakeley L. D.. Activation of ground blast-furnace slag by alkali-metal and alkaline-earth hydroxides. J. Am. Ceram. Soc, 1992, 75(12): 3233-3240.

[57] Song S., Sohn D., Jennings H. M., Mason T. O.. Hydration of alkali-activated ground granulated blast furnace slag. J. Mater. Sci, 2000, 35: 249-257.

[58] Fernández-Jiménez A., Palomo J. G., Puertas F.. Alkali-activated slag mortars. Mechanical strength behaviour. Cem. Concr. Res, 1999, 29: 1313-1321.

[59] Duran Ati C., Bilim C., Celik Ö., Karahan O.. Influence of activator on the strength and drying shrinkage of alkali-activated slag mortar. Constr. Build. Mater. 2009, 23(1): 548-555.

[60] Palacios M., Puertas F.. Effectiveness of mixing time on hardened properties of waterglass-activated slag pastes and mortars. ACI Mater. J. 2011, 108(1): 73-78.

[61] Gu J., Jin Z.. Quality control of the raw materials for alkali activated slag cement. Cement 1990, 8: 12-15.

[62] Wang P., Jin Z., Zhang Y.. Study on the composite activator for alkali activated slag cement. New Build. Mater. 2005, 8: 32-34.

[63] Milestone N. B.. Reactions in cement encapsulated nuclear wastes: need for toolbox of different cement types. Adv. Appl. Ceram, 2006, 105(1): 13-20.

[64] Bai Y., Collier N., Milestone N., Yang C.. The potential for using slags activated with near neutral salts as immobilisation matrices for nuclear wastes containing reactive metals. J. Nucl. Mater, 2011, 413(3): 183-192.

[65] Roy D.. Alkali-activated cements-opportunities and challenges. Cem. Concr. Res, 1999, 29(2): 249-254.

[66] Bernal S. A., Skibsted J., Herfort D.. Hybrid binders based on alkali sulfate-activated Portland clinker and metakaolin. In: Palomo, A. (ed.) XIII International Congress on the Chemistry of Cement, Madrid. CD-ROM proceedings, 2011.

[67] Richardson I. G., Brough A. R., Groves G. W., Dobson C. M.. The characterization of hardened alkali-activated blast-furnace slag pastes and the nature of the calcium silicate hydrate(C-S-H) paste. Cem. Concr. Res, 1994, 24(5): 813-829.

[68] Wang S. D., Scrivener K. L.. Hydration products of alkali-activated slag cement. Cem. Concr. Res, 1995, 25(3): 561-571.

[69] Rajaokarivony-Andriambololona Z. , Thomassin J. H. , Baillif, P. , Touray J. C. . Experimental hydration of two synthetic glassy blast furnace slags in water and alkaline solutions(NaOH and KOH 0. 1 N) at 40℃: structure, composition and origin of the hydrated layer. J. Mater. Sci, 1990, 25(5): 2399-2410.

[70] Richardson I. G. . Tobermorite/jennite and tobermorite/calcium hydroxide-based models for the structure of C-S-H: applicability to hardened pastes of tricalcium silicate, dicalcium silicate, Portland cement, and blends of Portland cement with blast-furnace slag, metakaolin, or silica fume. Cem. Concr. Res, 2004, 34(9): 1733-1777.

[71] Taylor R. , Richardson I. G. , Brydson R. M. D. . Nature of C-S-H in 20 year old neat ordinary Portland cement and 10% Portland cement-90% ground granulated blast furnace slag pastes. Adv. Appl. Ceram. 2007, 106(6): 294-301.

[72] Richardson I. G. . The calcium silicate hydrates. Cem. Concr. Res, 2008, 38(2): 137-158.

[73] Fernández-Jiménez A. . Cementos de escorias activadas alcalinamente: influencia de las variables y modelización del proceso. Thesis, Universidad Autónoma de Madrid, 2000.

[74] Pardal X. , Brunet F. , Charpentier T. , Pochard I. , Nonat A. . 27Al and 29Si solid-state NMR characterization of calcium-aluminosilicate-hydrate. Inorg. Chem, 2012, 51: 1827-1836.

[75] Renaudin G. , Russias J. , Leroux F. , Cau-dit-Comes C. , Frizon F. . Structural characterization of C-S-H and C-A-S-H samples-Part II: local environment investigated by spectroscopic analyses. J. Solid State Chem, 2009, 182(12): 3320-3329.

[76] Schneider J. , Cincotto M. A. , Panepucci H. . 29Si and 27 Al high-resolution NMR characterization of calcium silicate hydrate phases in activated blast-furnace slag pastes. Cem. Concr. Res, 2001, 31(7), 993-1001.

[77] Palacios M. , Puertas F. . Effect of carbonation on alkali-activated slag paste. J. Am. Ceram. Soc, 2006, 89(10): 3211-3221.

[78] Yang K. H. , Song J. K. , Ashour A. F. , Lee E. T. . Properties of cementless mortars activated by sodium silicate. Constr. Build. Mater, 2008, 22(9), 1981-1989.

[79] Yang K. H. , Song J. K. , Lee K. S. , Ashour A. F. . Flow and compressive strength of alkali-activated mortars. ACI Mater. J. 2009, 106(1): 50-58.

[80] Yang K. H. , Song J. K. . Workability loss and compressive strength development of cement-less mortars activated by combination of sodium silicate and sodium hydroxide. J. Mater. Civ. Eng, 2009, 21(3): 119-127.

[81] Rouseková I. , Bajza A. , Živica V. . Silica fume-basic blast furnace slag systems activated by an alkali silica fume activator. Cem. Concr. Res, 2007, 27(12): 1825-1828.

[82] Živica V. . High effective silica fume alkali activator. Bull. Mater. Sci, 2004, 27(2): 179-182.

[83] Živica V. . Effectiveness of new silica fume alkali activator. Cem. Concr. Comp, 2006, 28(1): 21-25.

[84] Bernal S. A., Rodríguez E. D., De Mejia Gutiérrez R., Provis J. L., Delvasto S.. Activation of metakaolin/slag blends using alkaline solutions based on chemically modified silica fume and rice husk ash. Waste Biomass Valor, 2012, 3(1): 99-108.

[85] Rodríguez E. D., Bernal S. A., Provis J. L., Paya J., Monzo J. M., Borrachero M. V.. Effect of nanosilica-based activators on the performance of an alkali-activated fly ash binder. Cem. Concr. Compos, 2013, 35(1): 1-11.

[86] Wang S. D., Scrivener K. L.. 29 Si and 27 Al NMR study of alkali-activated slag. Cem. Concr. Res, 2003, 33(5): 769-774.

[87] Bernal S. A., Provis J. L., Walkley B., San Nicolas R., Gehman J. D., Brice D. G., Kilcullen A., Duxson P., Van Deventer J. S. J.. Gel nanostructure in alkali-activated binders based on slag and fly ash, and effects of accelerated carbonation. Cem. Concr. Res, 2013, 53: 127-144.

[88] Gruskovnjak A., Lothenbach B., Holzer L., Figi R., Winnefeld F.. Hydration of alkali-activated slag: comparison with ordinary Portland cement. Adv. Cem. Res, 2006, 18(3): 199-128.

[89] Lloyd R. R., Provis J. L., Smeaton K. J., Van Deventer J. S. J.. Spatial distribution of pores in fly ash-based inorganic polymer gels visualised by Wood's metal intrusion. Microporous Mesoporous Mater. 2009, 126(1-2): 32-39.

[90] Ismail I., Bernal S. A., Provis J. L., Hamdan S., Van Deventer J. S. J.. Microstructural changes in alkali activated fly ash/slag geopolymers with sulfate exposure. Mater. Struct, 2013, 46(3): 361-373.

[91] Melo Neto A. A., Cincotto M. A., Repette W.. Drying and autogenous shrinkage of pastes and mortars with activated slag cement. Cem. Concr. Res, 2008, 38: 565-574.

[92] Häkkinen T.. The influence of slag content on the microstructure, permeability and mechanical properties of concrete: Part 1. Microstructural studies and basic mechanical properties. Cem. Concr. Res, 1993, 23(2): 407-421.

[93] Bakharev T., Sanjayan J. G., Cheng Y. B.. Effect of elevated temperature curing on properties of alkali-activated slag concrete. Cem. Concr. Res, 1999, 29(10): 1619-1625.

[94] Bernal S. A., Provis J. L., Brice D. G., Kilcullen A., Duxson P. Van Deventer J. S. J.. Accelerated carbonation testing of alkali-activated binders significantly underestimate the real service life: the role of the pore solution. Cem. Concr. Res, 2012, 42(10): 1317-1326.

[95] Faucon P., Delagrave A., Petit J. C., Richet C., Marchand J. M., Zanni H.. Aluminum incorporation in calcium silicate hydrates(C-S-H) depending on their Ca/Si ratio. J. Phys. Chem, 1999, B 103(37): 7796-7802.

[96] Andersen M. D., Jakobsen H. J., Skibsted J.. Incorporation of aluminum in the calcium silicate hydrate(C-S-H) of hydrated Portland cements: a high-field 27Al and 29 Si MAS NMR investigation. Inorg. Chem, 2003, 42(7): 2280-2287.

[97] Andersen M. D., Jakobsen H. J., Skibsted J.. A new aluminium-hydrate species in hydrated Portland cements characterized by 27Al and 29 Si MAS NMR spectroscopy. Cem. Concr. Res, 2006, 36(1): 3-17.

[98] Le Saoût G., Ben Haha M., Winnefeld F., Lothenbach B.. Hydration degree of alkali-activated slags: a 29 Si NMR study. J. Am. Ceram. Soc, 2011, 94 (12): 4541-4547.

[99] Sakulich A. R., Anderson E., Schauer C., Barsoum M. W.. Mechanical and microstructural characterization of an alkali-activated slag/limestone fine aggregate concrete. Constr. Build. Mater, 2009, 23: 2951-2959.

[100] Xu H., Provis J. L., Van Deventer J. S. J., Krivenko P. V.. Characterization of aged slag concretes. ACI Mater. J, 2008, 105(2): 131-139.

[101] Bakharev T., Sanjayan J. G., Cheng Y. B.. Alkali activation of Australian slag cements. Cem. Concr. Res, 1999, 29(1): 113-120.

[102] Fernández-Jiménez A., Puertas F.. Setting of alkali-activated slag cement. Influence of activator nature. Adv. Cem. Res, 2001, 13(3): 115-121.

[103] Sakulich A. R., Miller S., Barsoum M. W.. Chemical and microstructural characterization of 20-month-old alkali-activated slag cements. J. Am. Ceram. Soc, 2010, 93 (6): 1741-1748.

[104] Moseson A. J., Moseson D. E., Barsoum M. W.. High volume limestone alkali-activated cement developed by design of experiment. Cem. Concr. Compos, 2012, 34 (3): 328-336.

[105] Małolepszy J.. Activation of synthetic melilite slags by alkalies. In: Proceedings of the 8th International Congress on the Chemistry of Cement, Rio de Janeiro, Brazil, 1986, 4: 104-107.

[106] Collins F., Sanjayan J. G.. Early age strength and workability of slag pastes activated by NaOH and Na_2CO_3. Cem. Concr. Res, 1998, 28(5): 655-664.

[107] Collins F., Sanjayan J. G.. Workability and mechanical properties of alkali-activated slag concrete. Cem. Concr. Res, 1999, 29: 455-458.

[108] Hewlett P. C.. Lea's Chemistry of Cement and Concrete, 4th edn. Elsevier, Oxford, 1998.

[109] Douglas E., Brandstetr J.. A preliminary study on the alkali activation of ground granulated blast-furnace slag. Cem. Concr. Res, 1990, 20(5): 746-756.

[110] Fundi Y. S. A.. Alkaline pozzolana Portland cement. In: Krivenko P. V. (ed.) Proceedings of the First International Conference on Alkaline Cements and Concretes, Kiev, Ukraine, 1994, 1: 181-192. VIPOL Stock Company.

[111] Shi C., Day, R. L.. Chemical activation of blended cements made with lime and natural pozzolans. Cem. Concr. Res, 1993, 23(6): 1389-1396.

[112] Shi C., Day R. L.. Pozzolanic reaction in the presence of chemical activators: Part II. Reaction products and mechanism. Cem. Concr. Res, 2000, 30(4): 607-613.

[113] Shi C., Day R. L.. Pozzolanic reaction in the presence of chemical activators: Part I. Reaction kinetics. Cem. Concr. Res, 2000, 30(1), 51-58.

[114] Lloyd R. R., Provis J. L., Van Deventer J. S. J.. Pore solution composition and alkali diffusion in inorganic polymer cement. Cem. Concr. Res, 2010, 40 (9): 1386-1392.

[115] Oliveira C. T. A., John V. M., Agopyan V.. Pore water composition of clinker free granulated blast furnace slag cements pastes. In: Krivenko P. V. (ed.) Proceedings of the Second International Conference on Alkaline Cements and Concretes, Kiev, Ukraine, 1999: 109-119. ORANTA.

[116] Song S., Jennings H. M.. Pore solution chemistry of alkali-activated ground granulated blast-furnace slag. Cem. Concr. Res, 1999, 29: 159-170.

[117] Puertas F., Fernández-Jiménez A., Blanco-Varela M. T.. Pore solution in alkali-activated slag cement pastes. Relation to the composition and structure of calcium silicate hydrate. Cem. Concr. Res, 2004, 34(1): 139-148.

[118] Faucon P., Charpentier T., Nonat A., Petit J. C.. Triple-quantum two-dimensional 27Al magic angle nuclear magnetic resonance study of the aluminum incorporation in calcium silicate hydrates. J. Am. Chem. Soc, 1998, 120(46): 12075-12082.

[119] Glasser F. P.. Mineralogical aspects of cement in radioactive waste disposal. Miner. Mag, 2001, 65(5): 621-633.

[120] Roy A.. Sulfur speciation in granulated blast furnace slag: an X-ray absorption spectroscopic investigation. Cem. Concr. Res, 2009, 39: 659-663.

[121] Provis J. L., Duxson P., Van Deventer J. S. J.. The role of particle technology in developing sustainable construction materials. Adv. Powder Technol, 2010, 21(1): 2-7.

[122] Talling B., Brandstetr J.. Present state and future of alkali-activated slag concretes. In: Malhotra V. M. (ed.) 3rd International Conference on Fly Ash, Silica Fume, Slag and Natural Pozzolans in Concrete, ACI SP114, Trondheim, Norway. 1989, 2: 1519-1546. American Concrete Institute, Detroit, MI.

[123] Collins F., Sanjayan J. G.. Effects of ultra-fine materials on workability and strength of concrete containing alkali-activated slag as the binder. Cem. Concr. Res, 1999, 29(3): 459-462.

[124] Lim N. G., Jeong S. W., Her J. W., Ann K. Y.. Properties of cement-free concrete cast by finely grained nanoslag with the NaOH-based alkali activator. Constr. Build. Mater, 2012, 35: 557-563.

[125] Kumar R., Kumar S., Badjena S., Mehrotra S. P.. Hydration of mechanically activated granulated blast furnace slag. Metall. Mater. Trans, 2005, B 36(6): 873-883.

[126] ASTM International: Standard Test Method for Hydraulic Activity of Slag Cement by Reaction with Alkali(ASTM C1073-12). West Conshohocken, 2012.

[127] Osborn E. F., Roeder P. L., Ulmer G. C.. Part I-phase equilibria at solidus temper-

[128] Brown P. W.. The system Na_2O-CaO-SiO_2-H_2O. J. Am. Ceram. Soc, 1990, 73(11): 3457-3561.

[129] Kalousek G. L.. Studies of proportions of the quaternary system soda-lime-silica-water at 25℃. J. Res. Nat. Bur. Stand. 1944, 32: 285-302.

[130] Ilyukhin, V. V., Kuznetsov V. A., Lobatchov A. N., Bakshutov V. S.. Hydro-silicates of Calcium. Synthesis of Monocrystals and Crystal Chemistry. Nauka, Moscow, 1979.

[131] Taylor, H. F. W.. A method for predicting alkali ion concentrations in cement pore solutions. Adv. Cem. Res, 1987, 1(1): 5-16.

[132] Hong S. Y., Glasser F. P.. Alkali binding in cement pastes: Part I. The C-S-H phase. Cem. Concr. Res, 1999, 29(12): 1893-1903.

[133] Stade, H.. On the reaction of C-S-H (di, poly) with alkali hydroxides. Cem. Concr. Res, 1982, 19(5): 802-810.

[134] Atkins M., Bennett D., Dawes A., Glasser F., Kindness A., Read D.. A thermodynamic model for blended cements: Research Report for the Department of the Environment, DoE/HMIP/RR/92/005, 1991.

[135] Chen W., Brouwers H. J. H.. Alkali binding in hydrated Portland cement paste. Cem. Concr. Res, 2010, 40(5): 716-722.

[136] Hong S. Y., Glasser F. P.. Alkali sorption by C-S-H and C-A-S-H gels: Part II. Role of alumina. Cem. Concr. Res, 2002, 32(7): 1101-1111.

[137] Skibsted, J., Andersen, M. D.. The effect of alkali ions on the incorporation of aluminum in the calcium silicate hydrate(C-S-H) phase resulting from Portland cement hydration studied by 29 Si MAS NMR. J. Am. Ceram. Soc, 2013, 96(2): 651-656.

[138] García Lodeiro I., Macphee D. E., Palomo A., Fernández-Jiménez A.. Effect of alkalis on fresh C-S-H gels. FTIR analysis. Cem. Concr. Res, 2009, 39: 147-153.

[139] García-Lodeiro I.. Compatibility of cement gels C-S-H and N-A-S-H. Studies in real samples and in synthetic gels. Thesis, Universidad Autónoma de Madrid, 2008.

[140] Blakeman E. A., Gard J. A., Ramsay C. G., Taylor H. F. W.. Studies on the system sodium oxide-calcium oxide-silica-water. J. Appl. Chem. Biotechnol, 1974, 24(4-5): 239-245.

[141] Nelson E. B., Kalousek G. L.. Effects of Na_2O on calcium silicate hydrates at elevated temperatures. Cem. Concr. Res, 1997, 7(6): 687-694.

[142] Duxson P., Provis J. L., Lukey G. C., Van Deventer J. S. J., Separovic F., Gan Z. H.. 39KNMR of free potassium in geopolymers. Ind. Eng. Chem. Res, 2006, 45(26):

[143] Shi, C.. Steel slag-its production, processing, characteristics and cementitious prop-

erties. J. Mater. Civil Eng, 2004, 16(3): 230-236.

[144] Hu S., Wang H., Zhang G., Ding Q.. Bonding and abrasion resistance of geopolymeric repair material made with steel slag. Cem. Concr. Compos, 2008, 30(3): 239-244.

[145] Wang Y., Lin D.. The steel slag blended cement. Silic. Indus, 1983, 6: 121-126.

[146] Petropavlovsky O. N.. Slag alkaline binding systems and concretes based on steelmaking slag. Thesis, Kiev Civil Engineering Institute, 1987.

[147] Li D., Wu X.. Improvement of early strength of steel slag cement. Jiangsu Build. Mater, 1992, 4: 24-27.

[148] Shi C., Wu X., Tang M.. Research on alkali-activated cementitious systems in China. Adv. Cem. Res, 1993, 5(17): 1-7.

[149] Bin Q., Wu X., Tang M.. An investigation on alkali-BFS-steel slag cement. In: 2nd Beijing International Symposium on Cements and Concretes, Beijing, P. R. China, 1989: 288-294.

[150] Bin Q., Wu X., Tang M.. High strength alkali steel-iron slag binder. In: 9th International Congress on the Chemistry of Cement, New Delhi, India, 1992: 291-297.

[151] Shi C.. Corrosion resistant cement made with steel mill by-products. In: International Symposium on the Utilization of Metallurgical Slag, Beijing, P. R. China, 1999: 171-178.

[152] Shi C.. Characteristics and cementitious properties of ladle slag fines from steel production. Cem. Concr. Res, 2002, 32(3): 459-462.

[153] Kavalerova E., Petropavlovsky O., Krivenko P. V.. The role of solid-phase basicity on heat evolution during hardening of cements. In: Tammirinne, M. (ed.) International Conference on Practical Applications in Environmental Geotechnology(ecogeo*2000), Helsinki, Finland, 2000: 73-80. VTT Technical Research Centre of Finland.

[154] Natali Murri A., Rickard W. D. A., Bignozzi M. C., Van Riessen A.. High temperature behaviour of ambient cured alkali-activated materials based on ladle slag. Cem. Concr. Res, 2013, 43: 51-61.

[155] Bignozzi M. C., Manzi S., Lancellotti I., Kamseu E., Barbieri L., Leonelli C.. Mix-design and characterization of alkali activated materials based on metakaolin and ladle slag. Appl. Clay Sci, 2013, 73: 78-85.

[156] Shi, C.. Study on alkali activated phosphorous slag cement. J. Nanjing Inst. Chem. Technol, 1998, 10(2): 110-116.

[157] Shi C.. Influence of temperature on hydration of alkali activated phosphorous slag. J. Nanjing Inst. Chem. Technol, 1989, 11(1): 94-99.

[158] Fang Y., Mao Z., Wang C., Zhu Q.. Performance of alkali-activated phospho-r slagfly ash cement and the microstructure of its hardened paste. J. Chin. Ceram. Soc, 2007, 35(4): 451-455.

[159] Technologiya Metallov: Processing of slags of non-ferrous metallurgy. Chelyabinsk, Russia, 2008. http://www.technologiya-metallov.com/englisch/oekologie_4.htm.

[160] University of Wisconsin Recycled Materials Resource Center: Nonferrous slags-material description, Madison. 2013. http://rmrc.wisc.edu/ug-mat-nonferrous-slags.

[161] Małolepszy J., Deja J., Brylicki W.. Industrial application of slag alkaline concretes. In: Krivenko P. V. (ed.) Proceedings of the First International Conference on Alkaline Cements and Concretes, Kiev, Ukraine. 1994, 2: 989-1001. VIPOL Stock Company.

[162] Bin X., Yuan X.. Research of alkali-activated nickel slag cement. In: Krivenko P. V. (ed.) Proceedings of the Second International Conference on Alkaline Cements and Concretes, Kiev, Ukraine, 1999: 531-536. ORANTA.

[163] Zosin A. P., Priimak T. I., Avsaragov K. B.. Geopolymer materials based on magnesia-iron slags for normalization and storage of radioactive wastes. At. Energy, 1998, 85(1): 510-514.

[164] Narang K. C., Chopra S. K.. Studies on alkaline activation of BF, steel and alloy slags. Silic. Indus, 1983, 9: 175-182.

[165] Chen J.-X., Chen H.-B., Xiao P., Zhang L.-F.. A study on complex alkali-slag environmental concrete. In: Proceedings of the International Workshop on Sustainable Development and Concrete Technology, Beijing, China, 2004. 299-307. Center for Transportation Research and Education, Ames.

[166] Kalinkin A. M., Kumar S., Gurevich B. I., Alex T. C., Kalinkina E. V., Tyukavkina V. V., Kalinnikov V. T., Kumar R.. Geopolymerization behavior of Cu-Ni slag mechanically activated in air and in CO_2 atmosphere. Int. J. Miner. Proc, 2012, 112-113: 101-106.

[167] Komnitsas K., Zaharaki D.. Utilisation of low-calcium slags to improve the strength and durability of geopolymers. In: Provis J. L., Van Deventer J. S. J. (eds.) Geopolymers: Structure, Processing, Properties and Industrial Applications, 2009: 345-378. Woodhead, Cambridge.

[168] Komnitsas K., Zaharaki D., Perdikatsis V.. Effect of synthesis parameters on the compressive strength of low-calcium ferronickel slag inorganic polymers. J. Hazard. Mater, 2009, 161(2-3): 760-768.

[169] Komnitsas K., Zaharaki D., Bartzas G.. Effect of sulphate and nitrate anions on heavy metalimmobilisation in ferronickel slag geopolymers. Appl. Clay Sci, 2013, 73: 103-109.

[170] Komnitsas K., Zaharaki D., Perdikatsis V.. Geopolymerisation of low calcium ferronickel slags. J. Mater. Sci, 2007, 42(9): 3073-3082.

[171] Pontikes Y., Machiels L., Onisei S., Pandelaers L., Geysen D., Jones P. T., Blanpain B: Slags with a high Al and Fe content as precursors for inorganic polymers. Appl. Clay Sci, 2013, 73: 93-102.

[172] Sathonsaowaphak A., Chindaprasirt P., Pimraksa K.. Workability and strength of lignite bottom ash geopolymer mortar. J. Hazard. Mater, 2009, 168(1): 44-50.

[173] Chindaprasirt P., Jaturapitakkul C., Chalee W., Rattanasak U.. Comparative study on the characteristics of fly ash and bottom ash geopolymers. Waste Manag, 2009, 29(2): 539-543.

[174] Slavík R., Bednaík V., Vondruška M., Nemec A.. Preparation of geopolymer from fluidized bed combustion bottom ash. J. Mater. Proc. Technol, 2008, 200(1-3): 265-270.

[175] Xu H., Li Q., Shen L., Wang W., Zhai J.. Synthesis of thermostable geopolymer from circulating fluidized bed combustion(CFBC) bottom ashes. J. Hazard. Mater, 2010, 175(1-3): 198-204.

[176] Topçu I. B., Toprak M. U.. Properties of geopolymer from circulating fluidized bed combustion coal bottom ash. Mater. Sci. Eng. A, 2011, 528(3): 1472-1477.

[177] Luna Galiano Y., Fernández Pereira C., Vale J.. Stabilization/solidification of a municipal solid waste incineration residue using fly ash-based geopolymers. J. Hazard. Mater, 2011, 185(1): 373-381.

[178] Zheng L., Wang W., Shi Y.. The effects of alkaline dosage and Si/Al ratio on the immobilization of heavy metals in municipal solid waste incineration fly ash-based geopolymer. Chemosphere, 2010, 79(6): 665-671.

[179] Zheng L., Wang C., Wang W., Shi Y., Gao X.. Immobilization of MSWI fly ash through geopolymerization: Effects of water-wash. Waste Manag, 2011, 31(2): 311-317.

第4章 胶凝材料化学——低钙碱激发胶凝材料

John L. Provis，Ana Fernández-Jiménez，Elie Kamseu，
Cristina Leonelli and Angel Palomo

4.1 低钙碱激发胶凝材料

4.1.1 低钙碱激发胶凝材料的起步与发展

正如第2章所述，法国Davidovits教授的研究促进了早期低钙（包括无钙）碱激发胶凝材料的发展。最初设想这种材料作为一种阻燃材料取代有机聚合物材料，但随着对其潜在应用价值的认识，发现这种材料可作为一种混凝土生产过程中的胶凝材料[1]。然而地质聚合物混凝土领域的研究很快地集中在高钙体系。同时低钙体系应用的研究主要针对高温领域，如从黏土中提取类陶瓷性质的碱激发凝胶领域。早期在这一领域的研究仅仅为实现其商业目标，除了会议过程记录[2]和相应会议在签约杂志社[3]零星发表的论文等，所涉及的有价值科研信息很少。在1990年代早期[4,5]，偏高岭土碱激发胶凝材料的学术研究促进了最早发表的文献，Wastiels等[6-8]发表了第一篇关于描述由碱激发粉煤灰作为原料制备坚固耐用的胶凝材料的文章。粉煤灰基激发材料比黏土基胶凝材料具有更好的流变性，低钙AAM混凝土制品受到更多关注，自此工业界和学术界研究采用不同方法来提高流变性能。回顾胶凝材料体系中的低钙AAM胶凝材料的发展史，发现至2007年[9]在科学文献中的引用次数已经超过350次，说明人们对于了解和利用这种类型胶凝材料的热情高涨。

低钙碱激发体系中所形成的胶凝材料是一种高度无序、高度交联的铝硅酸盐凝胶。硅和铝存在于四面体网状结构中，四面体网络结构中铝的正电荷通过凝胶框架中的碱金属离子来平衡。在大量的文献中均指出这种凝胶结构和沸石的结构相似。Glukhovsky[10]和Davidovits[3]认为碱激发胶凝材料和古代罗马混凝土结构均为类沸石结构。Davidovits[3]根据沸石或类似结构（方沸石、方钠石、钙十字沸石、白榴石、六方钾霞石）描绘出分子片段的网络结构。随着研究的深入[11]，有学者发现水热合成沸石和碱铝硅酸盐的结构构造出纳米沸石的结构单元贯穿AAM凝胶，除了结晶沸石，在无序凝胶中可以看到包含许多这种结构，尤其是在高温养护时[12]。通过X射线和中子对泛函分析[13-16]表明偏高岭土制备的铝硅酸盐胶凝材料与在超过1000℃加热相同的凝胶形成长度为5~8Å的晶体结构非常相似。

这些信息提供了一个有用的结构模型，通过这个模型可以了解这些材料的化学性质和其他属性（如热力学），通过类水化硅酸钙和六水硅钙石模型理解由硅酸盐水泥水化形成的水化硅酸钙凝胶[17]。此时与研究和开发更广泛类水化硅酸钙C-S-H的结构的人数相比，关于"非沸石"胶凝材料的研究人员少了很多，包括在本报告的第3章讨论的由碱激发矿渣产生的铝取代凝胶[18]的研究。

4.1.2 碱激发胶凝材料之————地质聚合物

19 世纪 70 年代，Davidovits[1]第一次将"地质聚合物"这个词应用于煅烧黏土（尤其是偏高岭土）碱激发的产品中。这个术语采用了"聚硅酸"命名法[19]，在科技领域获得了一定的认可，经过三十多年的推广和引导终被地质聚合物研究所认可。第 1 章指出的"地质聚合物"，应用于各种碱激发胶凝材料，但当前主要应用于来源于粉煤灰和黏土低钙或无钙体系。关于"地质聚合物"一词是否可以（或应该）适用于富钙体系（如碱激发矿渣）的辩论此起彼伏；似乎解决这个问题主要基于营销（而不是科学）领域，在此不再深究。

关于低钙的 AAM 技术将被分为两个不同的领域在第 12、13 章分别进行详细介绍，分别作为替代水泥胶凝材料或烧制陶瓷的低成本原料[20]。设计、调节以及明确各领域配方间的差异，可扩展地质聚合物混凝土在其他领域的应用，如已经被证实的在 0.5< Si/Al< 300[21]范围内可以形成地质聚合物。尤其是在不需要抗侵蚀的环境中，在这个范围以外形成一些有其他性质的化合物可能适合其他的应用领域。然而贫硅（Si/Al<1）或富硅（Si/Al>5）的胶凝材料，其存在强度低、热稳定性差、耐化学腐蚀性差、易溶于水的缺点，使得此范围内的材料不适合于建筑应用领域。尽管这些材料可应用于这些特征不是问题（或者甚至是可取的）的领域，但是它们已超出本报告的范围。这种材料可以称为"地质聚合物"，但在建筑材料设计领域是否也应该归类为"碱激发胶凝材料"是另一码事。类似的评论可能也适用于其他领域，例如一些偏高岭土对水的需求量非常高，可能会形成过高孔隙率的胶凝材料，因此使得其成为唯一的原料。必须注意的是，在实验室环境中当定义和分析 AAM 类物质时，差的可塑性和/或其他的弱点会对碱激发材料替代传统水泥技术产生负面影响。

4.2 胶凝材料的结构特性和激发剂的影响

4.2.1 胶凝体系——氢氧化物激发

碱金属氢氧化物和铝硅酸盐基团在溶液中易结合，主要是由于固体碱氢氧化物的溶解释放大量的热[22]会导致其形成凝胶的变化。除此之外，也可将固体铝硅酸盐与碱金属氢氧化物一起煅烧形成前驱体，加水就能形成胶凝材料[23,24]，但是这个过程尚未像两部分（加固体前驱体的碱激发溶液）体系一样实现大规模利用，所以仍处于讨论和研究阶段。两百年前 Vicat[25] 和其他团队[26]对硝酸水泥的研究首次提出此类技术。碱的主要来源（KNO_3 或 K_2CO_3）为煅烧黏土，然后将生成的火山灰材料与石灰混合后形成水泥。然而，从那时起水泥的范围和发展规模一直受到限制。

低钙的胶凝材料（粉煤灰、偏高岭土和其他材料）由碱金属氢氧化物激发，其中碱铝硅酸盐凝胶、四面体铝和硅原子形成一个高度交联结构。碱阳离子[27,28]直接与框架中的氧离子（带负电）相连，而不是直接与（带正电）相邻四面体的离子连接，起电荷平衡作用。碱激发反应过程在一定程度上是离子交换的过程，主要取决于胶凝材料中碱阳离子的性质和渗透率（微米级和环状的凝胶）[29]。已知凝胶结构与胶凝材料的前驱体[11]密切相关，凝胶结构（前驱体凝胶）也是铝硅酸盐固体在水热合成沸石时的中间体[30-32]。随着时间的推移，会形成晶态沸石和相关材料，高温和高含水量下结晶度更高。碱激发粉煤灰胶凝材料在室温下强度发展缓慢，通常需要热养护或蒸汽养护[33,34]来帮助发展强度。

本章关于结晶相的形成的讨论主要基于钠盐作为碱源的激发剂，氢氧化钾激发铝硅酸盐的胶凝材料具有较低的结晶倾向[35]。氢氧化钠激发偏高岭土形成产物（整体 Si/Al 比率接近 1.0）主要是类方钠石体系中的长石，但此过程中会有少量的沸石（通常是沸石 Na-A 和/或二氧化硅八面沸石）作为瞬态或日后发展观察反应产物[27,36,37]。当偏高岭土进行同样的反应时需要使用热水提高液固比[38-40]。在胶凝材料中，氢氧化钠激发粉煤灰通常需要更高的 Si/Al 比，同时倾向于形成钠碱菱沸石（也称为碱菱沸石）和/或八面沸石（尽管这里面含有很高的 Si）以及富含 Al 的方钠石相和钙霞石相[41-43]。

核磁共振表明碱激发铝硅酸盐胶凝材料优越的 Q^4（nAl），n 分布在 1~4，但碱激发体系通常在这个范围内的高值一端[34,44,45]。在易固化的碱性胶凝材料或低/中等固化的硅酸盐激发剂[46]中，观察到胶凝材料中氢氧根的浓度（例如 Q^3 的硅酸盐部分）较低。

虽然在所有的硅/铝范围内，它们都会在早期出现固化，但是只有已经注明这些材料是易固化材料时，才会要求更高的硅酸盐浓度。在混合碱体系中，较大的阳离子优先形成凝胶，钠/钾比例 1∶1 混合胶凝材料孔溶液的凝胶优于钠/钾比例 1∶1 时的凝胶[47]。氢氧化钠对粉煤灰前驱体反应的加速程度优于氢氧化钾[48]。

Fernandez-Jimenez 等提出了碱激发粉煤灰的概念和微观结构模型[49]，如图 4-1 所示。碱激发剂攻击不完整球形玻璃壳空心粉煤灰颗粒的过程会导致在粒子内外有反应产物的形成，导致最终微观结构中嵌入不同反应程度的粉煤灰颗粒。其他课题组已经[43,50-52]开始从概念层面来研究反应模型，特别是通过添加可溶性二氧化硅催化剂[51,52]或镀一层高比表面积纳米颗粒[53]来讨论凝胶粒子表面的成核决定最终的反应程度。

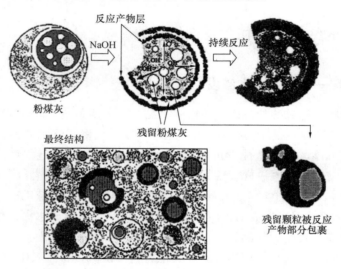

图 4-1 微观角度描述碱激发粉煤灰胶凝材料的形成
（改编自 Fernández-Jiménez 等[49]）

鉴于粉煤灰碱激发胶凝材料的高度异质性以及含铁材料呈现复杂的核磁共振相关特性，建立一个更全面的胶凝材料结构和碱激发粉煤灰化学性质的体系显得尤为重要。红外光谱的应用使其受益。Lee 和 Van Deventer[54]研究了氢氧化物或低模数硅酸盐溶液与粉煤灰颗粒

第 4 章 胶凝材料化学——低钙碱激发胶凝材料

之间的相互作用,为在粉煤灰颗粒表面上凝胶产物的结构演化提供了有用的见解。Fernandez-Jimenez 和 Palomo[55]通过红外光谱对碱激发胶凝材料进行了系统研究,实验结果表明凝胶进化分为两个阶段:凝胶 1,大量的 Si—O—Al 结合;凝胶 2,在较长反应时间时,随着凝胶程度的增加在内部形成硅交联结构(图 4-2)。该假设被后期的在线 ATR-FTIR 和离线分析验证,演示了凝胶结构早期到后期演变的化学机理[43,50],同步加速红外显微镜[56]结合在线 ATR-FTIR 可以获得空间分辨的信息。对偏高岭土的样品进行能量分散 X 射线衍射[57]和中子对泛函分析[16]得到基于凝胶 1/凝胶 2 过渡性能纳米结构信息。尽管这一过程的原位分析显然受到其发生的较长时间尺度(天/周)的限制。

低钙的物质如粉煤灰的掺入会产生许多潜在的影响,例如在碱激发的条件下,快速释放到溶液环境中的钙可能导致 $Ca(OH)_2$ 过度饱和,形成纳米微粒沉淀[58]。然而硅酸盐激发剂体系却并未如此,在最初的溶液中出现很多的硅将会导致形成硅酸盐络合钙而不是达到过饱和氢氧化钙阶段。即使相对少量的钙存在也会阻碍沸石结晶[59],从而抑制粉煤灰胶凝材料晶化,而基于偏高岭土制备的胶凝材料就不会发生这种情况。

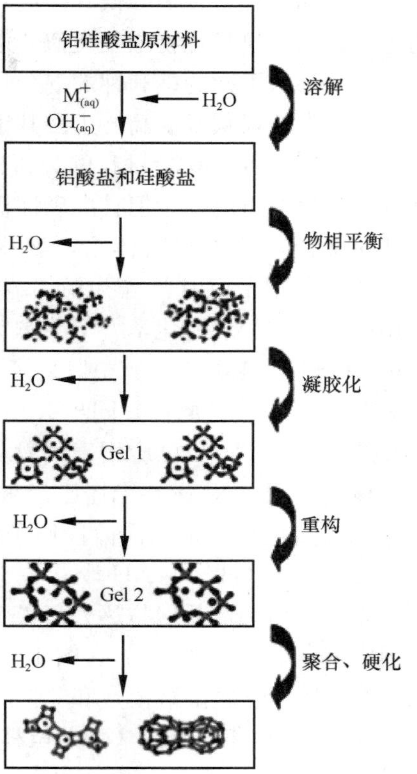

图 4-2 铝硅酸盐通过碱激发合成碱激发胶凝材料,多步凝胶反应概念模型(Duxson 等[9]版权归 Springer)

4.2.2 胶凝体系——硅酸盐激发

通过核磁共振、红外光谱等技术对低钙铝硅酸盐原料中不同激发剂进行研究,发现硅酸盐激发剂与碱激发的产物结构相类似。在原子尺度上观察到的主要差异是在各自的凝胶产品中的 Si/Al 比,通常稍有倾向沸石/似长石结晶物中的硅含量更高。微观结构上,硅酸盐激发的胶凝材料体系在微米或更大尺度上更加均匀[49,60],而不是作为渗透凝胶网络[61]保留,这被认为与较低硅胶随时间而致密化为不同颗粒的倾向相关。硅酸盐激发粉煤灰或偏高岭土往往提供更稳定的化学性质(即体系在比氢氧化物激发更宽的配合比设计和养护条件下开发出可接受的强度)。最早关于碱激发偏高岭土[5,19,62,63]和粉煤灰[6]应用的出版物中使用硅酸盐激发剂,从此很多研究一直致力于分析和了解这些材料。

通常激发低钙硅酸盐胶凝材料最有效的方法是将溶解碱硅酸盐加入固体前驱体。固体硅酸盐已经开始在粉煤灰胶凝材料中进行试验[64,65],但往往强度发展缓慢。有多种方法操纵硅酸钠溶液的组成,而且每种方法在胶凝材料结构中都有不同的影响。碱溶液加到硅酸盐溶液中调节模数(摩尔比率 SiO_2/M_2O,其中 M 是一种碱阳离子)在 1~2 之间,形成产生强度高和孔隙率低的胶凝材料[42,60,66-69]。最佳的配比也由胶凝材料中钠/铝的比值决定(例如,不包括在未反应的前驱体中的铝)大概为 1,虽然这在一定程度上也取决于 Si/Al 比[11,36]。使用高浓度的碱激发溶液也有优势,相当于降低了水灰比,可以为胶凝材料提供一个合适的碱含量，而稀释激发剂降低其碱度将导致效率降低[70]。碱金属阳离子对碱激发材料的影响与上文介绍的氢氧化物激发类似，但是应该[71]指出，根据粉煤灰组成、玻璃相含量和结构、粒度需要掺入不同的钠和钾的混合物实现最优性能。

碱激发粉煤灰或偏高岭土因其组成以及制备方法的差异性导致所形成的硅凝胶结构非常复杂。近年来各种胶凝材料的分析方法推陈出新，例如，量热法以及流变学[72-77]等为反应和硬化过程提供解释，但其准确性仍在讨论。然而，一些超越传统的实验室分析技术已应用到地质聚合物胶凝材料体系，该技术处于技术发展的最前沿。泛函理论方法结合X射线[13,14]以及中子衍射[15]作为局部结构的探针技术应用于碱激发偏高岭土胶凝材料早于首次应用研究硅酸盐水泥化学[78]，该技术应用于化学建筑材料（包括碱激发胶凝材料），最近已经被Meral等接受[79]。有史以来第一次基于分析碱激发偏高岭土使用原位中子对泛函实时描述反应体系[16]。同样，同步X射线荧光显微镜[58]、硬X射线纳米层析成像[80]和同步加速红外显微镜[56,81,82]已应用于碱激发胶凝材料，比应用于传统的水泥体系更早。

核磁共振（NMR）光谱已经被证明是一个分析碱激发硅酸盐胶凝材料的重要工具，因其能够直接探测未结晶阶段原子的配位结构[3,44,83-86]，进而分析硅和铝的连接环境，例如$^{23}Na^{[83,83,84]}$、$^1H^{[83,87]}$、$^2H^{[46]}$、$^{17}O^{[88,89]}$、$^{43}Ca^{[87,90]}$和$^{39}K^{[91]}$碱激发体系已被研究。解卷积核磁共振光谱给出大量数据与热力学模型表明其在进行分析和比较方面有很大价值[45]，但使用过程中需考虑其合理性[92]。图4-3显示了用^{29}Si MAS NMR光谱的值表征了碱激发硅酸盐的粉煤灰胶凝材料溶液模数和养护时间的函数关系。在所示的三个样品之间清楚可见结晶度（组分峰突出）和连通性（化学位移向负值偏移与更高的连接性和更低的Al含量相关）的差异。

正如上面所讨论的，红外光谱法可以应用于碱氢氧化物激发胶凝材料，也被应用于硅酸盐激发的体系。红外光谱结合核磁共振分析的结果可以表明，图4-2提出的一般反应机制也对硅酸盐激发胶凝材料有效。电阻率[93,94]也被用作反应动力学的表征手段，并且这些数据

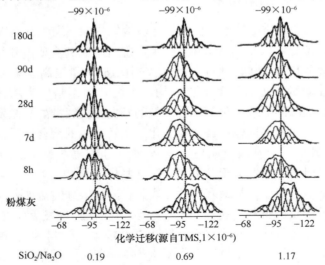

图4-3　AAFA浆体的^{29}Si MAS NMR-MAS，图中标记了溶液的SiO_2/Na_2O比和养护条件（密封于85℃的烘箱）

中通过原位高能 X 射线衍射获得的数据[57,95]与计算模型[94,96]相结合,为各个反应步骤的反应过程提供了一个更加详细的解释。还可以选择一些模拟方法解决以上问题——偏高岭土比粉煤灰更容易从概念上接受——包括使用半经验分子轨道计算[97]或密度泛函理论与蒙特卡洛模拟的多尺度粗晶方法[98,99]分析。

在硬化浆体结构的表征中,热导率测量[100]对凝胶的含水量(包括进行测试的相对湿度)特别敏感。在低钙胶凝材料中的凝胶水被认为主要是游离的并且在开放孔内可移动,与 NMR 光谱学的测试结果[46]以及通过气体吸附和正电子湮没寿命谱的表面干燥样品的分析完全一致[101]。膨胀计[102,103]还提供了碱激发胶凝材料中关于凝胶结构的有用信息,特别是与被确定为低连通性富含硅的凝胶相的作用相关,能够在温度升高时产生膨胀。起始温度和膨胀的程度已经用于表征这种材料的早期阶段;可能在 3~7d 后存在的少量膨胀相与高力学性能特别相关,这可能是由于其在基体中持续反应的原因,但是太多低连通性凝胶表现出成形性差、胶凝材料不成熟等特点[102,103]。

到目前为止,所有硅酸盐激发的低钙胶凝材料领域的文献介绍到此,可以与标准和先进的经验技术精心结合,来解决凝胶化学和结构的难题。没有一项分析技术可以提供所有必要的细节,并且分析方法(如在材料表征的所有领域)对于复杂材料的认识都是有限的,这些复合材料一般都是从纳米到厘米尺度的多相体系。在这些体系中的凝胶化学的许多细节需要使用自上而下和自下而上的方法进一步阐明,并且在未来几年,该领域的研究发展前景广阔。

4.2.3 胶凝体系——其他激发剂

使用除碱金属硅酸盐和氢氧化物之外的激发剂来激发低钙胶凝材料的研究已有报道。关于这些可能的替代品在第 1 章中已进行简要介绍。其中可能最有希望的是来自铝加工工业的副产品[104]的浓缩铝酸钠溶液激发,可形成强度超过 40MPa 的胶凝材料[105,106]和富含氧化铝的铝硅酸盐凝胶结构。碳酸盐激发剂由于其较弱的碱性使得在低钙体系中产生极缓慢的强度发展,这意味着需要加入 NaOH 以提供强度发展,导致生成类似部分碳化的 NaOH 激发凝胶产物[107]。硫酸盐激发类似于碳化。加入钙源如 CaO[108]或水泥熟料[109]来获得令人满意的强度发展是必要的,这些胶凝材料将在第 5 章进行详细的讨论。在以氢氧化物[110]或硅酸盐[94,110]作为激发剂的碱激发低 Ca 胶凝材料中添加碱硫酸盐,似乎对固化速率和强度发展总体上具有消极的影响,尽管硫酸盐没有明显地参与任何凝胶形成的过程。

4.3 粉煤灰及其与碱性溶液的相互作用

粉煤灰作为水泥胶凝材料的掺合料早已不是新鲜事。自 20 世纪 30 年代以来,粉煤灰就与硅酸盐水泥混合[111],包括世界上某些地区的实际生产中几乎无处不在使用(尽管在某些地区,这取决于管辖权和当地粉煤灰的可用性)。超过 50%粉煤灰的大体积粉煤灰混凝土,由于其特殊的施工性和养护条件而得到了更广泛的应用[112]。但是反应需要碱性环境,在没有熟料的情况下的激发需要更充分地理解粉煤灰玻璃化学,因为粉煤灰是用于形成胶凝材料中反应性组分的唯一来源。

在理解碱激发胶凝材料的形成和结构时，必须理解固体前驱体转化为最终胶凝材料结构的反应机理。在粉煤灰存在的情况下，地质聚合物反应机理复杂。来源不同的粉煤灰成分变化显著，即使源自同一发电站的粉煤灰，随时间变化其成分也并不相同，这使得问题变得非常复杂。对粉煤灰化学性质的探讨尤其重要。粉煤灰化学与碱激发产物的力学性能和耐久性能有着密切的关系。

反应体系中活性铝的含量对配制碱激发的粉煤灰胶凝材料是至关重要的，因为它影响铝硅酸盐凝胶的交联和化学稳定性。在没有足够反应性的铝的情况下，形成的凝胶呈现出可接受的机械强度发展，但是当暴露于湿气时强度发展是不稳定的，因此在大规模建筑应用中表现为有限的适用性。

Fernández-Jiménez[55,113-115]在具有相同的活性二氧化硅含量的情况下改变活性三氧化二铝的掺量，结果表明了反应性铝在凝胶形成和力学性能中的重要性。粉煤灰中含有大量的反应性氧化铝，氧化铝在反应初期释放到溶液中，表现出较好的性能。相反，具有低百分含量的活性氧化铝和/或其中所有活性氧化铝在反应的早期阶段消耗的粉煤灰通常表现出较差的性能。

为了共同解决这些问题，Duxson 和 Provis[120]基于二氧化硅、氧化铝的含量以及碱金属和碱土金属（即玻璃网络改性剂）含量的总和，开发了基于粉煤灰的类三元体系，如图4-4 所示。该图的基本概念是，碱金属阳离子和碱土金属阳离子的存在导致 Al 处于电荷平衡且形成具有反应性的铝硅酸盐玻璃体中，而不是生成低活性的莫来石嵌入富含玻璃状反应性较低的二氧化硅颗粒中。图 4-4 所示的强度分类非常相近，并且考虑了激发剂、养护条件和样品的龄期，而不是与数值强度值严格相关。同时还比较了来源不同[124]的几个 BFS 的组成。所研究的粉煤灰，根据 ASTM C618 判断大部分是 F 类；那些接近 BFS 区域的少数是 C 类。

一般来说，图 4-4 表明充足的铝和网状结构连接物是地质聚合物高强的必不可少的条

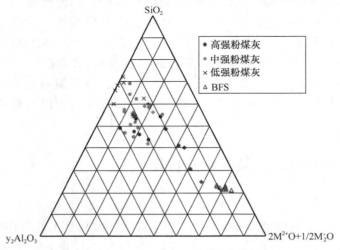

图 4-4　基于 Duxson 和 Provis[120]总结的浆料和灰浆的文献数据，
Diaz 和 Allouche[121]用 16 种不同的粉煤灰制备的混凝土的数据绘制的粉
煤灰氧化物组成与粉煤灰碱激发胶凝材料的抗压强度关联的类三元相图

件。另外，在本分析中未考虑粒度、铁含量、未燃烧碳的存在，以及硫酸盐或氯化物污染等潜在的因素。尽管如此，作为基于大量数据的预筛选的方法，该方法确实呈现出一些潜在的价值。已经注意到，基于大体积的粉煤灰的网络连通性分析看起来与反应性并不具有良好相关性[125]；这种半经验方法可以提供一种可能的替代方案。2003 年，Van Jaarsveld 等人[126]通过"化学溶解试验"、XRF 分析以及红外吸收光谱等检测手段，预测特定粉煤灰作为地质聚合物合成原材料时的作用。该预测似乎在目前还没有完全证实，并且由于玻璃相作为反应性的量度而在 X 射线衍射图谱中使用最大值的概念，也可以进一步增加对这一领域的理解，但很可能未来表征粉煤灰碱激发过程需要更详细的研究工作（但目前主要集中在数据量），这与单独的玻璃相分析相关[117,118,128-133]。

粉煤灰在水泥体系和碱激发环境的性能差异性、相关性以及水化动力学适用性等内容尚待解决[134-136]。目前，粉煤灰的碱激发反应模型仅从碱性角度出发未考虑钙化学的影响[96,137,138]，导致一些结果不同于水泥（或甚至碱激发的 BFS）在孔隙溶液中存在的环境的研究。这种模型在开发和应用中的重点在于，预测用于配合比设计为目的的碱激发粉煤灰胶凝材料的性能，因为它们在混凝土中的使用变得更加广泛，能更基础地表征（或甚定义）粉煤灰反应程度的能力[139]。

近年来，美国将粉煤灰列为危险废弃物，对其有毒组分的浸出和迁移做了大量的研究[140]。粉煤灰中铬的形态和迁移的研究[58]表明，使用硅酸盐激发剂可以降低 Cr 的迁移。同时氧化还原条件在控制 Cr 迁移的过程中具有重要作用[141]。此结论对共燃灰或城市固体废物焚烧灰的控制也特别重要[142]，因为这些灰更可能含有高危有害元素。应当注意，这不仅仅是碱激发研究界面临的问题，在硅酸盐水泥混凝土中使用大量的共燃灰或焚烧灰也引起了类似的问题，并且这些问题在整个行业和整个社会具有广泛的重要性。

粉煤灰的细度也被认为是碱激发活化的关键参数[123,143]，同时机械化学活化[144-146]已被证明能提高粉煤灰的性能。粉煤灰的表面化学在碱激发过程的早期起着重要作用[147]。虽然机械化学活化对粉体表面化学性质方面（与粒度和形状改变相反）的影响已经研究得较为深入，但硅酸盐材料中粉煤灰的影响仍然需要进一步研究。

最后，还应该注意的是，在碱激发过程中有价值的粉煤灰不仅是由烟煤的传统燃烧产生的灰。灰源还包括一些低 Ca 褐煤粉煤灰[148]、流化床粉煤灰[149,150]，以及由稻壳和树皮燃烧产生的富二氧化硅灰[151]，这些粉煤灰都具有一定的潜在价值。

4.4 天然矿产资源

4.4.1 偏高岭土

偏高岭土是高岭土经过 500～800℃的煅烧后脱羟基除去大部分化学结合水而未形成莫来石的矿物质[152-160]。高岭土来自天然矿床的开采、尾矿以及造纸工业废物，不同的来源将导致高岭土在粒度、纯度和结晶度方面的差异，造成其碱激发条件下具有不同的反应活性[161-163]。但是高岭土是否来源于废弃物，并非是控制其在碱激发胶凝材料合成中活性的根本。

偏高岭土在碱激发胶凝材料生产中的用途主要集中在"类陶瓷"应用和小规模实验室实

验，因其层状结构引起需水量增多从而导致混凝土的可施工性降低[164]。尽管如此，通过快速煅烧偏高岭土或使用机械压实的方法，可以获得球形度更好的颗粒形态来降低需水量，从而制备出低水胶比的碱激发偏高岭土混凝土[165-167]。研究需水量的解决方法，在未来该领域的发展中意义深远。

与此同时，偏高岭土的结构特性仍在讨论和研究中。除了晶态的层状黏土结构，由于偏高岭土在X射线下不显示结晶相，其结构分析存在挑战性，但也有残留块状/层状结构可直接观察到[168]，为潜在的进一步研究提供一些有用的信息。偏高岭土的煅烧温度对其在碱激发的反应性具有重要的影响[169-172]，因此了解和优化该过程至关重要。煅烧过程的直接分子模拟[160]和在500～750℃之间加热的产物之间显示出很小的纳米结构差异，这表明对于在该温度范围内加工的偏高岭土在煅烧期间进行一定程度的动力学控制，对反应性的差异有重要影响。在煅烧之前，特别是通过插入小的有机质对高岭土进行化学改性，能够进一步控制偏高岭土中的化学失常程度，从而影响反应性[173]。

偏高岭土的反应性由几何应变的Al位点决定，最初存在于高岭石的八面体层间[99]，经脱羟基连接后逐渐降低。这些位点的确切连接方式仍存在争议[159,160,174,175]，除了Ⅳ配位和Ⅴ配位为主的铝原子，还可能存在高应变的Ⅲ配位铝原子，这些位点无疑将会通过Al-O-Al键连接，这些事实意味着铝在碱性条件下非常容易发生反应。然而，这并不意味着在碱激发偏高岭土的反应程度可达到100%[92]，因为硅酸盐层很少被煅烧所破坏[174]，因此比铝酸盐层反应性低，所以胶凝材料微观结构中残留的颗粒在SEM下可见，其化学结构也可通过NMR来鉴别[44,45,60]。

偏高岭土可以与粉煤灰或各种炉渣混合形成复合前驱体，在反应过程中提供额外的Al[93,176,177]，调控反应速率和热性能[178,179]和/或控制碱-骨料反应[180]。这种利用方式还可以为上述可用性问题提供替代途径，并为未来该领域的发展提供更大的空间。

4.4.2　其他黏土

高岭土并非成功用于碱激发的唯一黏土原材料，其他1∶1和2∶1的黏土原材料也已经以这种方式增值。在葡萄牙，以钨尾矿[181-185]热处理后的黏土富集（主要是白云母）废弃物为原材料生产的胶凝材料已应用于混凝土和专用混凝土修补体系。伊利石、蒙脱石黏土在碱激发应用中也显示出强劲的潜力，并且黏土高硅含量的特点有利于低硅碱激发溶液的使用[186]。叶蜡石的热处理不能产生用于碱活性合适原材料，可能是由于通过该方法实现的矿物结构的化学破坏程度不足，尽管机械化学处理能够产生必要的反应性[187]。相反，用于碱激发的埃洛石可以通过热或机械方式进行预处理[188,189]。以高岭土和埃洛石的混合材料为原料，以碱和氢氧化钙作为复合激发剂制备出的Hwangtoh[190]已经用于生产施工性和力学性能满足结构应用的混凝土。在特定情况下，黏土中的铁含量对在确定碱激发期间的反应性和形成的反应产物的性质方面有重要的影响[191,192]。

4.4.3　长石和其他铝硅酸盐

作为地壳的主要成分的长石和天然沸石含有相应的可用于碱性激发的碱金属、铝和硅，故其用于碱性激发领域不足为奇。然而，此类矿物的高结晶度和良好的化学稳定性阻碍了碱激发过程，因此需要通过共混或化学改性来提高其反应活性。由钠长石、高岭土（煅烧或未煅烧的）和硅酸钠、添加或不添加粉煤灰的组合产生的碱激发胶凝材料具有可接受的（但不突出的）机械强度的发展。Xu 和 Van Deventer[197]对一组 15 种硅酸铝矿物与高岭石和低模数的混合 Na/K 硅酸盐溶液结合进行了试验，并提出了辉沸石的 5 元环结构是使其在矿物研究中表现较高反应性的原因[198,199]，发现矿物质在碱氢氧化物溶液中溶解的程度与形成的胶凝材料的抗压强度之间存在强烈的正相关性[197]。

最近，矿物的碱-热预处理已作为对大量矿物废物进行活化的手段。Pacheco-Torgal 和 Jalali[200]将富含黏土的钨矿开采废料与 Na_2CO_3 一起煅烧，产物因含过量碱导致风化问题。最近，Feng 等[24]开发了一种方法，通过钠长石和碱源（NaOH 或 Na_2CO_3）联合煅烧，将所得玻璃碾磨并直接与水混合形成一种混合碱激发水泥。其机械强度在 1d 达到 15MPa，28d 超过 40MPa。与硅酸盐水泥生产相比，其处理温度仅为 1000℃，可实现绿色节能减排。因此，诸如此类的方法应用在粉煤灰或 BFS 的供应困难区域，或者如果硅酸钠的世界市场由于未来更高的需求而变得更加受限，NaOH 或 Na_2CO_3 比硅酸钠的生产更容易和更廉价时，可以通过碱激发材料的途径实现。

4.5 火山灰和其他天然火山灰

如第 2 章所述，火山灰是古罗马胶凝材料和混凝土中使用的原始火山灰材料，由火山爆发产生的粉碎岩石和玻璃的小颗粒构成。从火山喷射时灰分的快速冷却导致相对高效的化学活性，特别是玻璃相和活性二氧化硅和氧化铝含量很高造成碱激发中的反应程度较高。从地质角度来看，火山灰颗粒也可随时间胶结在一起以形成称为凝灰岩的固体岩石，其也可被研磨和利用，并保留新鲜灰分的大部分反应性。来自欧洲、伊朗和非洲[201-211]地区的天然火山灰材料在迄今为止的研究中表现出良好的碱激发性能。

火山灰的化学成分主要由 SiO_2、Al_2O_3、Fe_2O_3 和 CaO 以及少量的 MgO、Na_2O、K_2O 和 TiO_2 以及微量元素组成。在晶体学上，斜长石、橄榄石和辉石等矿物嵌在玻璃状基质中；辉石使火山材料具有矿物特征，使火山灰显黑色。在火山灰中通常也发现闪石、云母和天然沸石，后两个组分在地质条件下为相变产物。

Ghukhovsky[10,212]提出，由于地质变换，一些火山岩成为沸石，然后在低温和低压下形成沉积岩，可能通过使用同一水泥体系前驱体与碱激发剂将这个过程应用于胶凝体系。因此，由胶凝体系形成的过程中直接合成碱硅铝酸盐矿物可保证形成的人造石材的强度和耐久性。火山灰中的无定形玻璃的性质，特别是高非结晶二氧化硅含量易在碱性溶液中的溶解，导致反应过程与观察到的偏高岭土或粉煤灰非常类似。对于具有低 CaO 含量的天然火山灰和含有富钠沸石和高可溶性硅酸盐含量的火山灰，最佳水玻璃模数（SiO_2/Na_2O 比）较低，但对于具有高 CaO 的天然火山灰或已经被煅烧[210]的火山灰则不同。高的温度下固化可以提高抗压强度（图 4-5）[207,209,213]和降低风化趋势[209]；添加供应的铝源如偏高岭土或铝酸钙水泥证明在这方面确实是有益的[204,209]。

火山灰的碱激发形成未反应的结晶相的铝硅酸盐凝胶。然而此相不具有粉煤灰中存在的莫来石或石英结晶相那样的惰性，在碱性溶液溶解颗粒表面的氧化铝和二氧化硅前驱体，同时溶解的硅酸盐和铝酸盐之间在颗粒表面发生反应。因此，在许多情况下，表面反应负责将未溶解的颗粒粘结到最终的地质聚合物结构中。与偏高岭土/粉煤灰不同，火山灰因其热解

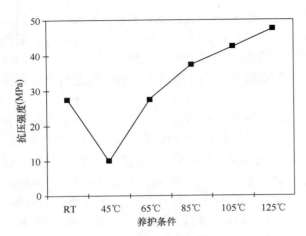

图 4-5 养护条件对碱激发的天然火山灰水泥（浮石型，源自 Taftan, 伊朗）抗压强度的影响

将样品 RT 在 25℃下密封 28d; 其他样品在 25℃下密封预养护 7d 后在所示温度的饱和水蒸气/蒸汽下养护 20h

（数据来自 Najafi Kani 等人[209]）

和地质特性以及环境诱导化学反应的差异（降低内在活性），导致溶解速率降低。这种低溶解率和缓慢的硬化行为，仍需要研究在火山灰的碱激发期间钙、铁和镁离子的溶解度，这些离子对反应的影响程度以及 Al 的作用及其可用性将由所研究的材料的具体化学性质决定。这些离子可能对天然火山灰胶凝材料结构的强化起到作用。但在此过程中，其存在形式多以氢氧化物的形式为主，对硬化浆体的力学性能和耐久性造成不利的影响。

因此，碱激发火山灰为实现火山灰高价值的充分利用提供了机会。由于火山灰的碱激发胶凝材料具有良好的致密性、良好的机械性能和较低孔隙率，使其适用于建筑领域。火山灰胶凝材料的强度被认为是由所形成的铝硅酸盐凝胶中化学键的强度，以及在凝胶中未反应的或部分反应的相和骨料之间发生的物理和化学相互作用所决定。考虑到许多前驱体颗粒是多孔的，也可以通过与凝胶的机械互锁而有助于强度发展，尽管多孔颗粒的高水需求有时是个问题。在这一领域进行的大多数工作都集中在生产小样品（浆料或砂浆）而不是混凝土，虽然已经发布的混凝土[211,213]在基于两种不同原料来源的可接受的低 w/b 比（0.42~0.45）的情况下似乎表现出良好的强度和可加工性。

4.6 低钙冶金渣

除用于硅酸盐水泥基和碱激发胶凝材料及混凝土生产中的高炉矿渣之外，还存在在碱激发条件下有显著反应性的各种可替代冶金渣。有一些可以归类为"低 Ca"，而其他"中钙"水平的冶金渣将在第 5 章中讨论。含有 2%~3% 的 CaO（和大于 30% 的 FeO）的氧化镁-铁渣在俄罗斯已经被碱激发用于废物固化，固化抗压强度高达 80MPa[214]。希腊的镍铁渣也被人们详细研究和开发[215-218]，一种情况是由碱硅酸盐单独激发[215,216]；另一种情况是与高岭土、偏高岭土、粉煤灰、赤泥和/或废玻璃等含有反应性 Si 和 Al 的原料复合激发[217,218]。上述体系中，低含量高活性的 Cr[218] 和高含量低活性的 Fe 已成为当前有色冶金废渣普遍存在的问题。然而，鉴于这些炉渣目前被视为危险废物，并且经常以对环境有危害的方式进行处置，例如倾倒入海洋。虽然不能应用到一般的建筑中，但是任何有益的利用形式都可能提供积极的方案。

4.7 合成体系

基于化学试剂（通常源自硝酸铝或铝酸钠与硅酸盐或胶态二氧化硅的组合）的碱激发胶

凝材料的直接合成方法主要用于开发陶瓷型碱激发材料,也用于研究没有因使用废物或天然材料合成碱激发胶凝材料而合成的模拟体系。而双粉末法则是通过将前驱体与碱性溶液中溶解的碱发生化学反应形成玻璃态物质,然后煅烧粉末得到纯铝硅酸盐前驱体粉末,随后在碱性溶液中与溶解的碱直接进行化学反应(尽管由该方法产生的凝胶通常含有较高的水含量)或者直接以独立的固体氧化铝和二氧化硅反应[219-216]。双粉末制备过程中,碱可以直接掺入固体前驱体之中得到"just-add-water"配合比设计,同时以碱金属氢氧化物或硅酸盐溶液的形式加入,可得到固体块状胶凝材料[81-82,227-230]。由于在大多数研究中使用的化学试剂的成本较高以及化学反应过程复杂,在大规模混凝土生产中立即使用这种途径的可能性显得很低。然而,通过研究和分析这些类型的材料可以获得宝贵的科学数据。这种加工途径还提供了产生相对纯的、低成本的类陶瓷结构的方法,或者说直接通过使用碱激发的单一制品,通过热处理破碎和使用这些非晶材料作为陶瓷的前驱体[231,232]。

4.8 小结

低钙碱激发胶凝材料(包括已描述为"地质聚合物"的那些胶凝材料)的开发通常基于对偏高岭土或粉煤灰的利用,尽管其他固体前驱体也已以更有限的方式得到使用。这些前驱体通常通过使用碱金属氢氧化物或硅酸盐溶液提供具有更高力学性能的产品,而碳酸盐或硫酸盐激发溶液通常在非高钙含量的情况下效果较差。在低钙碱激发胶凝材料体系中形成的凝胶类似于纳米结构的沸石,并且通常是高度交联的,因而有良好的力学性能和化学耐久性。这些凝胶中低含量的结合水确实导致耐久性混凝土的设计中存在一些连带问题,但是低钙碱激发胶凝材料似乎在一系列应用中具有很大的通用性,可在世界范围内对建筑材料生产中的一些未充分利用的资源进行评估。

参考文献

[1] Davidovits J.. Geopolymer Chemistry and Applications. Institut Géopolymère, Saint-Quentin, 2008.

[2] Davidovits J., Orlinski J. (eds.). Proceedings of Geopolymer'88-First European Conference on Soft Mineralurgy, Universite de Technologie de Compeigne, 1988.

[3] Davidovits J.. Geopolymers-inorganic polymeric new materials. J. Therm. Anal. 1991, 37(8): 1633-1656.

[4] Palomo A., Macias A., Blanco M. T., Puertas F.. Physical, chemical and mechanical characterisation of geopolymers. In: Proceedings of the 9th International Congress on the Chemistry of Cement, New Delhi, India, 1992, 5: 505-511. National Council for Cement and Building Materials.

[5] Palomo A., Glasser F. P.. Chemically-bonded cementitious materials based on metakaolin. Br. Ceram. Trans. J, 1992, 91(4): 107-112.

[6] Wastiels J., Wu X., Faignet S., Patfoort G.. Mineral polymer based on fly ash. In: Proceedings of the 9th International Conference on Solid Waste Management, Widener University, Philadelphia, PA, 8pp. 1993

[7] Wastiels J., Wu X., Faignet S., Patfoort G.. Mineral polymer based on fly ash. J. Resour. Manag. Technol, 1994, 22(3): 135-141.

[8] Patfoort G., Wastiels J., Bruggeman P., Stuyck L.. Mineral polymer matrix composites. In: Brandt A. M., Marshall I. H. (eds.) Proceedings of Brittle Matrix Composites 2(BMC 2), 1989: 587-592. Elsevier, Cedzyna.

[9] Duxson P., Fernández-Jiménez A., Provis J. L., Lukey, G. C., Palomo, A. Van Deventer J. S. J.. Geopolymer technology: the current state of the art. J. Mater. Sci, 2007, 42(9): 2917-2933.

[10] Glukhovsky V. D.. Ancient, modern and future concretes. In: Krivenko P. V., (ed.) Proceedings of the First International Conference on Alkaline Cements and Concretes, Kiev, Ukraine, 1994, 1: 1-9. VIPOL Stock Company.

[11] Provis J. L., Lukey G. C., Van Deventer J. S. J.. Do geopolymers actually contain nanocrystalline zeolites? -A reexamination of existing results. Chem. Mater, 2005, 17(12): 3075-3085.

[12] Fernández-Jiménez A., Monzó M., Vicent M., Barba A., Palomo A.. Alkaline activation of metakaolin-fly ash mixtures: obtain of zeoceramics and zeocements. Micropor. Mesopor. Mater. 2008, 108(1-3): 41-49.

[13] Bell J. L., Sarin P., Driemeyer P. E., Haggerty R. P., Chupas P. J., Kriven W. M.. X-ray pair distribution function analysis of a metakaolin-based, $KAlSi_2O_6 \cdot 5.5H_2O$ inorganic polymer(geopolymer). J. Mater. Chem, 2008, 18: 5974-5981.

[14] Bell J. L., Sarin P., Provis J. L., Haggerty R. P., Driemeyer P. E., Chupas P. J., Van Deventer J. S. J., Kriven, W. M.. Atomic structure of a cesium aluminosilicate geopolymer: a pair distribution function study. Chem. Mater. 2008, 20 (14): 4768-4776.

[15] White C. E., Provis J. L., Proffen T., Van Deventer J. S. J.. The effects of temperature on the local structure of metakaolin-based geopolymer binder: a neutron pair distribution function investigation. J. Am. Ceram. Soc, 2010. 93(10): 3486-3492.

[16] White C. E., Provis J. L., Llobet A., Proffen T., Van Deventer J. S. J.. Evolution of local structure in geopolymer gels: an in-situ neutron pair distribution function analysis. J. Am. Ceram. Soc, 2011, 94(10): 3532-3539.

[17] Richardson I. G.. Tobermorite/jennite- and tobermorite/calcium hydroxide-based models for the structure of C-S-H: applicability to hardened pastes of tricalcium silicate, β-dicalcium silicate, Portland cement, and blends of Portland cement with blast-furnace slag, metakaolin, or silica fume. Cem. Concr. Res, 2004, 34(9): 1733-1777.

[18] Puertas F., Palacios M., Manzano H., Dolado J. S., Rico A., Rodríguez J.. A model for the C-A-S-H gel formed in alkali-activated slag cements. J. Eur. Ceram. Soc, 2011, 31(12): 2043-2056.

[19] Davidovits J.. The need to create a new technical language for the transfer of basic scientific information. In: Gibb J. M., Nicolay D. (eds.) Transfer and Exploitation of

Scientific and Technical Information, EUR 7716, 1982: 316-320. Commission of the European Communities, Luxembourg.

[20] Duxson P., Provis J. L., Lukey G. C., Van Deventer J. S. J.. The role of inorganic polymer technology in the development of 'Green concrete'. Cem. Concr. Res, 2007, 37(12): 1590-1597.

[21] Fletcher R. A., MacKenzie K. J. D., Nicholson C. L., Shimada S.. The composition range of aluminosilicate geopolymers. J. Eur. Ceram. Soc, 2005, 25(9): 1471-1477.

[22] Provis J. L.. Activating solution chemistry for geopolymers. In: Provis J. L., Van Deventer, J. S. J. (eds.) Geopolymers: Structure, Processing, Properties and Industrial Applications, 2009: 50-71. Woodhead, Cambridge.

[23] Koloušek D., Brus J., Urbanova M., Andertova J., Hulinsky V., Vorel J.. P-reparation, structure and hydrothermal stability of alternative (sodium silicate-free) geopolymers. J. Mater. Sci, 2007, 42(22): 9267-9275.

[24] Feng D., Provis J. L., Van Deventer J. S. J.. Thermal activation of albite for the synthesis of one-part mix geopolymers. J. Am. Ceram. Soc, 2012, 95(2): 565-572.

[25] Vicat L.-J., Smith J. T.. A practical and scientific treatise on calcareous mortars and cements, artifi cial and natural; containing, directions for ascertaining the qualities of the different ingredients, for preparing them for use, and for combining them together in the most advantageous manner; with a theoretical investigation of their properties and modes of action. The whole founded upon an extensive series of original experiments, with examples of their practical application on the large scale. John Weale, Architectural Library, London, 1837.

[26] Treussart G.. On hydraulic and common mortars. Art. VII. Of artifi cial trass and puzzalona. J. Franklin Inst. 1838, 21(1),: 1-35.

[27] Provis J. L., Yong S. L., Duxson P.. Nanostructure/microstructure of metakaolin geopolymers. In: Provis J. L., Van Deventer J. S. J. (eds.). Geopolymers: Structure, Processing, Properties and Industrial Applications, 2009: 72-88. Woodhead, Cambridge.

[28] Fernández-Jiménez A., Palomo, A.. Nanostructure/microstructure of fly ash geopolymers. In: Provis J. L., Van Deventer J. S. J. (eds.). Geopolymers: Structure, Processing, Properties and Industrial Applications, 2009: 89-117. Woodhead, Cambridge.

[29] Bortnovsky O., Dědeček, J., Tvar užková Z., Sobalík Z., Šubrt J.. Metal ions as probes for characterization of geopolymer materials. J. Am. Ceram. Soc, 2008. 91(9): 3052-3057.

[30] Demortier A., Gobeltz N., Lelieur J. P., Duhayon C.. Infrared evidence for the formation of an intermediate compound during the synthesis of zeolite Na-A from metakaolin. Int. J. Inorg. Mater, 1999, 1(2): 129-134.

[31] Benharrats N., Belbachir M., Legrand A. P., D'Espinose de la Caillerie J.-B.. 29 Si

and 27 Al MAS NMR study of the zeolitization of kaolin by alkali leaching. Clay Miner, 2003, 38(1): 49-61.

[32] Slavík R., Bednařík V., Vondruška M., Skoba O., Hanzlíček T.. Proof of sodalite structures in geopolymers. Chem. Listy, 2005, 99: s471-s472.

[33] Criado M., Palomo A., Fernández-Jiménez A.. Alkali activation of fly ashes. Part 1: Effect of curing conditions on the carbonation of the reaction products. Fuel, 2005, 84(16): 2048-2054.

[34] Palomo A., Alonso S., Fernández-Jiménez A., Sobrados I., Sanz J.. Alkaline activation of fly ashes: NMR study of the reaction products. J. Am. Ceram. Soc, 2004, 87(6): 1141-1145.

[35] Duxson P., Mallicoat S. W., Lukey G. C., Kriven W. M., Van Deventer J. S. J.. The effect of alkali and Si/Al ratio on the development of mechanical properties of metakaolin-based geopolymers. Colloids Surf. A, 2007, 292(1): 8-20.

[36] Rowles M., O'Connor B.. Chemical optimisation of the compressive strength of aluminosilicate geopolymers synthesised by sodium silicate activation of metakaolinite. J. Mater. Chem, 2003, 13(5): 1161-1165.

[37] Zhang B., MacKenzie K. J. D., Brown I. W. M.. Crystalline phase formation in metakaolinite geopolymers activated with NaOH and sodium silicate. J. Mater. Sci, 2009, 44(17): 4668-4676.

[38] Heller-Kallai L., Lapides I.. Reactions of kaolinites and metakaolinites with NaOH - comparison of different samples(Part 1). Appl. Clay Sci, 2007, 35: 99-107.

[39] Rocha J., Klinowski J., Adams J. M.. Synthesis of zeolite Na-A from metakaolinite revisited. J. Chem. Soc. Faraday Trans, 1991, 87(18): 3091-3097.

[40] Barrer R. M., Mainwaring D. E.. Chemistry of soil minerals. Part XIII. Reactions of metakaolinite with single and mixed bases. J. Chem. Soc. Dalton Trans, 1972 22: 2534-2546.

[41] Oh J. E., Monteiro P. J. M., Jun S. S., Choi S., Clark S. M.. The evolution of strength and crystalline phases for alkali-activated ground blast furnace slag and fly ash-based geopolymers. Cem. Concr. Res, 2010, 40(2): 189-196.

[42] Criado M., Fernández-Jiménez A., de la Torre A. G., Aranda M. A. G., Palomo A.. An XRD study of the effect of the SiO_2/Na_2O ratio on the alkali activation of fly ash. Cem. Concr. Res, 2007, 37(5): 671-679.

[43] Rees C. A., Provis J. L., Lukey G. C., Van Deventer J. S. J.. Attenuated total reflectance Fourier transform infrared analysis of fly ash geopolymer gel aging. Langmuir, 2007, 23(15): 8170-8179.

[44] Duxson P., Provis J. L., Lukey G. C., Separovic F., Van Deventer J. S. J.. 29 Si NMR study of structural ordering in aluminosilicate geopolymer gels. Langmuir, 2005, 21(7): 3028-3036.

[45] Provis J. L., Duxson P., Lukey G. C., Van Deventer J. S. J.. Statistical thermody-

namic model for Si/Al ordering in amorphous aluminosilicates. Chem. Mater. 2005, 17 (11): 2976-2986.

[46] Duxson P., Lukey G. C., Separovic F., Van Deventer J. S. J.. The effect of alkali cations on aluminum incorporation in geopolymeric gels. Ind. Eng. Chem. Res, 2005, 44(4): 832-839.

[47] Duxson P., Provis J. L., Lukey G. C., Van Deventer J. S. J., Separovic F., Gan Z. H.. 39 K NMR of free potassium in geopolymers. Ind. Eng. Chem. Res, 2006, 45 (26): 9208-9210.

[48] Fernández-Jiménez A., Palomo A., Criado M.. Alkali activated fly ash binders. A comparative study between sodium and potassium activators. Mater. Constr, 2006, 56(281): 51-65.

[49] Fernández-Jiménez A., Palomo A., Criado M.. Microstructure development of alkali-activated fly ash cement: a descriptive model. Cem. Concr. Res, 2005, 35 (6): 1204-1209.

[50] Rees C. A., Provis J. L., Lukey G. C., Van Deventer J. S. J.. In situ ATR-FTIR study of the early stages of fly ash geopolymer gel formation. Langmuir, 2007, 23 (17): 9076-9082.

[51] Lloyd R. R., Provis J. L., Van Deventer J. S. J.. Microscopy and microanalysis of inorganic polymer cements. 1: Remnant fly ash particles. J. Mater. Sci, 2009, 44(2): 608-619.

[52] Lloyd R. R., Provis J. L., Van Deventer J. S. J.. Microscopy and microanalysis of inorganic polymer cements. 2: The gel binder. J. Mater. Sci, 2009, 44(2): 620-631.

[53] Rees C. A., Provis J. L., Lukey G. C., Van Deventer J. S. J.. The mechanism of geopolymer gel formation investigated through seeded nucleation. Colloids Surf. A, 2008, 318(1-3): 97-105.

[54] Lee W. K. W., Van Deventer J. S. J.. Use of infrared spectroscopy to study geopolymerization of heterogeneous amorphous aluminosilicates. Langmuir, 2003, 19(21): 8726-8734.

[55] Fernández-Jiménez A., Palomo A.. Mid-infrared spectroscopic studies of alkali-activated fly ash structure. Micropor. Mesopor. Mater. 2005, 86(1-3): 207-214.

[56] Hajimohammadi A., Provis J. L., Van Deventer J. S. J.. The effect of alumina release rate on the mechanism of geopolymer gel formation. Chem. Mater, 2010, 22(18): 5199-5208.

[57] Provis J. L., Van Deventer J. S. J.. Direct measurement of the kinetics of geopolymerisation by in-situ energy dispersive X-ray diffractometry. J. Mater. Sci, 2007, 42(9): 2974-2981.

[58] Provis J. L., Rose V., Bernal S. A., Van Deventer J. S. J.. High resolution nanoprobe X-ray fluorescence characterization of heterogeneous calcium and heavy metal distributions in alkali activated fly ash. Langmuir, 2009, 25(19): 11897-11904.

[59] Catalfamo P., Di Pasquale S., Corigliano F., Mavilia L.. Influence of the calcium content on the coal fly ash features in some innovative applications. Resourc. Conserv. Recyc, 1997, 20(2): 119-125.

[60] Duxson P., Provis J. L., Lukey G. C., Mallicoat S. W., Kriven W. M., Van Deventer J. S. J.. Understanding the relationship between geopolymer composition, microstructure and mechanical properties. Colloids Surf. A. 2005, 269(1-3): 47-58.

[61] Lloyd R. R.. Accelerated ageing of geopolymers. In: Provis J. L., Van Deventer J. S. J. (eds.) Geopolymers: Structure, Processing, Properties and Industrial Applications, 2009: 139-166. Woodhead, Cambridge.

[62] Davidovits J.. Mineral polymers and methods of making them. Compiler: U. S, 4349386[P]. 1982.

[63] Davidovits J.. Synthetic mineral polymer compound of the silicoaluminates family and preparation process. Compiler: U. S, 4472199[P]. 1984.

[64] Yang K.-H., Song J.-K., Ashour A. F., Lee E.-T.. Properties of cementless mortars activated by sodium silicate. Constr. Build. Mater, 2008, 22(9): 1981-1989.

[65] Yang K. H., Song J. K.. Workability loss and compressive strength development of cementless mortars activated by combination of sodium silicate and sodium hydroxide. J. Mater. Civ. Eng, 2009, 21(3): 119-127.

[66] Criado M., Fernández-Jiménez A., Palomo A.. Alkali activation of fly ash. Effect of the SiO_2/Na_2O ratio. Part I: FTIR study. Micropor. Mesopor, Mater. 2007, 106(1-3): 180-191.

[67] Steveson M., Sagoe-Crentsil K.. Relationships between composition, structure and strength of inorganic polymers. Part I-Metakaolin-derived inorganic polymers. J. Mater. Sci, 2005, 40(8): 2023-2036.

[68] Steveson M., Sagoe-Crentsil K.. Relationships between composition, structure, and strength of inorganic polymers. Part 2. Fly ash-derived inorganic polymers. J. Mater. Sci. 2005, 40(16): 4247-4259.

[69] Lloyd R. R., Provis J. L., Smeaton K. J., Van Deventer J. S. J.. Spatial distribution of pores in fly ash-based inorganic polymer gels visualised by Wood's metal intrusion. Micropor. Mesopor. Mater, 2009, 126(1-2): 32-39.

[70] Phair J. W., Van Deventer J. S. J.. Effect of the silicate activator pH on the microstructural characteristics of waste-based geopolymers. Int. J. Miner. Proc, 2002, 66(1-4): 121-143.

[71] Van Jaarsveld J. G. S., Van Deventer J. S. J.. Effect of the alkali metal activator on the properties of fly ash-based geopolymers. Ind. Eng. Chem. Res, 1999, 38(10): 3932-3941.

[72] Rahier H., Simons W., Van Mele B., Biesemans M.. Low-temperature synthesized aluminosilicate glasses. 3. Influence of the composition of the silicate solution on production, structure and properties. J. Mater. Sci, 1997, 32(9): 2237-2247.

[73] Rahier H., Van Mele B., Biesemans M., Wastiels J., Wu X.. Low-temperature synthesized aluminosilicate glasses. 1. Low-temperature reaction stoichiometry and structure of a model compound. J. Mater. Sci, 1996, 31(1): 71-79.

[74] Rahier H., Van Mele B., Wastiels J.. Low-temperature synthesized aluminosilicate glasses. 2. Rheological transformations during low-temperature cure and high-temperature properties of a model compound. J. Mater. Sci, 1996, 31(1): 80-85.

[75] Yao X., Zhang Z., Zhu H., Chen Y.. Geopolymerization process of alkali-metakaolinite characterized by isothermal calorimetry. Thermochim. Acta, 2009, 493(1-2): 49-54.

[76] Granizo M. L., Blanco M. T.. Alkaline activation of metakaolin-an isothermal conduction calorimetry study. J. Therm. Anal, 1998, 52(3): 957-965.

[77] Palomo A., Banfi ll P. F. G., Fernandéz-Jiménez A., Swift D. S.. Properties of alkali-activated fly ashes determined from rheological measurements. Adv. Cem. Res, 2005, 17(4): 143-151.

[78] Skinner L. B., Chae S. R., Benmore C. J., Wenk H. R., Monteiro P. J. M.. N-anostructure of calcium silicate hydrates in cements. Phys. Rev. Lett, 2010, 104: 195-202.

[79] Meral C., Benmore C. J., Monteiro P. J. M.. The study of disorder and nanocrystallinity in C-S-H, supplementary cementitious materials and geopolymers using pair distribution function analysis. Cem. Concr. Res, 2011, 41(7): 696-710.

[80] Provis J. L., Rose V., Winarski R. P., Van Deventer J. S. J.. Hard X-ray nanotomography of amorphous aluminosilicate cements. Scripta Mater. 2011, 65 (4): 316-319.

[81] Hajimohammadi A., Provis J. L., Van Deventer J. S. J.. The effect of silica availability on the mechanism of geopolymerisation. Cem. Concr. Res, 2011, 41(3): 210-216.

[82] Hajimohammadi A., Provis J. L., Van Deventer J. S. J.. Time-resolved and spatially-resolved infrared spectroscopic observation of seeded nucleation controlling geopolymer gel formation. J. Colloid Interf. Sci, 2011, 357(2): 384-392.

[83] Rowles M. R., Hanna J. V., Pike K. J., Smith M. E., O'Connor B. H.. 29 Si, 27 Al, 1 H and 23 Na MAS NMR study of the bonding character in aluminosilicate inorganic polymers. Appl. Magn. Reson, 2007, 32: 663-689.

[84] Barbosa V. F. F., MacKenzie K. J. D., Thaumaturgo C.. Synthesis and characterisation of materials based on inorganic polymers of alumina and silica: sodium polysialate polymers. Int. J. Inorg. Mater, 2000, 2(4): 309-317.

[85] Criado M., Fernández-Jiménez A., Palomo A., Sobrados I., Sanz J.. Alkali activation of fly ash. Effect of the SiO_2/Na_2O ratio. Part II: 29 Si MASNMR survey. Micropor. Mesopor, Mater, 2008, 109(1-3): 525-534.

[86] Brus J., Kobera L., Urbanová M., Koloušek D., Kotek J.. Insights into the structural transformations of aluminosilicate inorganic polymers: a comprehensive solid-

state NMR study. J. Phys. Chem. C, 2012, 116(27): 14627-14637.

[87] Vance E. R., Perera D. S., Hanna J. V., Pike K. J., Aly Z., Blackford M. G., Zhang Y., Zhang Z., Rowles M., Davis J., Uchida O., Yee P., Ly L.. Solid state chemistry phenomena in geopolymers with Si/Al~2. In: International Workshop on Geopolymers and Geopolymer Concrete, Perth, Australia. CDROM proceedings, 2005.

[88] Duxson P.. Structure and thermal conductivity of metakaolin geopolymers. Ph. D. Thesis, University of Melbourne, 2006.

[89] Gehman J. D., Provis J. L.. Generalized biaxial shearing of MQMAS NMR spectra. J. Magn. Reson, 2009, 200(1): 167-172.

[90] MacKenzie K., Rahner N., Smith M., Wong, A.. Calcium-containing inorganic polymers as potential bioactive materials. J. Mater. Sci, 2010, 45(4): 999-1007.

[91] Barbosa V. F. F., MacKenzie K. J. D.. Synthesis and thermal behaviour of potassium sialate geopolymers. Mater. Lett, 2003, 57(9-10): 1477-1482.

[92] Provis J. L., Van Deventer J. S. J.. Discussion of "Synthesis and microstructural characterization of fully-reacted potassium-poly(sialate-siloxo) geopolymeric cement matrix", by Y. Zhang et al. ACI Mater. J, 2009, 106(1): 95-96.

[93] Zhang Y. S., Li Z. J., Sun W., Li W. L. Setting and hardening of geopolymeric cement pastes incorporated with fly ash. ACI Mater. J, 2009, 106(5): 405-412.

[94] Provis J. L., Walls P. A., Van Deventer J. S. J.. Geopolymerisation kinetics. 3. Effects of Cs and Sr salts. Chem. Eng. Sci, 2008, 63(18): 4480-4489.

[95] Provis J. L., Van Deventer J. S. J.. Geopolymerisation kinetics. 1. Insitu energy dispersive X-ray diffractometry. Chem. Eng. Sci, 2007, 62(9), 2309-2317.

[96] Provis J. L., Van Deventer J. S. J.. Geopolymerisation kinetics. 2. Reaction kinetic modelling. Chem. Eng. Sci, 2007, 62(9): 2318-2329.

[97] Zhang Y., Sun W.. Semi-empirical AM1 calculations on 6-memebered alumino-silicate rings model: implications for dissolution process of metakaoline in alkaline solutions. J. Mater. Sci, 2007, 42(9): 3015-3023.

[98] White C. E., Provis J. L., Kearley G. J., Riley D. P., Van Deventer J. S. J.. Density functional modelling of silicate and aluminosilicate dimerisation solution chemistry. Dalton Trans. 2011, 40(6): 1348-1355.

[99] White C. E., Provis J. L., Proffen T., Van Deventer J. S. J.. Molecular mechanisms responsible for the structural changes occurring during geopolymerization: multiscale simulation. AIChE J, 2012, 58(7): 2241-2253.

[100] Duxson P., Lukey G. C., Van Deventer J. S. J.. Thermal conductivity of metakaolin geopolymers used as a first approximation for determining gel interconnectivity. Ind. Eng. Chem. Res, 2006, 45(23): 7781-7788.

[101] Vance E. R., Hadley J. H., Hsu F. H., Drabarek E.. Positron annihilation lifetime spectra in a metakaolin-based geopolymer. J. Am. Ceram. Soc, 2008, 91(2): 664-666.

[102] Provis J. L., Yong C. Z., Duxson P., Van Deventer J. S. J.. Correlating mechanical

and thermal properties of sodium silicate-fly ash geopolymers. Colloids Surf. A, 2009, 336(1-3): 57-63.

[103] Provis J. L., Duxson P., Harrex R. M., Yong C. Z., Van Deventer J. S. J.. Valorisation of fly ashes by geopolymerisation. Global NEST J, 2009, 11(2): 147-154.

[104] Van Riessen A., Jamieson E., Kealley C. S., Hart R. D., Williams R. P.. Bayer-Geopolymers: an exploration of synergy between the alumina and geopolymer industries. Cem. Concr. Compos, 2013, 41, 29-33.

[105] Phair J. W., Van Deventer J. S. J.. Characterization of fly-ash-based geopolymeric binders activated with sodium aluminate. Ind. Eng. Chem. Res, 2002, 41(17): 4242-4251.

[106] Nugteren H., Ogundiran M. B., Witkamp G. -J., Kreuzer M. T.. Coal fly ash activated by waste sodium aluminate as an immobilizer for hazardous waste. In: 2011 World of Coal Ash Conference, Denver, CO. CD-ROM proceedings. ACAA/CAER, 2011.

[107] Fernández-Jiménez A., Palomo A.. Composition and microstructure of alkali activated fly ash binder: effect of the activator. Cem. Concr. Res, 2005, 35(10): 1984-1992.

[108] Shi C., Day R. L.. Acceleration of the reactivity of fly ash by chemical activation. Cem. Concr. Res, 1995, 25(1): 15-21.

[109] Bernal S. A., Skibsted J., Herfort D.. Hybrid binders based on alkali sulfate-activated Portland clinker and metakaolin. In: Palomo A. (ed.) 13th International Congress on the Chemistry of Cement, Madrid. CD-ROM proceedings, 2011

[110] Criado M., Fernández-Jiménez A., Palomo A.. Effect of sodium sulfate on the alkali activation of fly ash. Cem. Concr. Compos, 2010, 32(8): 589-594.

[111] Davis R. E., Carlson R. W., Kelly J. W., Davis H. E.. Properties of cements and concretes containing fly ash. J. Am. Concr. Inst, 1937, 33, 577-612.

[112] Bilodeau A., Malhotra V. M.. High-volume fly ash system: concrete solution for sustainable development. ACI Mater. J, 2000, 97(1): 41-48.

[113] Fernández-Jiménez A., de la Torre A. G., Palomo A., López-Olmo G., Alonso M. M., Aranda M. A. G.. Quantitative determination of phases in the alkaline activation of fly ash. Part I. Potential ash reactivity. Fuel, 2006, 85(5-6): 625-634.

[114] Fernández-Jiménez A., de la Torre A. G., Palomo A., López-Olmo G., Alonso M. M., Aranda M. A. G.. Quantitative determination of phases in the alkaline activation of fly ash. Part II: Degree of reaction. Fuel, 2006, 85(14-15): 1960-1969.

[115] Fernández-Jiménez A., Palomo A., Sobrados I., Sanz J.. The role played by the reactive alumina content in the alkaline activation of fly ashes. Micropor. Mesopor. Mater, 2006, 91(1-3): 111-119.

[116] Winnefeld F., Leemann A., Lucuk M., Svoboda P., Neuroth M.. Assessment of

phase formation in alkali activated low and high calcium fly ashes in building materials. Constr. Build. Mater, 2010, 24(6): 1086-1093.

[117] Keyte L. M.. Fly ash glass chemistry and inorganic polymer cements. In: Provis J. L., Van Deventer J. S. J. (eds.) Geopolymers: Structure, Processing, Properties and Industrial Applications, 2009: 15-36. Woodhead, Cambridge.

[118] Keyte L. M.. What's wrong with Tarong? The importance of fly ash glass chemistry in inorganic polymer synthesis. Ph. D. Thesis, University of Melbourne, Australia, 2008.

[119] Lloyd R. R., Provis J. L., Van Deventer J. S. J.. Pore solution composition and alkali diffusion in inorganic polymer cement. Cem. Concr. Res, 2010, 40 (9): 1386-1392.

[120] Duxson P., Provis J. L.. Designing precursors for geopolymer cements. J. Am. Ceram. Soc, 2008, 91(12): 3864-3869.

[121] Diaz E. I., Allouche E. N.. Recycling of fly ash into geopolymer concrete: Creation of a database. In: Green Technologies Conference 2010, IEEE, Grapevine, TX, USA. CD-ROM proceedings, 2010.

[122] Diaz E. I., Allouche E. N., Eklund S.. Factors affecting the suitability of fly ash as source material for geopolymers. Fuel, 2010, 89: 992-996.

[123] Diaz-Loya E. I., Allouche E. N., Vaidya S.. Mechanical properties of fly-ash-based geopolymer concrete. ACI Mater. J, 2011, 108(3): 300-306.

[124] Shi C., Krivenko P. V., Roy D. M.. Alkali-Activated Cements and Concretes. Taylor & Francis, Abingdon, 2006.

[125] Towler M. R., Stanton K. T., Mooney P., Hill R. G., Moreno N., Querol X.. Modelling of the glass phase in fly ashes using network connectivity theory. J. Chem. Technol. Biotechnol, 2002, 77: 240-245.

[126] Van Jaarsveld J. G. S., Van Deventer J. S. J., Lukey G. C.. The characterisation of source materials in fly ash-based geopolymers. Mater. Lett, 2003, 57(7): 1272-1280.

[127] Diamond S.. On the glass present in low-calcium and in high-calcium fly ashes. Cem. Concr. Res, 1983, 13(4): 459-464.

[128] Chancey R. T., Stutzman P., Juenger M. C. G., Fowler D. W.. Comprehensive phase characterization of crystalline and amorphous phases of a Class F fly ash. Cem. Concr. Res, 2010, 40(1): 146-156.

[129] Gustashaw K., Chancey R., Stutzman P., Juenger M.. Quantitative characterization of fly ash reactivity for use in geopolymer cements. In: Palomo A(ed.) 13th International Congress on the Chemistry of Cement, Madrid, Spain. CD-ROM proceedings, 2011.

[130] Chen-Tan N. W., Van Riessen A., Ly C. V., Southam D. C.. Determining the reactivity of a fly ash for production of geopolymer. J. Am. Ceram. Soc, 2009, 92(4): 881-887.

[131] Bumrongjaroen W., Muller I. S., Pegg I. L.. Characterization of glassy phase in fly ash from Iowa State University, Vitreous State Laboratory, Catholic University of America, Report VSL-07R520X-1, 2007.

[132] Valcke S. L. A., Saraber A. J., Pipilikaki P., Fischer H. R., Nugteren H. W.. Screening coal combustion fly ashes for application in geopolymers. Fuel, 2013, 106: 490-497.

[133] Zhang Z., Wang H., Provis J. L.. Quantitative study of the reactivity of fly ash in geopolymerization by FTIR. J. Sust. Cem. -Based Mater, 2012, 1(4): 154-166.

[134] Brouwers H. J. H., Van Eijk R. J.. Fly ash reactivity: extension and application of a shrinking core model and thermodynamic approach. J. Mater. Sci, 2002, 37(10): 2129-2141.

[135] Brouwers H. J. H., Van Eijk R. J.. Reactivity of fly ash: extension and application of a shrinking core model. Concr. Sci. Eng, 2002, 4, 106-113.

[136] Das S. K., Yudhbir. A simplified model for prediction of pozzolanic characteristics of fly ash, based on chemical composition. Cem. Concr. Res, 2006, 36(10): 1827-1832.

[137] Li C., Li Y., Sun H., Li L.. The composition of fly ash glass phase and its dissolution properties applying to geopolymeric materials. J. Am. Ceram. Soc, 2011, 94(6): 1773-1778.

[138] Chen C., Gong W., Lutze W., Pegg I., Zhai J.. Kinetics of fly ash leaching in strongly alkaline solutions. J. Mater. Sci, 2011, 46(3): 590-597.

[139] Ben Haha M., De Weerdt K., Lothenbach B.. Quantification of the degree of reaction of fly ash. Cem. Concr. Res, 2010, 40(11): 1620-1629.

[140] Sear L. K. A.. Coal fired power station ash products and EU regulation. Coal Comb. Gasif. Prod, 2009, 1: 63-66.

[141] Zhang J., Provis J. L., Feng D., Van Deventer J. S. J.. The role of sulfide in the immobilization of Cr(VI) in fly ash geopolymers. Cem. Concr. Res, 2008, 38(5): 681-688.

[142] Álvarez-Ayuso E., Querol X., Plana F., Alastuey A., Moreno N., Izquierdo M., Font O., Moreno T., Diez S., Vázquez E., Barra M.. Environmental, physical and structural characterisation of geopolymer matrixes synthesised from coal(co-)combustion fly ashes. J. Hazard. Mater, 2008, 154(1-3): 175-183.

[143] Rickard W. D. A., Williams R., Temuujin J., Van Riessen A.. Assessing the suitability of three Australian fly ashes as an aluminosilicate source for geopolymers in high temperature applications. Mater. Sci. Eng. A, 2011, 528(9): 3390-3397.

[144] Kumar R., Kumar S., Mehrotra S. P.. Towards sustainable solutions for fly ash through mechanical activation. Resourc. Conserv. Recyc, 2007, 52(2): 157-179.

[145] Kumar S., Kumar R., Alex T. C., Bandopadhyay A., Mehrotra S. P.. Influence of reactivity of fly ash on geopolymerisation. Adv. Appl. Ceram, 2007, 106(3):

120-127.

[146] Kumar S., Kumar R.. Mechanical activation of fly ash: effect on reaction, structure and properties of resulting geopolymer. Ceram. Int, 2011, 37(2): 533-541.

[147] Lee W. K. W., Van Deventer J. S. J.. Structural reorganisation of class F fly ash in alkaline silicate solutions. Colloids Surf. A, 2002, 211(1): 49-66.

[148] Škvára F., Kopecky L. Šmilauer V., Bittnar Z.. Material and structural characterization of alkali activated low-calcium brown coal fly ash. J. Hazard. Mater, 2009, 168(2-3): 711-720.

[149] Topçu I. B., Toprak M. U.. Properties of geopolymer from circulating fluidized bed combustion coal bottom ash. Mater. Sci. Eng. A, 2011, 528(3): 1472-1477.

[150] Xu H., Li Q., Shen L., Zhang M., Zhai J.. Low-reactive circulating fluidized bed combustion(CFBC) fly ashes as source material for geopolymer synthesis. Waste Manag, 2010, 30(1): 57-62.

[151] Songpiriyakij S., Kubprasit T., Jaturapitakkul C., Chindaprasirt P.. Compressive strength and degree of reaction of biomass and fly ashbased geopolymer. Constr. Build. Mater, 2011, 24(3): 236-240.

[152] Brindley G. W., Nakahira M.. The kaolinite-mullite reaction series. 2. Metakaolin. J. Am. Ceram. Soc, 1959, 42(7): 314-318.

[153] MacKenzie K. J. D., Brown I. W. M., Meinhold R. H., Bowden M. E.. Outstanding problems in the kaolinite-mullite reaction sequence investigated by 29Si and 27Al solid-state nuclear magnetic resonance. 1. Metakaolinite. J. Am. Ceram. Soc, 1985, 68(6): 293-297.

[154] Collins D. R., Fitch A. N., Catlow C. R. A.. Time-resolved powder neutron diffraction study of the thermal reactions in clay minerals. J. Mater. Chem, 1991, 1(6): 965-970.

[155] Gualtieri A., Bellotto M.. Modelling the structure of the metastable phases in the reaction sequence kaolinite-mullite by Xray scattering experiments. Phys. Chem. Miner, 1998, 25: 442-452.

[156] McConville C. J., Lee W. E., Sharp J. H.. Microstructural evolution in fired kaolinite. Br. Ceram. Trans, 1998, 97(4): 162-168.

[157] Brindley G. W., Nakahira M.. The kaolinite-mullite reaction series. 1. A survey of outstanding problems. J. Am. Ceram. Soc, 1959, 42(7): 311-314.

[158] Lee S., Kim Y. J., Lee H. J., Moon H.-S.. Electron-beam-induced phase transformations from metakaolinite to mullite investigated by EFTEM and HRTEM. J. Am. Ceram. Soc, 2001, 84(9): 2096-2098.

[159] Sperinck S., Raiteri P., Marks N., Wright K.. Dehydroxylation of kaolinite to metakaolin- a molecular dynamics study. J. Mater. Chem, 2011, 21(7): 2118-2125.

[160] White C. E., Provis J. L., Proffen T., Riley D. P., Van Deventer J. S. J.. Density functional modeling of the local structure of kaolinite subjected to thermal dehydrox-

ylation. J. Phys. Chem. A, 2010, 114(14): 4988-4996.

[161] Granizo M. L., Blanco-Varela M. T., Palomo A.. Influence of the starting kaolin on alkali- activated materials based on metakaolin. Study of the reaction parameters by isothermal conduction calorimetry. J. Mater. Sci, 2000, 35(24): 6309-6315.

[162] Zibouche F., Kerdjouj H., d'Espinose de la Caillerie J.-B., Van Damme H.. Geopolymers from Algerian metakaolin. Influence of secondary minerals. Appl. Clay Sci, 2009, 43(3-4): 453-458.

[163] Zhang Z. H., Yao X., Zhu H. J., Hua S. D., Chen Y.. Activating process of geopolymer source material: Kaolinite. J. Wuhan Univ. Technol.- Mater Sci. Ed, 2009, 24(1): 132-136.

[164] Provis J. L., Duxson P., Van Deventer J. S. J.. The role of particle technology in developing sustainable construction materials. Adv. Powder Technol, 2010, 21(1): 2-7.

[165] Marín-López C., Reyes Araiza J., Manzano-Ramírez A., Rubio Avalos J., Perez-Bueno J., Muñiz-Villareal M., Ventura-Ramos E., Vorobiev Y.. Synthesis and characterization of a concrete based on metakaolin geopolymer. Inorg. Mater, 2009, 45(12): 1429-1432.

[166] San Nicolas R.. Approche performantielle des bétons avec métakaolins obtenus par calcination fly ash. Ph. D. Thesis, Université de Toulouse, France, 2011.

[167] Živica V., Balkovic S., Drabik M.. Properties of metakaolin geopolymer hardened paste prepared by high-pressure compaction. Constr. Build. Mater, 2011, 25(5): 2206-2213.

[168] Lee S., Kim Y. J., Moon H. S.. Energy-filtering transmission electron microscopy (EF-TEM) study of a modulated structure in metakaolinite, represented by a 14 Å modulation. J. Am. Ceram. Soc, 2003, 86(1): 174-176.

[169] Wang M. R., Jia D. C., He P. G., Zhou Y.. Influence of calcination temperature of kaolin on the structure and properties of final geopolymer. Mater. Lett, 2010, 64(22): 2551-2554.

[170] Cioffi R., Maffucci L., Santoro L.. Optimization of geopolymer synthesis by calcination and polycondensation of a kaolinitic residue. Resour. Conserv. Recyc, 2003, 40(1): 27-38.

[171] Elimbi A., Tchakoute H. K., Njopwouo D.. Effects of calcination temperature of kaolinite clays on the properties of geopolymer cements. Constr. Build. Mater, 2011, 25(6): 2805-2812.

[172] Medri V., Fabbri S., Dedecek J., Sobalik Z., Tvaruzkova Z., Vaccari A.. Role of the morphology and the dehydroxylation of metakaolins on geopolymerization. Appl. Clay Sci, 2010, 50(4): 538-545.

[173] White C. E., Provis J. L., Gordon L. E., Riley D. P., Proffen T., Van Deventer J. S. J.. The effect of temperature on the local structure of kaolinite intercalated with

potassium acetate. J. Am. Ceram. Soc, 2011, 23(2): 188-199.

[174] White C. E. , Provis J. L. , Proffen T. , Riley D. P. , Van Deventer J. S. J. . Co-mbining density functional theory(DFT) and pair distribution function(PDF) analysis to solve the structure of metastable materials: the case of metakaolin. Phys. Chem. Chem. Phys, 2010, 12(13): 3239-3245.

[175] White C. E. , Perander L. M. , Provis J. L. , Van Deventer J. S. J. . The use of XANES to clarify issues related to bonding environments in metakaolin: a discussion of the paper S. Sperinck et al. , "Dehydroxylation of kaolinite to metakaolin-a molecular dynamics study", J. Mater. Chem, 21, 2118-2125. J. Mater. Chem, 2011, 21(19): 7007-7010.

[176] Van Jaarsveld J. G. S. , Van Deventer J. S. J. , Lukey G. C. . The effect of composition and temperature on the properties of fly ash- and kaolinite-based geopolymers. Chem. Eng. J, 2002, 89(1-3): 63-73.

[177] Van Jaarsveld J. G. S. , Van Deventer J. S. J. , Lukey G. C. . A comparative study of kaolinite versus metakaolinite in fly ash based geopolymers containing immobilized metals. Chem. Eng. Commun, 2004, 191(4): 531-549.

[178] Bernal S. A. , Rodríguez E. D. , Mejía de Gutierrez R. , Gordillo M. , Provis J. L. . Mechanical and thermal characterisation of geopolymers based on silicate-activated metakaolin/slag blends. J. Mater. Sci, 2011, 46(16): 5477-5486.

[179] Bernal S. A. , Provis J. L. , Mejía de Gutierrez R. , Rose V. . Evolution of binder structure in sodium silicate-activated slag-metakaolin blends. Cem. Concr. Compos, 2011, 33(1): 46-54.

[180] Krivenko P. V. , Petropavlovsky O. , Gelevera A. , Kavalerova E. . Alkali-aggregate reaction in the alkali-activated cement concretes. In: Bilek V. , Keršner Z. (eds.) Proceedings of the 4th International Conference on Non-Traditional Cement & Concrete, Brno, Czech Republic. ZPSV, a. s. 2011.

[181] Pacheco-Torgal F. , Castro-Gomes J. , Jalali S. . Investigations about the effect of aggregates on strength and microstructure of geopolymeric mine waste mud binders. Cem. Concr. Res, 2007, 37(6): 933-941.

[182] Pacheco-Torgal F. , Castro-Gomes J. , Jalali S. . Adhesion characterization of tungsten mine waste geopolymeric binder. Infl uence of OPC concrete substrate surface treatment. Constr. Build. Mater, 2008, 22(3): 154-161.

[183] Pacheco-Torgal F. , Castro-Gomes J. , Jalali S. . Tungsten mine waste geopolymeric binder: preliminary hydration products investigations. Constr. Build. Mater, 2009, 23(1): 200-209.

[184] Pacheco-Torgal F. , Castro-Gomes J. P. , Jalali S. . Investigations on mix design of tungsten mine waste geopolymeric binder. Constr. Build. Mater, 2008, 22 (9): 1939-1949.

[185] Pacheco-Torgal F. , Castro-Gomes J. P. , Jalali S. . Investigations of tungsten mine

waste geopolymeric binder: Strength and microstructure. Constr. Build. Mater, 2008, 22(11): 2212-2219.

[186] Buchwald A., Hohmann M., Posern K., Brendler E.. The suitability of thermally activated illite/ smectite clay as raw material for geopolymer binders. Appl. Clay Sci, 2009, 46(3): 300-304.

[187] MacKenzie K. J. D., Komphanchai S., Vagana R.. Formation of inorganic polymers (geopolymers) from 2:1 layer lattice aluminosilicates. J. Eur. Ceram. Soc, 2008, 28(1): 177-181.

[188] MacKenzie K. J. D., Brew D. R. M., Fletcher R. A., Vagana R.. Formation of aluminosilicate geopolymers from 1:1 layer-lattice minerals pre-treated by various methods: a comparative study. J. Mater. Sci, 2007, 42(12): 4667-4674.

[189] MacKenzie K. J. D.. Utilisation of non-thermally activated clays in the production of geopolymers. In: Provis J. L., Van Deventer J. S. J. (eds.) Geopolymers: Structure, Processing, Properties and Industrial Applications, 2009: 296-316. Woodhead, Cambridge.

[190] Yang K.-H., Hwang H.-Z., Lee S.. Effects of water-binder ratio and fine aggregate-total aggregate ratio on the properties of hwangtoh-based alkali-activated concrete. J. Mater. Civil Eng, 2010, 22(9): 887-896.

[191] Gomes K. C., Torres S. M., De Barros S., Vasconcelos I. F., Barbosa N. P.. Mechanical properties of geopolymers with iron rich precursors. In: Palomo A. (ed.) 13th International Congress on the Chemistry of Cement, Madrid, Spain. CD-ROM proceedings, 2011.

[192] Gomes K. C., Lima G. S. T., Torres S. M., De Barros S., Vasconcelos I. F., Barbosa N. P.. Iron distribution in geopolymer with ferromagnetic rich precursor. Mater. Sci. Forum, 2010, 643: 131-138.

[193] Xu H., Van Deventer J. S. J.. Geopolymerisation of multiple minerals. Miner. Eng, 2002, 15(12): 1131-1139.

[194] Xu H., Van Deventer J. S. J.. Effect of source materials on geopolymerization. Ind. Eng. Chem. Res, 2003, 42(8): 1698-1706.

[195] Xu H., Van Deventer J. S. J.. The effect of alkali metals on the formation of geopolymeric gels from alkali-feldspars. Colloids Surf. A, 2003, 216(1-3): 27-44.

[196] Xu H., Van Deventer J. S. J.. Factors affecting the geopolymerization of alkali-feldspars. Miner. Metall. Proc, 2002, 19(4): 209-214.

[197] Xu H., Van Deventer J. S. J.. The geopolymerisation of alumino-silicate minerals. Int. J. Miner. Proc, 2000, 59(3): 247-266.

[198] Xu H., Van Deventer J. S. J., Roszak S., Leszczynski J.. Ab initio study of dissolution reactions of 5-membered aluminosilicate framework rings. Int. J. Quant. Chem, 2004, 96(4): 365-373.

[199] Xu H., Van Deventer J. S. J.. Microstructural characterisation of geopolymers syn-

thesised from kaolinite/stilbite mixtures using XRD, MAS-NMR, SEM/EDX, TEM/EDX, and HREM. Cem. Concr. Res, 2002, 32(11): 1705-1716.

[200] Pacheco-Torgal F., Jalali S.. Influence of sodium carbonate addition on the thermal reactivity of tungsten mine waste mud based binders. Constr. Build. Mater, 2010, 24(1): 56-60.

[201] Leonelli C., Kamseu E., Boccaccini D. N., Melo U. C., Rizzuti A., Billong N., Piselli P.. Volcanic ash as alternative raw materials for traditional vitrified ceramic products. Adv. Appl. Ceram, 2007, 106(3): 141-148.

[202] Kamseu E., Leonelli C., Perera D. S., Melo U. C., Lemougna P. N.. Investigation of volcanic ash based geopolymers as potential building materials. Interceram, 2009, 58(2-3): 136-140.

[203] Allahverdi A., Mehrpour K., Najafi Kani E.. Investigating the possibility of utilizing pumicetype natural pozzolan in production of geopolymer cement. Ceram.-Silik, 2008, 52(1): 16-23.

[204] Bondar D., Lynsdale C. J., Milestone N. B., Hassani N., Ramezanianpour A. A.. Effect of adding mineral additives to alkali-activated natural pozzolan paste. Constr. Build. Mater, 2011, 25(6): 2906-2910.

[205] Bondar D., Lynsdale C. J., Milestone N. B., Hassani N., Ramezanianpour A. A.. Effect of heat treatment on reactivity-strength of alkali-activated natural pozzolans. Constr. Build. Mater, 2011, 25(10): 4065-4071.

[206] Najafi Kani E., Allahverdi A.. Effect of chemical composition on basic engineering properties of inorganic polymeric binder based on natural pozzolan. Ceram.-Silik, 2009, 53(3): 195-204.

[207] Najafi Kani E., Allahverdi A.. Effects of curing time and temperature on strength development of inorganic polymeric binder based on natural pozzolan. J. Mater. Sci, 2009, 44: 3088-3097.

[208] Chávez-García, M. L., García T. A., de Pablo L.. Synthesis and characterization of geopolymers from clinoptilolite tuff. In: Palomo A. (ed.) 13th International Congress on the Chemistry of Cement, Madrid, Spain. CD-ROM proceedings, 2011.

[209] Najafi Kani E., Allahverdi A., Provis J. L.. Efflorescence control in geopolymer binders based on natural pozzolan. Cem. Concr. Compos, 2012, 34(1): 25-33.

[210] Bondar D., Lynsdale C. J., Milestone N. B., Hassani N., Ramezanianpour A. A.. Effect of type, form, and dosage of activators on strength of alkali-activated natural pozzolans. Cem. Concr. Compos, 2011, 33(2): 251-260.

[211] Bondar D., Lynsdale C. J., Milestone N. B., Hassani N.. Oxygen and chloride permeability of alkali-activated natural pozzolan concrete. ACI Mater. J, 2012, 104(1): 53-62.

[212] Glukhovsky V. D.. Gruntosilikaty(Soil Silicates). Gosstroyizdat, Kiev, 1959.

[213] Bondar D., Lynsdale C. J., Milestone N. B., Hassani N., Ramezanianpour A. A..

Engineering properties of alkali-activated natural pozzolan concrete. ACI Mater. J, 2011, 108(1): 64-72.

[214] Zosin A. P., Priimak T. I., Avsaragov K. B.. Geopolymer materials based on magnesia-iron slags for normalization and storage of radioactive wastes. At. Energy, 1998, 85(1): 510-514.

[215] Komnitsas K., Zaharaki D., Perdikatsis V.. Geopolymerisation of low calcium ferronickel slags. J. Mater. Sci, 2007, 42(9): 3073-3082.

[216] Komnitsas K., Zaharaki D., Perdikatsis V.. Effect of synthesis parameters on the compressive strength of low-calcium ferronickel slag inorganic polymers. J. Hazard. Mater, 2009, 161(2-3): 760-768.

[217] Zaharaki D., Komnitsas K., Perdikatsis V.. Use of analytical techniques for identification of inorganic polymer gel composition. J. Mater. Sci, 2010, 45(10): 2715-2724.

[218] Komnitsas K., Zaharaki D.. Utilisation of low-calcium slags to improve the strength and durability of geopolymers. In: Provis J. L., Van Deventer J. S. J. (eds.) Geopolymers: Structure, Processing, Properties and Industrial Applications, 2009: 345-378. Woodhead, Cambridge.

[219] Hos J. P., McCormick P. G., Byrne, L. T.. Investigation of a synthetic aluminosilicate inorganic polymer. J. Mater. Sci, 2002, 37(11): 2311-2316.

[220] Gordon M., Bell J. L., Kriven W. M.. Comparison of naturally and synthetically derived, potassium-based geopolymers. Ceram. Trans, 2005, 165: 95-106.

[221] Cui X.-M., Zheng G.-J., Han Y.-C., Su F., Zhou J.. A study on electrical conductivity of chemosynthetic Al_2O_3-$2SiO_2$ geopolymer materials. J. Power Sourc, 2008, 184(2): 652-656.

[222] Zheng G., Cui X., Zhang W., Tong, Z.. Preparation of geopolymer precursors by sol-gel method and their characterization. J. Mater. Sci, 2009, 44: 3991-3996.

[223] Fernández-Jiménez A., Vallepu R., Terai T., Palomo A., Ikeda K.. Synthesis and thermal behavior of different aluminosilicate gels. J. Non-Cryst. Solids, 2006: 352: 2061-2066.

[224] García-Lodeiro I., Fernández-Jiménez A., Blanco M. T., Palomo A.. FTIR study of the sol-gel synthesis of cementitious gels: C-S-H and N-A-S-H. J. SolGel Sci. Technol, 2008, 45(1): 63-72.

[225] Vallepu R., Fernández-Jiménez A. M., Terai T., Mikuni A., Palomo A., MacKenzie K. J. D., Ikeda K.. Effect of synthesis pH on the preparation and properties of K-Al-bearing silicate gels from solution. J. Ceram. Soc. Japan, 2006, 114(7): 624-629.

[226] Phair J. W., Smith J. D., Van Deventer J. S. J.. Characteristics of aluminosilicate hydrogels related to commercial "Geopolymers". Mater. Lett, 2003, 57(28): 4356-4367.

[227] Hajimohammadi A., Provis J. L., Van Deventer J. S. J.. One-part geopolymer mixes from geothermal silica and sodium aluminate. Ind. Eng. Chem. Res, 2008, 47(23): 9396-9405.

[228] O'Connor S. J., MacKenzie K. J. D.. Synthesis, characterisation and thermal behaviour of lithium aluminosilicate inorganic polymers. J. Mater. Sci, 2010 45, (14): 3707-3713.

[229] O'Connor S. J., MacKenzie K. J. D.. A new hydroxide-based synthesis method for inorganic polymers. J. Mater. Sci, 2010, 45(12): 3284-3288.

[230] Brew D. R. M., MacKenzie K. J. D.. Geopolymer synthesis using silica fume and sodium aluminate. J. Mater. Sci, 2007, 42(11): 3990-3993.

[231] Bell J. L., Driemeyer P. E., Kriven W. M.. Formation of ceramics from metakaolin-based geopolymers: Part I-Cs-based geopolymer. J. Am. Ceram. Soc, 2009, 92(1): 1-8.

[232] Bell J. L., Driemeyer P. E., Kriven W. M.. Formation of ceramics from metakaolin-based geopolymers. Part II: K-based geopolymer. J. Am. Ceram. Soc, 2009, 92(3): 607-615.

第 5 章　胶凝材料化学——复合体系和中钙体系

J. L. Provis，S. A. Bernal

5.1　引言

前两章分别对高钙含量与低钙含量的碱激发胶凝体系进行了探讨，本章将简要讨论由中钙含量前驱体和复合前驱体在复合胶凝体系中的研究进展。近年来，以低碳高强耐久性优异的可替代胶凝体系的迫切需要和第 3、4 章中现有的对高钙和低钙含量碱激发胶凝材料（AAMs）强度发展和耐久性的较成熟的理解，激发了人们对复合体系的关注。此类胶凝材料利用硅酸盐水泥熟料和碱激发 BFS（主要是 C-S-H 凝胶）或碱激发铝硅酸盐（地质聚合物凝胶）[1-3]之间的水化反应产物共存的特性，发挥其机械强度和耐久性之间的协同作用。混合富含活性钙的高铝硅酸盐水泥（包含硅酸盐水泥熟料）和碱为利用铝硅酸盐固废提供可能，这提供了一种增值固废材料的途径。但单独激发时获得的工业产品没有良好的强度发展。

本章将简要讨论存在于 AAM 体系构成中钙的凝胶相化学，同时，对不同前驱体材料的组合在碱激发胶凝体系中的发展进行了展望。这些主要集中在科研方面而非仅仅应用于工业发展上，在大量发表的相关技术论文中都只描述，在研究较为成熟的 AAM 体系中以低体积百分率掺杂一种新型废弃物所获得的强度发展及显微结构（主要通过显微镜学、衍射学和红外线以及光谱分析）。这些刊物在碱激发化学稳步发现的可靠性研究中发挥了重要作用。由于科学研究数据有限，在一本著作中详尽地描述研究成果是不可能的。

5.2　复合胶凝体系中的共生凝胶

第 3 章和第 4 章探讨碱激发材料中凝胶相形成的化学过程，高钙体系（导致形成托勃莫来石类 C-S-H 凝胶）和低钙体系（导致形成类沸石 N-A-S-H 凝胶）这两种类型凝胶具有良好的性能，使得在整个材料中发挥最大化效应的利用变得合理可靠[4]。在碱性介质中，高活性铝硅酸盐和钙源水化反应后，C-S-H 和 N-A-S-H 凝胶是共存体系中的主要结构。在此过程中需控制 pH 值，以防止活性钙离子通过 $Ca(OH)_2$ 的形式沉淀下来[4-9]。

两种纯凝胶和复合方式的配合比设计、材料制备都已有研究[10-13]。与 AAM 体系相似，pH 值在合成相形成中扮演着决定其稳定性的重要角色。当 pH>11 时，C-S-H 类型凝胶富存于 SiO_2 中，比传统的硅酸盐水泥具有更高的聚合度。当 pH>12.5 时，N-A-S-H 凝胶的显微结构与碱激发胶凝材料反应所得生成物相似[10]。

在高 Ca/Si 比以及高 Al 和低 Al 的条件下，二次反应产物 $Ca(OH)_2$ 大量生成[11]。不同的碱性条件会导致 C-S-H 凝胶变为含 Ca 的 N-S-H 凝胶[11]。Bernal[14,15]和 Ben Haha[16]等人也得到相同的结果。此外，在高碱性条件下，胶凝体系中铝和碱的同时反应会导致

C-S-H 相和类水化钙黄长石晶型相的相互交联[12]。研究表明，Na_2O-CaO-SiO_2-Al_2O_3-H_2O 共沉淀胶凝体系中，高碱性环境下 Ca^{2+} 会取代固相中的 Na^+ 形成富含高浓度 Ca 和 Na 型的 C-A-S-H 凝胶，从而破坏 N-A-S-H 凝胶结构[13]。当生成的 C-A-S-H 和 N-A-S-H 凝胶在水中混合时（图 5-1），通过 TEM-EDX 对两种凝胶后期产物的化学特性观察分析，它们并非显示两者混合性质，这意味着在适宜的条件下形成的共存混合物显示出其真实的热力学相。

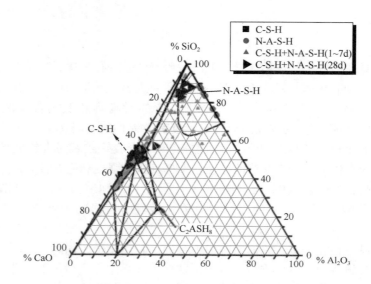

图 5-1　在超纯水中混合预先合成的 C-S-H 和 N-A-S-H 凝胶沉淀后 1～7d（小灰色三角形）和 28d（大三角形）的 TEM-EDX 数据得到类三元相图

不同区域对应不同相，龄期越长，其化学组成区别越明显（改编自文献[13]）

5.3　中钙体系中的激发剂

中钙 AAM 胶凝体系中，最佳激发剂的选择主要依据使用的前驱体的性质。正如第 3、4 章所述，氢氧化物激发剂对低 Ca 胶凝体系的效果较好，硅酸盐和硫酸盐类适用于高钙胶凝体系。硅酸盐类激发剂是使用范围最为广泛的胶凝合成物。这都为中钙胶凝体系研究奠定了基础，该体系在硫酸盐和碳酸盐碱激发作用下形成高 pH 值的 Ca 体系，为铝硅酸盐混合物提供了有利的反应条件。

当复合体系中的钙含量较高时，可通过在前驱体粉末中加水获得合适的凝结时间和强度发展；比如 C 类粉煤灰基混凝土已在美国落基山脉生产[17,18]，尽管生产工艺由于缺少碱源而超出碱激发反应类别。含有熟料的材料比如 Pyrament 复合水泥体系[19]仅加水就可以被激发。然而，在 AAMs 体系中为不同常规的特例。一般而言，在第 3、4 章的激发剂选择用于中钙体系也可以，读者们可以参考这些章节内容。

在本章中，以下部分将简要探讨用于生产中钙含量 AAMs 体系的单一前驱体或复合前驱体或适用于每种类型的前驱体激发剂。

5.4 单一原料

5.4.1 高钙粉煤灰

根据 ASTMC618[20]标准,高钙粉煤灰被定义为 C 类灰,通常含有超过 20%的 CaO(有时高达 40%),类似于欧洲标准中的钙质灰。在 AAM 合成中,高钙灰分布在世界上煤炭资源丰富的地区,尤其是在美国西部和亚洲、欧洲地区的褐煤或次烟煤,但其未像第 4 章所讨论的低钙灰(F 类)那样受到重视。

C 类粉煤灰的 AAMs 最早出现在专利里,而非研究论文里。1991 年,Lone Star 公开的专利是由柠檬酸和碱氢氧化物或碳酸盐与 C 类粉煤灰混合的体系[21];1996 年,路易斯安那州立大学发明一种不含柠檬酸的化合物[22]。源自新西兰亨利特的 C 类粉煤灰,具有很好的强度发展性能,它与硅酸盐水泥混凝土相比,徐变和收缩性能相当[27]。

利用 C 类粉煤灰制得的 AAM 凝胶最大优点是凝胶结构致密,并且小孔多为有序排列且孔径是纳米级的。如图 5-2 所示,分别以相同方法制备 F 类粉煤灰和 C 类粉煤灰样品,这些胶凝孔道中都填充着可以在实验条件下进入大于 11nm 孔径的合金(伍德合金),在显微镜下可以明显地观察到孔径特征。F 类粉煤灰基胶凝材料被金属大量渗透,而 C 类粉煤灰基则显示被伍德合金微量填充,这表明 C 类粉煤灰基胶凝材料具有较低的渗透孔隙率。

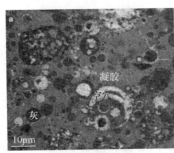
(a)F类粉煤灰

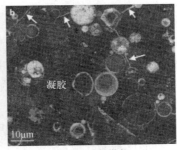
(b)C类粉煤灰

图 5-2 由粉煤灰与偏硅酸钠激发剂反应产生的 AAM 样品的背散射电子图像。伍德合金侵入孔中,因此通过大于 11nm 并与外部相通的孔在两个图像中都显示为亮区域。(a)中的凝胶明显比(b)亮,表明凝胶孔隙率高得多。(b)中的箭头表示由未反应的粉煤灰颗粒导致凝胶相裂纹,使液态金属覆盖这些区域(改编自文献[28])

Diaz-Loya 等人[29]对来自北美不同燃煤发电厂的 25 种灰进行碱激发实验,发现 C 类灰具有最好的强度发展。也并不是所有报道都是高钙灰比低钙灰能够产生强度较高的 AAM 胶凝材料,Winnefeld 等人[30]和 Oh 等人[31]发现与 F 类灰相比,各种高钙灰在强度方面的性能发展和凝胶相演化不太理想。Winnefeld 等人的研究表明[30],所测试的德国高钙灰中的 SO_3 含量非常高(在氧化物基础上高达 14%,超过 ASTM C618 Class C 粉煤灰中的允许最大值 5%)和 Al_2O_3 含量低;四种灰中两种具有极低的高聚物组分($SiO_2+Al_2O_3+Fe_2O_3$),只满足 ASTM C618[20]分类为 C 级所需最少组分的 50%。因此,AAM 胶凝材料强度发展较低。综上所述,较高钙灰而言,低钙(F 级)灰可提供更好的强度发展。类似地,在 Oh 等人的研究中[31]，C类灰在有限的激发剂作用下14d抗压强度为35MPa，但受激发剂模数的变化而不稳定，在较高或较低激发剂模数下强度都非常低。Guo等人[32]也发现了类似的情况。

高钙灰的主要缺点之一是同一等级不同灰之间的差异性（无论是否符合C类要求ASTM C618）比低钙灰大，导致产品稳定性比低钙灰基AAM体系所得产品差。这就要求对每种灰进行更仔细的配合比设计工作。Díaz-Loyae等人[29]研究表明，C类灰需水量高、初凝时间短（小于5min），是对大批量生产混凝土的一大挑战。

5.4.2 其他单一原料

在AAMs体系中，中钙含量的单一原料在文献中很少提到。在Fares和Tagnit-Hamou的研究中提到[33]，炼铝底渣（质量分数14.6%CaO）用NaOH激发形成胶凝材料，蒸养28d后强度可达60MPa。Tashima等人[34]利用废铝硅酸钙玻璃通过碱激发得到的砂浆强度超过80MPa，这表明对某些地区废弃材料进行合理商业定位，极有可能在不久的将来可以用碱激发技术生产出高强度材料。

5.5 复合原料

5.5.1 铝硅酸盐+$Ca(OH)_2$+碱

胶凝体系中包括不含钙或含少量钙的铝硅酸盐，还有氢氧化钙和碱，这些体系是典型的且经常用来分析火山灰反应的模型，但这并非本章的重点。本章主要分析高碱在体系中的碱诱导反应活性。Alonso等人[7,8]发现在强碱作用下，$Ca(OH)_2$作为激发剂与偏高岭土会形成无定形铝硅酸钠凝胶，与缺少$Ca(OH)_2$的情况相似。这可能由于$Ca(OH)_2$在高pH值下的溶解度过低，对反应无影响。C-S-H型凝胶也被报道是在弱碱即NaOH环境下MK和$Ca(OH)_2$激发所致的工业副产物[7,8,35]。Dombrowski等人[9]在相同的碱激发条件下研究FA/$Ca(OH)_2$比率对形成碱激发胶凝材料的影响，发现胶凝材料中$Ca(OH)_2$含量的增加促进了C-S-H型凝胶的形成，力学强度和胶凝结构得到改善。

Shi发现Na_2SO_4激发粉煤灰与石灰混合物会有C-S-H和AFt相（高硫型水化硫铝酸钙）水化产物的生成[36]。而Williams等[37]用$Ca(OH)_2$和NaOH激发粉煤灰，其产物主要以C-S-H为主和少量排列有序的钙铝榴石。在天然火山灰中添加$Ca(OH)_2$和Na_2SO_4作为激发剂，发现C-S-H凝胶中产生AFm（单硫型水化硫铝酸钙）和AFt相[38-40]，以4:1混合火山灰和$Ca(OH)_2$制得此化学类激发剂。无论在机械过程还是热处理过程，这是一种相对有效的控制凝胶行程的方法[41]。

5.5.2 煅烧黏土+矿渣+碱

在上文的基础上，本文对BFS和偏高岭土混合形成碱激发胶凝材料的进展进行详细介绍。Yip和Van Deventer[5]提出了BFS与偏高岭土的碱激发高度依赖于碱激发剂的碱性和固体前驱体的共混比。BFS在强碱条件下其颗粒（粒化矿渣）被快速反应所覆盖。低溶解度的Ca^{2+}离子易形成$Ca(OH)_2$沉淀而非生成C-S-H型凝胶。这种体系的特征在于通过偏高岭土的激发形成铝硅酸盐凝胶,而钙通常沉淀为 $Ca(OH)_2$。

另一方面,在弱碱条件下(通过稀释的激发剂溶液或具有超过 2.0 的激发剂模数),GBFS 中的钙在早期溶解,形成 C-S-H 型凝胶,铝硅酸盐反应形成高 Al 凝胶,进而导致 N-A-S-H 和 C-S-H 型共存凝胶。在这些体系中形成的反应产物的性质与强度发展也被报道过[4,6]。在强碱条件下形成的 N-A-S-H 凝胶承担主要强度;而在弱碱条件下,则以 C-S-H 凝胶为主,且在 C-S-H 和 N-A-S-H 凝胶共存体系中含量最高[6]。这些结果与 Buchwald 等人[42]在复合胶凝体系中观察到的结果一致。当凝胶共存发生时,外加 Al 嵌入 C-S-H 型凝胶中,导致该产品中平均链长度增加,地质聚合物凝胶呈现较低的交联度。

BFS 和偏高岭土的复合可有效改善反应进程[43-45]。同时激发剂碱度的增加会加速缩合反应[43,44],提高孔溶液中的铝浓度,促进 C-S-H 凝胶的冷凝[43],改变反应历程。然而,这种关系取决于偏高岭土前驱体复合物的含量和溶液模数。Bernal 等人[14,44,46]仅用激发 BFS(高模数和相对低 Na_2O 浓度的激发剂)的胶凝材料,在相同条件下激发 BFS/偏高岭土复合物,发现随着偏高岭土含量增加,凝固时间增加,反应产物减少。在此阶段,BFS 和偏高岭土的溶解-冷凝过程同时发生,可用的相对含量低的碱快速形成铝硅酸盐凝胶,抑制 MK 之后的溶解。

当 KOH 作为激发剂时,偏高岭土含量高会降低 BFS/偏高岭土胶凝体系的反应程度[47]。在这种情况下,需要更高的 SiO_2/K_2O 比来延长凝固时间,同时会降低机械强度。Lecomte 等人[48]确定了混合物中的细小 BFS 颗粒与煅烧黏土是初始硬化和早期强度发展的主要来源。他们确定 KOH 和高模数硅酸钠复合激发的 BFS 和煅烧黏土为 2∶1 的胶凝材料中的高配位 [$Q^4(nAl)$] 和低配位(Q^1 和 Q^2)Si。这些作者都是先混合黏土和激发剂,30min 后加入 BFS。这使得容易反应的黏土溶解,但提出了关于凝胶相在动力学上是稳定的,而非在热力学上接近长期稳定的问题。Zhang 等人[49]在固化重金属废物中确定了一个 BFS/偏高岭土以 1∶1 时最优的强度和固化效果,而 Bernal 等人[14,15,46,50,51]发现更高含量的偏高岭土需要使用更高的激发剂模数,抗压强度的增加与激发剂模数呈现函数关系。Burciaga-Díaz 等人[52]也发现了类似的结果。

当受热导致收缩和开裂过程被更好地控制且复合体系中的损害更小时,混合 BFS-偏高岭土样品在 1000℃ 煅烧后的残余强度也比不含偏高岭土 BFS 基材料更好[53]。

5.5.3 粉煤灰+矿渣+碱

长期以来,粉煤灰、BFS 和碱的组合是一个有前途的 AAM 胶凝配方。第一篇关于此领域的论文可追溯到 1997 年[54],其研究重点是与 Trief 过程进行比较,用于制造钙含量低于硅酸盐的替代性凝胶水泥,其中胶凝体系作为研究的一部分。学者克服高碱含量下风化、养护条件导致的水损失以及反应进程等问题,制备出抗压强度在 28d 时约为 100MPa 并且在 180d 时增加至 120MPa 的性能优异的材料[25]。Bijen 和 Waltje[55]也指出了 NaOH 激发材料的风化和需水量大的问题,并发现对 NaOH 用量超过干燥胶凝质量的 4% 时,对反应几乎没有影响。虽然并未说明养护条件,但是这些结果也符合由于碱浸出在水下固化 AAMs 时观察到的难题。

在许多研究中,BFS 的加入有利于粉煤灰基胶凝体系的凝固和强度发展;例如,Li 和 Liu[56]发现 4% 的 BFS 使组成为 90% 粉煤灰/10% 偏高岭土混合物在 14d 时固化强度增加

40%。Kumar 等人[57]发现在较低温度下固化时,BFS 对体系的化学性质有重要影响,而当固化温度升高时,粉煤灰则发挥更大作用。这可能是由于粉煤灰、BFS 这两种物质与硅酸盐水泥混合时的表面活化能不同所导致的[58,59]。

20 世纪 90 年代,以 BFS-粉煤灰、偏硅酸钠激发剂以及少量硅酸盐熟料作为原材料制备的名为 Diabind 的耐酸性材料,已用于管材生产进行销售[60]。最近,BFS-粉煤灰混合物经过技术、商业和监管部门的审查,已经成为一种主要的商用 AAM 体系材料(在澳大利亚由 Zeobond 作为"E-Crete"销售)[2]。在 AAM 体系中,前驱体的独特复配设计展现了巨大的商业价值。

在中国,碱激发材料有重大发展。正如 Shi 和 Qian[61]以及 Pan 和 Yang[62]所述,中国具有丰富的粉煤灰和 BFS 资源。目前,中国致力于通过复合激发剂提高强度和改善耐久性的研究,其他的碱源如碳酸盐与 NaOH 的复合激发剂的研究发现取得重大进展[62]。

Puertas 等人[63]研究了激发剂掺量、BFS/粉煤灰配合比以及固化温度对力学性能的影响,发现配合比是最重要因素。其主要反应产物是 C-S-H 型凝胶,还有具有高浓度四面体的结合铝和嵌入其结构中钠离子[63]。水化碱-铝硅酸盐凝胶相类似于那些碱激发粉煤灰的特征也被证实[64]。这可能是由于粉煤灰在中钙含量的碱性介质中的溶解度得到改善,并且粉煤灰在碱激发反应过程的参与程度相应增加[65]。NMR 结果作为一部分的碳化数据[66]证明了凝胶共存理论。在复合粉煤灰/BFS 胶凝体系中发现加速碳化改变了 C-S-H 型凝胶的结构而非改变 N-A-S-H 型凝胶。复合胶凝体系中添加 BFS 会增加 Na^+ 的孔溶液浓度,表明该离子与 C-S-H 型凝胶的结合度高于 N-A-S-H 型凝胶,这分别与基于链和基于框架结构微观化学相符合[67]。

Escalante García 等人[68]也进行了激发条件和 BFS-粉煤灰 AAM 胶凝体系配合比的参数研究,发现(类似于如在前面部分中讨论的 BFS-煅烧黏土复合物,其化学机理相同)当粉煤灰含量较高时,低的激发剂模数和较高的激发剂用量对良好强度的发展是必要的。

Lloyd 等人[28,69,70]通过扫描电子显微镜中的元素组成图和背散射电子成像发现反应边界存在部分反应的 BFS 颗粒,其形态和化学特征与不含粉煤灰的体系中观察到的不同,特别是 BFS 反应边界中的 Si 耗尽区[69,70],表明 BFS 的钙会影响粉煤灰的反应,而且粉煤灰的铝和硅导致 BFS 颗粒反应途径的差异。精确控制这个过程仍然略有难度,仍需不断研究。

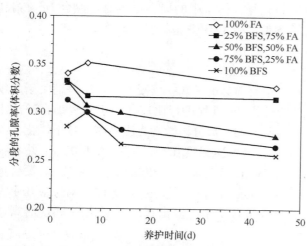

图 5-3 复合 BFS-粉煤灰胶凝材料的孔隙率随组成与养护时间的变化图,用 X 射线显微层析测定(改编自文献[3])

在保证耐久性前提下,每种前驱体对复合胶凝体系中孔隙率和孔结构的影响尤为重要。测定密度时的孔隙率并没有显示与矿渣含量相关[28],可能是由于 C-S-H 型凝胶在高温干燥操作中引起脱水反应[71],从 X 射线断层扫描获得的数据表明,矿渣含量增加,孔隙率减小(图 5-3)[3]。有趣的是,断层扫描数据显示含量为 50% 矿渣或以上的样品有明显区别，孔隙率随时间降低，相比之下，含量低于50%的矿渣则没有这种区别。虽然这个矿粉比例可能会因使用的粉煤灰和矿渣不同造成反应活性不同（特别是钙含量），但一般认为，类似于碱激发矿渣的 AAM 胶凝体系结构在矿渣含量低时是可以实现的，在世界上一些矿渣供应受限的地方尤为重要，或者添加另一种铝硅酸盐组分以控制流变性或凝结时间。Lloyd 等人[72]通过直接将这些结果与耐酸性进行比较，显示出较好的相关性，断层扫描数据发现酸渗透程度随矿渣量从0%到100%之间增大而增加，虽然这两个研究中测试的矿渣和粉煤灰的来源不同。但1∶1比例构成的复合物与全由矿渣组成的样品有相似的性能。

5.5.4 赤泥复合胶凝体系

赤泥是拜耳法提取氧化铝的碱性副产物，其中铝土矿矿石在 NaOH 溶液中被消化。提铝后固体残渣中含有大量的 NaOH。当前，赤泥经脱水后直接送到尾矿大坝。材料因为含有 Fe 的氧化物所以显红色，通常在不纯的铝土矿中也富含杂质 Si，如果工艺操作效率低于100%效率（最常见的情况），也常含一些残留的 Al。这种材料产量估计为 3500 万 t[73]，是完全没有利用的并且能够提供碱激发材料（AAM）生产所需物质，特别是潜在的活性碱、Ca、Al 和 Si，其原材料单一便宜。许多赤泥中活性 Al 含量低是 AAMs 使用面临的最大困难；因为拜耳法是为了提取 Al，并不是把 Al 残留在废物中。因此，除了少量直接由硅酸钠激发的富铝的赤泥研究工作以外，通常用赤泥作为唯一的胶凝组分研究尚未成功。

高岭土[73]、煅烧油页岩[75]和粉煤灰[76]等作为补充 Al 源的材料，已成功地用于含赤泥的碱激发复合物的研究中。然而，在上文讨论的高 Ca 胶凝体系中，涉及最多的是使用赤泥和矿渣结合而非只用矿渣进行低成本生产高性能胶凝材料，这种方法制得的胶凝产品性能值得期待。Pan 等人[77,78]研究了使用固体硅酸钠来激发矿渣和赤泥组成的复合物，发现了赤泥本身并没有产生好的强度发展，但矿渣和赤泥一起被激发，产生了很高强度。通过固体硅酸钠和硅铝酸钠复合激发矿渣-赤泥混合物，抗压强度在 28d 时超过 55MPa，并且冻融破坏、碳化速率和碳化加速后的强度损失低于控制下的硅酸盐水泥[79,80]。煅烧相对富铝的赤泥也显示出，通过增强氧化铝的活性，地质聚合物性能也会得到改善。

5.5.5 硅酸盐水泥＋铝硅酸盐＋碱：复合水泥

为更好地利用硅酸盐水泥熟料和碱激发铝硅酸盐为组分的复合胶凝材料，应充分考虑两种组分结构发展的机理差异，控制反应生成条件和优化配合比设计，有助于胶凝体系中强度和耐久性的提高。近年来，硅酸盐水泥在氢氧化钠、碳酸钠、硅酸钠和硫酸钠等不同混合碱性条件下水化的研究已进行大量工作，并在很大程度上超出了本文的范围。本章将具体地分析碱与添加的铝硅酸盐相互作用形成的产品的强度发展以及硅酸盐熟料或水泥发挥的次要作用。这种复合水泥体系，特别是在生产中添加含有铝硅酸盐的粉煤灰，得到的产品具有高性能、低 CO_2 特点，最近在国际会议研究和应用中越来越受关注。

硫铝酸钙（钙矾石 AFt）和单硫型水化硫铝酸钙（AFm）相因复合 AAM 胶凝体系中的普通硅酸盐水泥（OPC）组分产生的水化产物，明显不同于标准普通硅酸盐水泥水化形成的水化产物。高浓度的 Na^+ 作为氢氧化物或硫酸盐溶液有利于 AFm 结构的生成，碱取代导致钙矾石成"U 相"[86-91]，是通过添加大量硫酸盐来实现的。反应过程也受到强碱条件下

Ca(OH)$_2$溶解度低的影响。硅酸钠在硅酸盐水泥熟料的水化中具有强烈的促进作用。Blazhis 和 Rostovskaya[96]向硅酸钠/氟化钾混合物掺入硅酸盐水泥熟料中，生产的胶凝材料的早期强度明显升高（1d 固化后抗压强度高达 86MPa）。氟化钾控制凝结时间，复合激发剂有利于机械强度的发展和形成结晶及无序的硅酸盐与铝硅酸盐胶凝相产物。

AAM-硅酸盐水泥体系方面，关于铝硅酸盐物质与硅酸盐熟料结合[97]或硅酸盐熟料本身[4,83,92,98-100]的研究表明，其对胶凝材料耐久性、机械性能的研究产生深远影响。Krivenko 等人[101]发现硅酸钠激发粉煤灰-硅酸盐水泥复合胶凝体系具有优异的固化危险废物（Pb）能力，同时蒸汽养护有利于其机械强度的发展。

Gelevera 和 Munzer[102]通过对碱硅酸盐激发的硅酸盐熟料/BFS 复合物的研究发现，高含量 BFS 与硅酸盐碱性溶液反应，抗压强度会升高。硅酸盐熟料/BFS 复合物通过碱激发具有优异的抗冻性[102]。Fundi[103]研究了由 50%～70%天然火山灰（火山岩）和 30%～50%硅酸盐水泥被不同的钠和钾复合物激发。在 50%普通硅酸盐水泥配置的样品中，K$_2$CO$_3$ 相对无效，而硅酸钠（特别是）和硫酸钠被验证为最有效的化学激发剂，当以低至 3%质量的剂量加入时，其硬化浆体的力学性能提高多达 100%。向 70%硅酸盐水泥中掺入硫酸钠的天然火山灰可得到高性能的硅酸盐水泥（在 12 个月后抗压强度为 74MPa）。在这些条件下，用硅藻土、方解石和硫铝酸盐且不存在硬化胶凝材料情况下，形成低硅酸钙水化产物并逐渐促进碱硅铝酸盐的结晶。

Sanitskii[98,99]发现碱金属碳酸盐和硫酸盐能高效激发熟料和源自黏土的铝硅酸盐或粉煤灰的复合物。硅酸盐熟料-粉煤灰-硫酸钠复杂体系通过研究 Ca(OH)$_2$-Na$_2$SO$_4$ 和 Ca(OH)$_2$-Na$_2$SO$_4$-粉煤灰悬浮液的相互作用来评价的。在 Ca(OH)$_2$-Na$_2$SO$_4$-粉煤灰的悬浮液体系中，pH 值随时间增加。在铝硅酸盐物质存在时，Ca(OH)$_2$ 和 Na$_2$SO$_4$ 反应生成 NaOH 是由石膏和提供 Al$_2$O$_3$ 的粉煤灰反应生成的钙矾石所控制（图 5-4）。硅酸盐熟料煅烧黏土-Na$_2$SO$_4$ 相关的机理也被注意到。特别是生成钙矾石形成过程中碱性的重要性也被提及，同样也在硅酸盐水泥-粉煤灰-Na$_2$CO$_3$ 胶凝体系中提到[92]，其中 CaCO$_3$ 沉淀是在高 pH 值条件下发生的。在这两个体系中，观察到铝硅酸盐型凝胶作为二级凝胶相，和 C-S-H 唯一凝胶相相比，改善了产品性能。

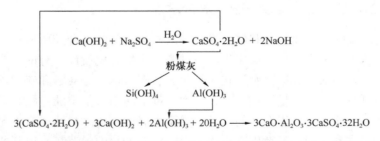

图 5-4 氢氧化钙-硫酸钠-粉煤灰体系相互反应模型图（改编自文献[98]）

5.5.6 铝硅酸盐＋其他含钙原料＋碱

钙的硅酸盐[4]、碳酸盐矿物[104-107]和铝硅酸盐联合可改善碱激发胶凝体系的性质。含钙物质可有效平衡钙源中钙的可用性和激发溶液的碱性。在中碱激发条件下，相比于硅酸盐水泥或者 BFS，硅酸盐中的钙溶解度更低，这就导致微量形成甚至不形成 C-S-H 类型的凝胶。未反应的矿物粒子会破坏铝硅酸盐凝胶的网状结构，导致总强度的降低。碳酸盐似乎提供更多有效钙和承担胶凝体系的强度，但白云石对胶凝体系性质的变化影响小于方解石，因为镁不能像钙一样形成具有强度的凝胶相[104]。由 BFS、石灰岩和硅酸钠或碳酸钠混合而成的胶凝材料都被证明有很好的强度发展和力学性能[105-107]。同时还要用硅藻土作为补充 SiO_2 的原料。这种特别的研究工作一般被发展中国家所关注，以用于低成本建筑材料，所以优先使用安全环保的碳酸钠作为激发剂；抗压强度 3d 内就能达到 40MPa；二氧化碳排放和耗能比硅酸盐水泥少了 97%，同时比计算的成本少 50%。

在碱激发复合胶凝材料中，第二重要的钙质成分是铝酸钙水泥。这种材料在各种碱性溶液中如何避免转换反应被广泛讨论，同时它也是 AAM 复合胶凝体系中的有用成分。铝酸钙水泥复合天然沸石[109]、煅烧的黏土[110]或者天然火山灰[111]，能提高胶凝材料的强度，除此之外，还能控制风化。当掺量低于 20%的时候，铝酸钙水泥替代铝硅酸盐作为原料，而不是形成常规的透明状水化产物，其实人们对胶凝材料的物相演化至今仍缺乏系统的研究。

5.6 小结

富钙前驱体和主流的铝硅酸盐前驱体碱激发反应形成胶凝材料，不同配合比设计和不同前驱体存在巨大差异，只要用各种合理的方式制备不同的胶凝体系，就能控制其性能、成本和环境影响。在很多体系中能观察到高钙和低钙胶凝材料具有各自明显不同的结构，潜在表明胶凝体系的热力学性能和长期稳定性。由中钙原料形成的胶凝产品的性质不能仅仅由前驱体或其他原料决定，钙的利用对优化产品性能有影响，当含钙物质与激发剂结合时，它们之间存在显著的协同作用。这一领域需要大量实践经验，目前仅是在很小范围内做了一定的理论研究。该领域未来在科学和技术方面的发展潜力巨大。

参考文献

[1] Shi C., Fernández-Jiménez A., Palomo A.. New cements for the 21st century: the pursuit of an alternative to Portland cement. Cem. Concr. Res, 2011, 41(7): 750-763.

[2] Van Deventer J. S. J., Provis J. L., Duxson P.. Technical and commercial progress in the adoption of geopolymer cement. Miner. Eng, 2012, 29: 89-104.

[3] Provis J. L., Myers R. J., White C. E., Rose V., Van Deventer J. S. J.. X-ray microtomography shows pore structure and tortuosity in alkali-activated binders. Cem. Concr. Res, 2012, 42(6): 855-864.

[4] Yip C. K., Lukey G. C., Provis J. L., Van Deventer J. S. J.. Effect of calcium silicate sources on geopolymerisation. Cem. Concr. Res, 2008, 38(4): 554-564.

[5] Yip C. K., Van Deventer J. S. J.. Microanalysis of calcium silicate hydrate gel formed within a geopolymeric binder. J. Mater. Sci, 2003, 38(18): 3851-3860.

[6] Yip C. K., Lukey G. C., Van Deventer J. S. J.. The coexistence of geopolymeric gel and calcium silicate hydrate at the early stage of alkaline activation. Cem. Concr.

Res,2005,35(9):1688-1697.

[7] Alonso S., Palomo A.. Calorimetric study of alkaline activation of calcium hydroxidemetakaolin solid mixtures. Cem. Concr. Res, 2001, 31(1): 25-30.

[8] Alonso S., Palomo A.. Alkaline activation of metakaolin and calcium hydroxide mixtures: influence of temperature, activator concentration and solids ratio. Mater. Lett, 2001, 47(1-2): 55-62.

[9] Dombrowski K., Buchwald A., Weil M.. The influence of calcium content on the structure and thermal performance of fly ash based geopolymers. J. Mater. Sci, 2007, 42(9): 3033-3043.

[10] García-Lodeiro I., Fernández-Jiménez A., Blanco M. T., Palomo, A.. FTIR study of the sol-gel synthesis of cementitious gels: C-S-H and N-A-S-H. J. Sol-Gel Sci. Technol, 2008, 45(1): 63-72.

[11] García Lodeiro I., Macphee D. E., Palomo A., Fernández-Jiménez A.. Effect of alkalis on fresh C-S-H gels. FTIR analysis. Cem. Concr. Res, 2009, 39: 147-153.

[12] García Lodeiro I., Fernández-Jimenez A., Palomo A., Macphee D. E.. Effect on fresh C-S-H gels of the simultaneous addition of alkali and aluminium. Cem. Concr. Res, 2010, 40(1): 27-32.

[13] García-Lodeirov I., Palomo A., Fernández-Jiménez A., Macphee D. E.. Compatibility studies between N-A-S-H and C-A-S-H gels. Study in the ternary diagram Na_2O-CaO-Al_2O_3-SiO_2-H_2O. Cem. Concr. Res, 2011, 41(9): 923-931.

[14] Bernal S. A., Mejía de Gutiérrez R., Rose V., Provis J. L.. Effect of silicate modulus and metakaolin incorporation on the carbonation of alkali silicate-activated slags. Cem. Concr. Res, 2010, 40(6): 898-907.

[15] Bernal S. A., Provis J. L., Rose V., Mejía de Gutiérrez R.. High-resolution X-ray diffraction and fluorescence microscopy characterization of alkali-activated slag-metakaolin binders. J. Am. Ceram. Soc, 2013, 96(6): 1951-1957.

[16] Ben Haha M., Le Saout G., Winnefeld F., Lothenbach B.. Influence of activator type on hydration kinetics, hydrate assemblage and microstructural development of alkali activated blast-furnace slags. Cem. Concr. Res, 2011, 41(3): 301-310.

[17] Cross D., Stephens J., Vollmer J.. Structural applications of 100 percent fly ash concrete. In: World of Coal Ash 2005, Lexington. 2005: 131.

[18] Berry M., Stephens J., Cross D.. Performance of 100 % fly ash concrete with recycled glass aggregate. ACI Mater. J, 2011, 108(4): 378-384.

[19] Husbands T. B., Malone P. G., Wakeley L. D.. Performance of Concretes Proportioned with Pyrament Blended Cement, U. S. Army Corps of Engineers Construction Productivity Advancement Research Program, Report CPAR-SL-94-2, 1994.

[20] ASTM International: Standard Specification for Coal Fly Ash and Raw or Calcined Natural Pozzolan for Use in Concrete (ASTM C618-12). West Conshohocken, 2012.

[21] Gravitt B. B., Heitzmann R. F., Sawyer J. L.. Hydraulic cement and composition employing the same. Compiler: U. S, 4997, 484[P]. 1991.

[22] Roy A., Schilling P. J., Eaton H. C.. Alkali activated class C fly ash cement. Compiler: U. S, 5565028[P]. 1996.

[23] Nicholson C. L., Fletcher R. A., Miller N., Stirling C., Morris J., Hodges S., MacKenzieK. J. D., Schmücker M.. Building innovation through geopolymer technology. Chem. N. Z, 2005, 69(3): 10-12.

[24] Perera D. S., Nicholson C. L., Blackford M. G., Fletcher R. A., Trautman R. A.. Geopolymers made using New Zealand flyash. J. Ceram. Soc. Jpn, 2004, 112(5): S108-S111.

[25] Lloyd R. R.. The durability of inorganic polymer cements. Ph. D. thesis, University of Melbourne, Australia, 2008.

[26] Keyte L. M.. What's wrong with Tarong? The importance of fly ash glass chemistry in inorganic polymer synthesis. Ph. D. thesis, University of Melbourne, Australia, 2008.

[27] Lee N. P.. Creep and Shrinkage of Inorganic Polymer Concrete, BRANZ Study Report 175, BRANZ, 2007.

[28] Lloyd R. R., Provis J. L., Smeaton K. J., Van Deventer J. S. J.. Spatial distribution of pores in fly ash-based inorganic polymer gels visualised by Wood's metal intrusion. Micropor. Mesopor. Mater, 2009, 126(1-2): 32-39.

[29] Diaz-Loya E. I., Allouche E. N., Vaidya S.. Mechanical properties of fly-ash-based geopolymer concrete. ACI Mater. J, 2011, 108(3): 300-306.

[30] Winnefeld F., Leemann A., Lucuk M., Svoboda P., Neuroth M.. Assessment of phase formation in alkali activated low and high calcium fly ashes in building materials. Constr. Build. Mater, 2010, 24(6): 1086-1093.

[31] Oh J. E., Monteiro P. J. M., Jun S. S., Choi S., Clark S. M.. The evolution of strength and crystalline phases for alkali-activated ground blast furnace slag and fly ash-based geopolymers. Cem. Concr. Res, 2010, 40(2): 189-196.

[32] Guo X. L., Shi H. S., Dick W. A.. Compressive strength and microstructural characteristics of class C fly ash geopolymer. Cem. Concr. Compos, 2010, 32(2): 142-147.

[33] Fares G., Tagnit-Hamou A.. Chemical and thermal activation of sodium-rich calcium alumino-silicate binder. In: Beaudoin J. J., (ed.) 12th International Congress on the Chemistry of Cement, Montreal, Canada. Cement Association of Canada, Ottawa, Canada. CD-ROMproceedings, 2007.

[34] Tashima M. M., Soriano L., Monzó J., Borrachero M. V., Payà J.. Novel geopolymeric material cured at room temperature. Adv. Appl. Ceram, 2013, 112(4): 179-183.

[35] Granizo M. L., Alonso S., Blanco-Varela M. T., Palomo A.. Alkaline activation

of metakaolin: effect of calcium hydroxide in the products of reaction. J. Am. Ceram. Soc. 2002, 85(1): 225-231.

[36] Shi C.. Early microstructure development of activated lime-fly ash pastes. Cem. Concr. Res, 1996, 26(9): 1351-1359.

[37] Williams P. J., Biernacki J. J., Walker L. R., Meyer H. M., Rawn C. J., Bai J.. Microanalysis of alkali-activated fly ash-CH pastes. Cem. Concr. Res, 2002, 32(6): 963-972.

[38] Shi C., Day R. L.. Chemical activation of blended cements made with lime and natural pozzolans. Cem. Concr. Res, 1993, 23(6): 1389-1396.

[39] Shi C., Day R. L.. Pozzolanic reaction in the presence of chemical activators: Part II Reaction products and mechanism. Cem. Concr. Res, 2000, 30(4): 607-613.

[40] Shi C., Day R. L.. Pozzolanic reaction in the presence of chemical activators: Part I. Reaction kinetics. Cem. Concr. Res, 2000, 30(1): 51-58.

[41] Shi C., Day R. L.. Comparison of different methods for enhancing reactivity of pozzolans. Cem. Concr. Res, 2001, 31(5): 813-818.

[42] Buchwald A., Hilbig H., Kaps C.. Alkali-activated metakaolin-slag blends-performance and structure in dependence on their composition. J. Mater. Sci, 2007, (9): 3024-3032.

[43] Buchwald A., Tatarin R., Stephan D.. Reaction progress of alkaline-activated metakaolinground granulated blast furnace slag blends. J. Mater. Sci, 2009, 44(20): 5609-5617.

[44] Bernal S. A., Provis J. L., Mejía de Gutierrez R., Rose, V.. Evolution of binder structure in sodium silicate-activated slag-metakaolin blends. Cem. Concr. Composm, 2001, 33(1): 46-54.

[45] White C. E., Page K., Henson N. J., Provis J. L.. Insitu synchrotron X-ray pair distribution function analysis of the early stages of gel formation in metakaolin-based geopolymers. Appl. Clay. Sci, 2013, 73: 17-25.

[46] Bernal S. A., Mejía de Gutiérrez R., Provis J. L.. Engineering and durability properties of concretes based on alkali-activated granulated blast furnace slag/metakaolin blends. Constr. Build Mater, 2012, 33: 99-108.

[47] Cheng T. W., Chiu J. P.. Fire-resistant geopolymer produced by granulated blast furnace slag. Miner. Eng, 2003, 16(3): 205-210.

[48] Lecomte I., Liégeois M., Rulmont A., Cloots R., Maseri F.. Synthesis and characterization of new inorganic polymeric composites based on kaolin or white clay and on groundgranulated blast furnace slag. J. Mater. Res, 2003, 18(11): 2571-2579.

[49] Zhang Y., Sun W., Chen, Q., Chen L.. Synthesis and heavy metal immobilization behaviors of slag based geopolymer. J. Hazard. Mater, 2007, 143(1-2): 206-213.

[50] Bernal López S., Gordillo M., Mejía de Gutiérrez R., Rodríguez Martínez E., Delvasto Arjona S., Cuero R.. Modeling of the compressive strength of alternative con-

cretes using the response surface methodology. Rev. Fac. Ing. -Univ. Antioquia, 2009, 49: 112-123.

[51] Bernal S. A.. Carbonatación de Concretos Producidos en Sistemas Binarios de una Escoria Siderúrgica yun Metacaolín Activados Alcalinamente. Ph. D. thesis, Universidad del Valle, 2009.

[52] Burciaga-Díaz O., Escalante-García J. I., Arellano-Aguilar R., Gorokhovsky A.. Statistical analysis of strength development as a function of various parameters on activated metakaolin/slag cements. J. Am. Ceram. Soc, 2010, 93(2): 541-547.

[53] Bernal S. A., Rodríguez E. D., Mejía de Gutiérrez R., Gordillo M., Provis J. L.. Mechanical and thermal characterisation of geopolymers based on silicate-activated metakaolin/slag blends. J. Mater. Sci, 2011, 46(16): 5477-5486.

[54] Smith M. A., Osborne G. J.. Slag/fly ash cements. World Cem. Technol. 1977, 1(6): 223-233.

[55] Bijen J., Waltje H.. Alkali activated slag-fly ash cements. In: Malhotra, V. M. (ed.) 3rd International Conference on Fly Ash, Silica Fume, Slag and Natural Pozzolans in Concrete, ACI SP114, Trondheim, Norway, Detroit, MI. 1989, 2: 1565-1578. American Concrete Institute.

[56] Li Z., Liu S.. Influence of slag as additive on compressive strength of fly ash-based geopolymer. J. Mater. Civil Eng, 2007, 19(6): 470-474.

[57] Kumar S., Kumar R., Mehrotra S. P.. Influence of granulated blast furnace slag on the reaction, structure and properties of fly ash based geopolymer. J. Mater. Sci, 2010, 45(3): 607-615.

[58] Ma W., Sample D., Martin R., Brown P. W.. Calorimetric study of cement blends containng fly ash, silica fume, and slag at elevated temperatures. Cem. Concr. Aggr, 1994, 16(2): 93-99.

[59] Schindler A. K., Folliard K. J.. Heat of hydration models for cementitious materials. ACI Mater. J, 2005, 102(1): 24-33.

[60] Blaakmeer J.. Diabind: An alkali-activated slag fly ash binder for acid-resistant concrete. Adv. Cem. Based Mater, 1994, 1(6): 275-276.

[61] Shi C., Qian J.. Increasing coal fly ash use in cement and concrete through chemical activation of reactivity of fly ash. Energy Sources, 2003, 25(6): 617-628.

[62] Pan Z., Yang N.. Updated review on AAM research in China. In: Shi C., Shen X. (eds.)First International Conference on Advances in Chemically-Activated Materials, Jinan, China. Bagneux, France. 2010: 45-55. RILEM.

[63] Puertas F., Martínez-Ramírez S., Alonso S., Vázquez E.. Alkali-activated fly ash/slag cement. Strength behaviour and hydration products. Cem. Concr. Res, 2000, 30: 1625-1632.

[64] Puertas F., Fernández-Jiménez A.. Mineralogical and microstructural characterisation of alkali-activated fly ash/slag pastes. Cem. Concr. Compos, 2003, 25(3):

287-292.

[65] Temuujin J., Van Riessen A., Williams R.. Influence of calcium compounds on the mechanical properties of fly ash geopolymer pastes. J. Hazard. Mater, 2009, 167(1-3): 82-88.

[66] Bernal S. A., Provis J. L., Walkley B., San Nicolas R., Gehman J. D., Brice D. G., Kilcullen A., Duxson P., Van Deventer J. S. J.. Gel nanostructure in alkali-activated binders based on slag and fly ash, and effects of accelerated carbonation. Cem. Concr. Res, 2013, 53: 127-144.

[67] Lloyd R. R., Provis J. L., Van Deventer J. S. J.. Pore solution composition and alkali diffusion in inorganic polymer cement. Cem. Concr. Res, 2010, 40(9): 1386-1392.

[68] Escalante García J. I., Campos-Venegas K., Gorokhovsky A., Fernández A.. Cementitious composites of pulverised fuel ash and blast furnace slag activated by sodium silicate: Effect of Na_2O concentration and modulus. Adv. Appl. Ceram, 2006, 105(4): 201-208.

[69] Lloyd R. R., Provis J. L., Van Deventer J. S. J.. Microscopy and microanalysis of inorganic polymer cements. 1: Remnant fly ash particles. J. Mater. Sci, 2009, 44(2): 608-619.

[70] Lloyd R. R., Provis J. L., Van Deventer J. S. J.. Microscopy and microanalysis of inorganic polymer cements. 2: The gel binder. J. Mater. Sci, 2009, 44(2): 620-631.

[71] Ismail I., Bernal S. A., Provis J. L., Hamdan S., Van Deventer J. S. J.. Drying-induced changes in the structure of alkali-activated pastes. J. Mater. Sci, 2013, 48(9): 3566-3577.

[72] Lloyd R. R., Provis J. L., Van Deventer J. S. J.. Acid resistance of inorganic polymer binders. 1. Corrosion rate. Mater. Struct, 2012, 45(1-2): 1-14.

[73] Dimas D., Giannopoulou I. P., Panias D.. Utilization of alumina red mud for synthesis of inorganic polymeric materials. Miner. Proc. Extr. Metall. Rev, 2009, 30(3): 211-239.

[74] Wagh A. S., Douse V. E.. Silicate bonded unsintered ceramics of Bayer process waste. J. Mater. Res, 1991, 6(5): 1094-1102.

[75] Sun W.-b., Feng X.-p., Zhao G.-x.. Effect of distortion degree on the hydration of red mud base cementitious material. J. Coal Sci. Eng, 2009, 15(1): 88-93.

[76] Kumar A., Kumar S.. Development of paving blocks from synergistic use of red mud and fly ash using geopolymerization. Constr. Build. Mater, 2013, 38: 865-871.

[77] Pan Z., Fang Y., Pan Z., Chen Q., Yang N., Yu, J., Lu J.. Solid alkali-slag-red mud cementitious material. J. Nanjing Univ. Chem. Technol, 1998, 20(2): 34-38.

[78] Pan Z., Fang Y., Zhao C., Yang N.. Research on alkali activated slag-red mud cement. Bull. Chin. Ceram. Soc, 1999, 18(3): 34-39.

[79] Pan Z. H., Li D. X., Yu J., Yang N. R.. Properties and microstructure of the hardened alkaliactivated red mud-slag cementitious material. Cem. Concr. Res, 2003, 33(9): 1437-1441.

[80] Pan Z., Cheng L., Lu Y., Yang N.. Hydration products of alkali-activated slag-red mud cementitious material. Cem. Concr. Res, 2002, 32(3): 357-362.

[81] Ye N., Zhu J., Liua J., Li, Y., Ke X., Yang J.. Influence of thermal treatment on phase transformation and dissolubility of aluminosilicate phase in red mud. Mater. Res. Soc. Symp. Proc, 2012: 1488. doi 10. 1557/opl. 2012. 1546.

[82] Shi C., Day R. L.. Acceleration of the reactivity of fly ash by chemical activation. Cem. Concr. Res, 1995, 25(1): 15-21.

[83] Palomo A., Fernández-Jiménez A., Kovalchuk G. Y., Ordoñez L. M., Naranjo M. C.. OPCfly ash cementitious systems. Study of gel binders formed during alkaline hydration. J. Mater. Sci, 2007, 42(9): 2958-2966.

[84] Ruiz-Santaquiteria C., Fernández-Jiménez A., Skibsted J., Palomo A.. Clay reactivity: production of alkali activated cements. Appl. Clay Sci, 2013, 73: 11-16.

[85] Donatello S., Fernández-Jimenez A., Palomo A.. Very high volume fly ash cements. Early age hydration study using Na_2SO_4 as an activator. J. Am. Ceram. Soc, 2013, 96(3): 900-906.

[86] Way S. J., Shayan A.. Early hydration of a Portland cement in water and sodium hydroxide solutions: composition of solutions and nature of solid phases. Cem. Concr. Res, 1989, 19, 759-769.

[87] Li G., Le Bescop P., Moranville M.. The U phase formation in cement-based systems containing high amounts of Na_2SO_4. Cem. Concr. Res, 1996, 26(1): 27-33.

[88] Li G., Le Bescop P., Moranville M.. Expansion mechanism associated with the secondary formation of the U phase in cement-based systems containing high amounts of Na_2SO_4. Cem. Concr. Res, 1996, 26(2): 195-201.

[89] Li G., Le Bescop P., Moranville-Regourd M.. Synthesis of the U phase ($4CaO \cdot 0.9Al_2O_3 \cdot 1.1SO_3 \cdot 0.5Na_2O \cdot 16H_2O$). Cem. Concr. Res, 1997, 27(1): 7-13.

[90] Martínez-Ramírez S., Palomo A.. OPC hydration with highly alkaline solutions. Adv. Cem. Res, 2001, 13(3): 123-129.

[91] Martínez-Ramírez S., Palomo A.. Microstructure studies on Portland cement pastes obtained in highly alkaline environments. Cem. Concr. Res, 2001, 31, 1581-1585.

[92] Fernández-Jiménez A., Sobrados I., Sanz J., Palomo A.. Hybrid cements with very low OPC content. In: Palomo A. (ed.) 13th International Congress on the Chemistry of Cement, Madrid, Spain. CSIC, Madrid, Spain. CD-ROM proceedings, 2011.

[93] Scheetz B. E., Hoffer J. P.. Characterization of sodium silicate-activated Portland cement: 1. Matrices for low-level radioactive waste forms. In: Al-Manaseer A. A., Roy D. M. (eds.) Concrete and Grout in Nuclear and Hazardous Waste Disposal (ACI SP 158). American Concrete Institute, Detroit, MI. 1995: 91-110.

[94] Brykov A. S., Danilov B. V., Korneev V. I., Larichkov A. V.. Effect of hydrated sodium silicates on cement paste hardening. Russ. J. Appl. Chem, 2002, 75 (10): 1577-1579.

[95] Brykov A. S., Danilov B. V., Larichkov A. V.. Specific features of Portland cement hydration in the presence of sodium hydrosilicates. Russ. J. Appl. Chem, 2006, 79(4): 521-524.

[96] Blazhis A. R., Rostovskaya G. S.. Super quick hardening high strength alkaline clinker and clinker-free cements. In: Krivenko P. V. (ed.) Proceedings of the First International Conference on Alkaline Cements and Concretes, Kiev, Ukraine. 1994, 1: 193-202. VIPOL Stock Company.

[97] Tailby J., MacKenzie K. J. D.. Structure and mechanical properties of aluminosilicate geopolymer composites with Portland cement and its constituent minerals. Cem. Concr, Res, 2010, 40(5): 787-794.

[98] Sanitskii M. A.. Alkaline Portland cements. In: Krivenko P. V. (ed.) Proceedings of the Second International Conference on Alkaline Cements and Concretes, Kiev, Ukraine. 1999: 315-333. ORANTA.

[99] Sanitskii M. A., Khaba P. M., Pozniak O. R., Zayats B. J., Smytsniuk R. V., Gorpynko A. F.. Alkali-activated composites cements and concretes with fly ash additive. In: Krivenko P. V. (ed.) Proceedings of the Second International Conference on Alkaline Cements and Concretes, Kiev, Ukraine, 1999, 472-479. ORANTA.

[100] Bernal S. A., Skibsted J., Herfort D.. Hybrid binders based on alkali sulfate-activated Portland clinker and metakaolin. In: Palomo A. (ed.) 13th International Congress on the Chemistry of Cement, Madrid. CSIC, Madrid, Spain. CD-ROM proceedings, 2011.

[101] Krivenko P. V., Kovalchuk G., Kovalchuk O.. Fly ash based geocements modified with calcium- containing additives. In: Bílek V., Keršner Z. (eds.) Proceedings of the 3rd International Symposium on Non-traditional Cement and Concrete. 2008: 400-409. ZPSV A. S., Brno.

[102] Gelevera A. G., Munzer K.. Alkaline Portland and slag Portland cements. In: Krivenko P. V. (ed.) Proceedings of the First International Conference on Alkaline Cements and Concretes, Kiev, Ukraine. 1994. 1: 173-179. VIPOL Stock Company.

[103] Fundi Y. S. A.. Alkaline pozzolana Portland cement. In: Krivenko P. V. (ed.) Proceedings of the First International Conference on Alkaline Cements and Concretes, Kiev, Ukraine. 1994, 1: 181-192. VIPOL Stock Company.

[104] Yip C. K., Lukey G. C., Provis J. L., Van Deventer J. S. J.. Carbonate mineral addition to metakaolin-based geopolymers. Cem. Concr. Compos, 2008, 30 (10): 979-985.

[105] Sakulich A. R., Anderson E., Schauer C., Barsoum M. W.. Mechanical and microstructural characterization of an alkali-activated slag/limestone fine aggregate

[106] Sakulich A. R., Anderson E., Schauer C. L., Barsoum M. W.. Influence of Si: Al ratio on the microstructural and mechanical properties of a fine-limestone aggregate alkali-activated slag concrete. Mater. Struct. 2010, 43(7): 1025-1035.

[107] Moseson A. J., Moseson D. E., Barsoum M. W.. High volume limestone alkali-activated cement developed by design of experiment. Cem. Concr. Compos, 2012, 34(3): 328-336.

[108] Miller S. A., Sakulich A. R., Barsoum M. W., Jud Sierra E.. Diatomaceous earth as a pozzolan in the fabrication of an alkali-activated fine-aggregate limestone concrete. J. Am. Ceram. Soc, 2010, 93(9): 2828-2836.

[109] Ding J., Fu Y., Beaudoin J. J.. Effect of different inorganic salts/alkali on conversion-prevention in high alumina cement products. Adv. Cem. Based Mater, 1996, 4(2): 43-47.

[110] Fernández-Jiménez A., Palomo A., Vazquez T., Vallepu, R., Terai T., Ikeda K.. Alkaline activation of blends of metakaolin and calcium aluminate cement. Part I: Strength and microstructural development. J. Am. Ceram. Soc, 2008, 91(4): 1231-1236.

[111] Najafi Kani E., Allahverdi A., Provis J. L.. Efflorescence control in geopolymer binders based on natural pozzolan. Cem. Concr. Compos, 2012, 34(1): 25-33.

第6章 外 加 剂

Francisca Puertas, Marta Palacios and John L. Provis

6.1 碱激发体系的外加剂定义

在开始讨论碱激发胶凝材料中的外加剂之前，首先给外加剂下一个定义是非常必要的。外加剂是碱激发胶凝材料配比中必不可少的组分，有时候被说成是硅酸盐水泥的外加剂（矿物外加剂或化学外加剂）。在碱化学激发的背景下，碱激发剂和固体（铝）硅酸盐都不应该被认为是一种外加剂，它们均为碱激发胶凝材料的组分，同样，添加的熟料化合物或相关材料（如水泥窑灰）已经超出了本文的范围。虽然也会引入作为促凝剂和缓凝剂的无机组分，但是本章主要讨论有机外加剂。

研究表明，有机外加剂和无机外加剂（常用于硅酸盐水泥混凝土技术）对碱激发水泥混凝土、砂浆、水泥浆体性能的影响还没有被完全研究清楚。此外，也许是因为条件的变化，不同作者报告的结果往往是矛盾（对立的）的，例如：被激发材料（炉渣、粉煤灰、偏高岭土）的自然性质不同、碱激发剂的性质和浓度的差异也和外加剂的种类及用量有关。然而，这些研究都一致显示，有机外加剂和无机外加剂在硅酸盐水泥和碱激发水泥体系环境下的行为非常不同，虽然在这方面进行深入研究仍然是必要的，但是目前已经提出了很多关于这些差异的原因。专利文献中确实有大量碱激发领域的各种外加剂的参考文献，然而大多数专利中提供的科学背景资料有限，也出于对专利保护的策略，所以从这些资源中获取详细的信息是非常困难的。

有些点是值得研究的，比如通常在 AAM 混凝土中外加剂的使用情况。

在 AAM 混凝土中添加一种液体外加剂，外加剂中水的含量应该被考虑在混凝土的配合比设计范围之内，因为大多数配合比设计对水灰比都十分敏感。

粉煤灰中未燃尽的碳组分通常在确定所需外加剂掺量时存在问题。由于外加剂对有机组分的选择性吸附，碳含量稍微增加，外加剂所需的量就会急剧增加。

添加外加剂的研究结果表明，当决定使用哪种外加剂时，其预期效果和副作用都应明确，方便对结果进行判断。尽管作为基本常识，但在具体环境下仍然是非常重要的。例如，疏水聚合物链越多，引气性越好。这个就最终使用材料的预期属性而言，可能是优点亦可能是缺点。疏水性有助于使表面防水达到抗冻融性所需的含气量，但太多的空气对耐久性和渗透性不利。另外，AAMs 中的一些塑化剂会增加坍落度，如果是引气造成的，而非由有效的增塑作用造成，那么这是不可取的。

在碱激发混凝土中使用养护组分（内部或外部）是比较有意思的，有机内养护化合物应该被认为是外加剂，但目前在该领域的所有文献中似乎没有相关报道。

在此基础上，以下是根据性质进行分类的外加剂对碱激发混凝土、砂浆和浆体性能的影响的回顾。

6.2 引气剂

Douglas 等人[1]发现,在硅酸钠/石灰浆激发 BFS 混凝土中添加磺化烃型引气剂,能有效地提高含气量到 6%。然而,Rostami 和 brendley[2]发现,使用单一引气剂并不会提高硅酸钠激发粉煤灰混凝土的抗冻融性能。抗冻融性将在本书第 10 章介绍,不仅涉及夹带的含气量而且涉及它的分布,所以它可能是一种可以给 AAM 和 OPC 混凝土中引入相同含量空气的外加剂,但是对孔隙率的影响不同,因此对抗冻融性能的影响也不同。这个仍然是当前文献中所详细记载的。

Bakharev 等人[3]研究了烷基芳基磺酸盐引气剂对用水玻璃或 NaOH 和 Na_2CO_3 激发 BFS 混凝土的工作性、收缩性能和机械强度的影响。这些作者的结论是,该外加剂会提高工作性,不影响机械强度,并减少收缩。

其他的研究表明,在碱激发砂浆和混凝土中使用无机外加剂,含气量会被改善,这其实是因为气泡由内部形成而不是由空气引入,胶凝材料经过化学反应产生气体,使孔隙变大从而形成泡沫胶凝材料化学反应。例如,Arellano Aguilar 等人[4]通过给偏高岭土砂浆和混凝土,或粉煤灰/偏高岭土比例为 25%/75% 复合的砂浆和混凝土中添加铝粉得到了轻质混凝土,在这两种情况下都使用 Na_2O 浓度为 15.2% 的硅酸钠作为激发剂,发现高碱性环境中添加铝粉会产生氢气,使其体积增大,密度减小。为了制备性能优良的加气混凝土,就必须平衡导热性能和强度。更多关于轻质和发泡碱激发材料会在本书第 12 章中介绍。

6.3 促凝剂和缓凝剂

有大量的外加剂被用来加速或者减缓碱激发水泥的反应进程,尽管它们的性能变化范围很广,而且到现在为止作用机理也没有被完全解释清楚。

Chang[5]在硅酸钠激发 BFS 混凝土中加入 H_3PO_4 作为缓凝剂,实验结果表明,其具有强的浓度敏感性:当 H_3PO_4 浓度低于 0.78mol/L 时影响不大,浓度在 0.8~0.84mol/L 之间有轻微的影响,但是在浓度为 0.87mol/L 时有很强的缓凝作用(设定时间为 6h)。同时 0.87mol/L 的磷酸也会降低早期的抗压强度,增大干缩。磷酸和石膏同时使用会降低磷酸的缓凝效果,也会影响强度的发展(和单独使用磷酸或单独使用石膏一样),未能减小干缩。这种行为潜在的化学机理或许和 Gong 与 Yang[6]发现的现象有关,在碱硅酸盐激发 BFS-赤泥共混物中加入高浓度磷酸钠时,有很强的缓凝作用,这是因为磷酸根离子会在反应体系中结合钙离子形成磷酸钙沉淀,因此具有缓凝作用。Lee 与 Van Deventer[7]证实磷酸钾是碱金属硅酸盐激发粉煤灰中的一种非常有效的缓凝剂,而且不会损失 28d 强度。然而 Shi and Li[8]发现当在碱激发磷渣中加入 Na_3PO_4 并没有明显的缓凝作用,也有报道称[9]硼酸盐和磷酸盐可作为中等强度的碱激发 BFS 矿渣水泥的促凝剂。

Bilim 等人[10]发现缓凝剂对碱硅酸盐激发 BFS 凝结时间的影响比对普通水泥砂浆的凝结时间的影响低得多。随着激发剂中硅酸盐模数的增加,该聚合物的缓凝作用下降,同时在开始 60min 内对工作性有积极的影响。该外加剂减缓了早期强度的发展,但是对后期砂浆的机械强度没有影响。

使用硼酸盐作为硅酸盐水泥缓凝剂是众所周知的[11]；在碱激发C类粉煤灰中加入硼酸盐会导致其出现快速凝结的现象[12]。因此要想改变新拌浆体的凝结时间要求硼酸盐质量浓度应大于7%。但是在该浓度下对胶凝材料的强度有不利影响。Tailby和Mackenzie[13]发现了一种创新应用，将硼酸盐作为硅酸钠和煅烧黏土与熟料矿物混合反应生成铝硅酸盐胶凝材料的缓凝剂，硼酸盐延缓铝酸盐胶凝材料的形成使熟料组分能够充分水化，因此提高了混合胶凝材料强度的发展。

与硅酸盐水泥体系相似，4%的NaCl可加快碱激发反应。而在1.5mol/L的$Na_2Si_2O_5$溶液作为激发剂激发BFS混凝土体系中[9]，添加8%的盐会发现有缓凝作用，反应几乎停止以及机械强度的发展也会延迟甚至停止。Brough等人[14]发现，在硅酸盐激发BFS浆体中，当NaCl的含量达到4%时会有促凝作用，但是高于4%会有缓凝作用；然而苹果酸是一种更有效的缓凝剂。在其他的硅酸盐激发BFS混凝土中[15]，NaCl的掺量达到20%时对凝结时间才有很小的影响，超过这个掺量才有显著的缓凝效果；NaCl对最终强度发展的影响有限。然而，长远来看，在钢筋混凝土中加入如此大量的氯化物的影响是存在潜在危害的[16]；这些掺合料应主要考虑在无钢筋加固建筑中，或者钢不是主要的增强材料（例如合成纤维增强）。

Lee和Van Deventer[7,17]测试了一系列盐在碱硅酸盐激发粉煤灰体系中的作用，Mg和Ca盐结果总结于图6-1；钙盐一般表现出促凝的效果，而镁盐的影响不大，KCl和KNO_3都能延缓凝结时间。Provis等人[18]研究了钙盐和锶盐对偏高

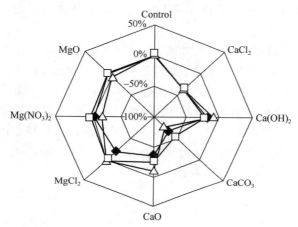

图6-1 三种不同粉煤灰/高岭土比例的共混物中加入0.09mol/L钙盐和镁盐，它们对维卡终凝时间的影响（每一组符号反映了不同的粉煤灰/高岭土比），共混物用Na-K混合硅酸盐溶液激发

岭土和硅酸钠体系的动力学的影响，发现硝酸盐和硫酸盐在初始阶段表现出缓凝作用，但是会被后期凝胶（胶凝材料）的形成和它的结构所干预，导致在没有它们的地方会形成更加多孔和渗透性更好的凝胶。

硝酸盐的作用比硫酸盐要强。相反，关于硫酸盐激发的讨论已在第3、4、5章进行，包含Douglas和Brandstetr在内的许多作者[19]都表示Na_2SO_4在碱硅酸盐激发BFS胶凝材料中有显著的促凝作用，再次突出了高钙和低钙胶凝材料体系之间在强度发展和不同组分的影响方面的差异。有钙离子的地方就可能形成AFm和AFt相，在讨论硫酸盐对碱激发胶凝材料的作用时，铝和硫都非常重要[20]；在钙离子浓度不是很高的情况下，这些相的形成是不常见的。

酒石酸和各种硝酸盐等也可作为调节碱激发胶凝材料凝结时间的物质[21,22]，尽管会对后期强度造成相应的影响。Pu等人[23]开发出了专有的碱激发BFS胶凝材料无机调凝剂，命名为YP-8，能有效地延长凝结时间70～120min，强度不损失。Rattanasak等人[24]研究各种有机和无机外加剂对硅酸钠激发高钙粉煤灰的强度影响，研究表明蔗糖对初凝时间没有影

响，但是会延缓终凝时间，还会提高材料的抗压强度。最后，$C_{12}A_7$ 会减少粉煤灰激发胶凝材料的凝结时间，但是它会降低 28d 的机械强度[25]。

6.4 减水剂和高效减水剂

Finland 发明了"F-Concrete"并注册了专利[26,27]，这是首次使用减水剂的碱激发胶凝材料。在这种情况下，对木质素磺酸盐进行了尝试。然而，在这方面进行的研究并没有解释这些激发剂在激发体系中的行为。Douglas 和 Brandstetr[19]也报道说，木质素磺酸盐和萘磺酸盐型减水剂在碱激发 BFS 体系中都不是很有效，Wang 等人[28]也观察到木质素磺酸盐会降低抗压强度而不会改善和易性。

Bakharev 等人[3] 报道说，木质素磺酸盐的混合物在硅酸盐水泥混凝土和水玻璃、NaOH 或 Na_2CO_3 激发 BFS 体系中的作用相似，即提高混凝土的和易性，与此同时能够延缓凝结时间和提高机械强度的发展。这些外加剂也会稍微地减少干燥收缩。在同一研究中，萘系外加剂提高了水玻璃或 NaOH、Na_2CO_3 激发 BFS 混凝土前几分钟的和易性，但随后会引起快凝同时降低后期的机械强度和强化收缩。因此，在碱激发矿渣混凝土中使用这些外加剂是有害的。

Collins 和 Sanjayan[29]研究了葡萄糖酸钙和葡萄糖酸钠对 NaOH、Na_2CO_3 混合激发 BFS 砂浆和混凝土的影响。结果表明，砂浆的和易性得到了改善，并观察到 120min 流动度保持水泥效果要好，但 1d 强度降低，随其掺量增大，强度下降的趋势变得平缓；此外，缓凝剂会随着水的渗出被带出来，这样会软化混凝土的表面。

Puertas 等人[30]研究了乙烯基共聚物和聚羧酸系减水剂对水玻璃激发 BFS 和粉煤灰砂浆以及净浆的影响。实验结果表明，2%掺量的乙烯基共聚物净浆的流动性并未改善，会延缓激发反应导致其 2～28d 之间的强度（图 6-2）降低。相比之下，尽管聚羧酸减水剂不会提高

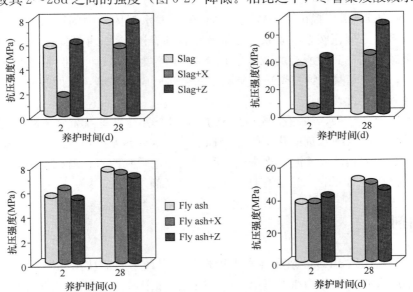

图 6-2 添加乙烯基共聚物(X)和聚羧酸共聚物减水剂(Z)的碱硅酸盐激发粉煤灰和矿渣砂浆的 2d 和 28d 强度
（由 Puertas 等人改编[30]）

净浆的流动性，也不会对砂浆的机械强度产生影响。聚羧酸减水剂的种类对激发过程和碱激发 BFS 水泥行为有明显的影响，然而这些影响不如对碱激发粉煤灰体系那么明显。在聚羧酸减水剂对碱激发粉煤灰浆体流变学的影响方面，Criado 等人[31]也得出了相似的结论。

Kashani 等人[32]发明了一种梳状聚羧酸减水剂，可以减小硅酸钠激发 BFS 胶凝材料高达 40% 的屈服应力。但此实验仅在实验室进行设计并未得到市场化推广。相似化学分子是降低屈服应力还是增加屈服应力取决于分子结构的详细信息，然而主链同时带有正电荷和负电荷的分子表现出有效的增塑效果。

Sathonsaowaphak 等人[33]报道了在 0.01～0.03 的萘/灰比下使用萘系外加剂，可以轻微提高和易性，同时能够保持碱活性底灰砂浆的机械强度。当掺加量达到 0.03～0.09 之间时，其硬化浆体机械强度就会降低。Kong 和 Sanjayan[34]研究了添加聚羧酸和磺酸基减水剂激发的偏高岭土净浆、砂浆和混凝土的工作性，发现拌合物的和易性提高，混凝土的耐火性有所下降。然而，Hardjito 和 Rangan[35]发现当萘系磺酸盐减水剂的掺量达到 2% 时，可以改善粉煤灰/硅酸钠碱激发胶凝材料混凝土的坍落度，但是不会损失 3d 和 7d 抗压强度。

Palacios 等人[36]研究了几种减水剂与碱激发矿渣水泥之间的相互作用，发现三聚氰胺基、萘系基和乙烯基共聚物外加剂对碱激发矿渣浆体的吸附比普通硅酸盐浆体要低 3～10 倍，使用的外加剂溶液都有不同的 pH 值。研究发现 pH=11.7 的 NaOH 激发矿渣悬浮液的 Zata 电位（大约 -2mV）比普通硅酸盐悬浮液的 Zata 电位（大约 +0.5mV）要低一点，而且偏向负电位，这个可以作为减水剂在两种水泥上吸附行为的差异的一部分解释。

这些作者还发现外加剂对碱激发矿渣流变参数的影响直接取决于减水剂的类型和掺量，以及碱激发剂溶液的 pH 值。在 pH=13.6 的 NaOH 激发矿渣体系中，唯一观察到降低了流变参数的外加剂是萘系减水剂。掺量低至每克矿渣加 1.26mg 的萘系减水剂都能减小 98% 的屈服应力。然而，当使用水玻璃溶液作为激发剂时，却没有哪种减水剂能提高流动性[36,38]。

Palacios 和 Puertas[39]提出了高效减水剂在碱激发水泥体系中的现象。这些作者研究了不同类型减水剂（三聚氰胺系、萘系、乙烯共聚物和聚羧酸系）在强碱性介质下的化学稳定性，除了萘系减水剂，其他所有的减水剂在 NaOH 环境且 pH>13 时都是稳定的。在如此高的 pH 值下，乙烯基共聚物与聚羧酸类减水剂会进行碱性水解，从而改变它们的结构，因此具有分散性和流变性。

鉴于文献报道的结果，明确在高 pH 值的 AAM 体系中用来提高 AAM 水泥、砂浆、混凝土流动性的新型高效减水剂的化学性质是稳定的。这些减水剂应该进行改进，使通常用在硅酸盐水泥减水剂加到 AAM 混凝土中的一些缓凝作用得到改善；就减小坍落度损失方面而言，这些缓凝作用是需要的，但是有时会使凝结时间额外延长。

6.5 减缩剂

许多碱硅酸盐激发 BFS 砂浆和混凝土出现的一个最重要的技术问题，是会发生较大的化学收缩和干燥收缩，这些将在本书第 10 章中进行详细讨论。在相同的制备、养护、储存环境下，碱激发胶凝材料的收缩率是硅酸盐水泥的 4 倍[40]。虽然使用多种外加剂不会完全地消除收缩，但是基于使用不同纤维（腈纶、聚丙烯、碳纤维或玻璃纤维，包括一些为减缩所特别设计的纤维）的许多方法已经作为物理方式来减缓这个问题[41-44]。纤维增韧碱激发胶凝材料的产品将在本书第 12 章中进行详细介绍。

几乎没有任何关于减缩剂对碱激发 BFS 体系影响的研究报道，Bakharev 等人[3]报道了标准的减缩剂（化学信息不详），和引气剂一块使用，可以使干缩降低到 OPC 之下（无外加剂）。同时研究表明 BFS 混凝土添加石膏可以减小收缩，这个是由于形成的钙矾石和单硫铝酸钙所引起的膨胀，这样能够补偿收缩，但是这并不是大多数 AAM 体系的机理。

Palacios 和 Puertas[37,45]发现在水玻璃激发 BFS 砂浆中使用基于聚丙二醇的减缩剂（SRAs）可以减小 85% 的自收缩和 50% 的干缩（图 6-3），所观察到的 SRAs 对收缩有益的影响主要是由于孔隙水表面张力的下降和外加剂对孔改变的结果。具体来说，外加剂会导致大孔的生成，提高孔直径在 $0.1 \sim 1.0~\mu m$ 范围内的孔的百分比，这些孔的毛细管压力要比不加外加剂的砂浆中主要的小毛细管压力低得多。SRAs 也观察到可以减缓矿渣的激发过程，并且掺量越大，减缓作用越强，但是它不会改变浆体的矿物组成。

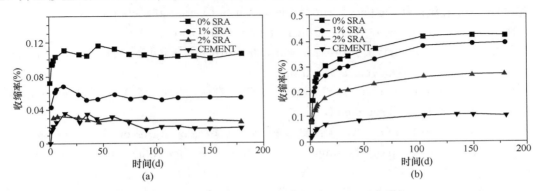

图 6-3 添加 SRAs 和不添加 SRAs，不同的养护条件下的水玻璃激发 BFS 砂浆的收缩
(a) RH = 99%；(b) RH = 50% （由 Puertas 等人得出的结果[43]）

6.6 小结

综上所述，硅酸盐水泥的外加剂历经几十年的发展已能精确控制其流变性质和水化过程。然而，其在碱激发胶凝材料体系中的作用要么无效要么有害。这主要是由于碱激发过程与硅酸盐水泥的水化过程相比有不同的化学作用，特别是大部分的碱激发胶凝材料合成过程中存在很高的 pH 值环境。虽然一些高质量的科学研究在过去的十几年均已发表，然而这篇文献中科学信息的深度和广度都不够，而且许多关键领域仍需要从根本上进行详细的分析。在未来这个领域里，为碱激发这种特别的化学环境设计专用有机外加剂将变得非常必要，进而能释放碱激发胶凝材料外加剂的应用潜力。

参考文献

[1] Douglas E., Bilodeau A., Malhotra V. M.. Properties and durability of alkali-activated slag concrete. ACI Mater. J, 1992, 89 (5): 509-516.

[2] Rostami H., Brendley W.. Alkali ash material: a novel fly ash-based cement. Envi-

[3] Bakharev T., Sanjayan J. G., Cheng Y. B.. Effect of admixtures on properties of alkaliactivated slag concrete. Cem. Concr. Res, 2000, 30(9): 1367-1374.

[4] Arellano Aguilar R., Burciaga Díaz O., Escalante García J. I.. Lightweight concretes of activated metakaolin-fly ash binders, with blast furnace slag aggregates. Constr. Build. Mater, 2010, 24 (7): 1166-1175.

[5] Chang J. J.. A study on the setting characteristics of sodium silicate-activated slag pastes. Cem. Concr. Res, 2003, 33(7): 1005-1011.

[6] Gong C., Yang N.. Effect of phosphate on the hydration of alkali-activated red mud-slag cementitious material. Cem. Concr. Res, 2000, 30 (7): 1013-1016.

[7] Lee W. K. W., Van Deventer J. S. J.. Effects of anions on the formation of aluminosilicate gel in geopolymers. Ind. Eng. Chem. Res, 2002, 41(18): 4550-4558.

[8] Shi C., Li Y.. Investigation on some factors affecting the characteristics of alkali-phosphorus slag cement. Cem. Concr. Res, 1989, 19 (4): 527-533.

[9] Talling B., Brandstetr J.. Present state and future of alkali-activated slag concretes. In: Malhotra V. M. (ed.) 3rd International Conference on Fly Ash, Silica Fume, Slag and Natural Pozzolans in Concrete, ACI SP114, 1989, 2: 1519-1546. American Concrete Institute, Trondheim, Norway.

[10] Bilim C., Karahan O., Atis C. D.. Infl uence of admixtures on the properties of alkali-activated slag mortars subjected to different curing conditions. Mater. Design, 2013, 44: 540-547.

[11] Bensted J., Callaghan I. C., Lepre A.. Comparative study of the effi ciency of various borate compounds as set-retarders of class G oilwell cement. Cem. Concr. Res, 1991, 21 (4): 663-668.

[12] Nicholson C. L., Murray B. J., Fletcher R. A., Brew D. R. M. MacKenzie K. J. D., Schmücker M.. Novel geopolymer materials containing borate structural units. In: Davidovits J. (ed.) World Congress Geopolymer, 2005, 31-33. Geopolymer Institute, Saint-Quentin.

[13] Tailby J., MacKenzie K. J. D.. Structure and mechanical properties of aluminosilicate geopolymer composites with Portland cement and its constituent minerals. Cem. Concr. Res, 2010, 40(5): 787-794.

[14] Brough A. R., Holloway M., Sykes J., Atkinson A.. Sodium silicate-based alkali-activated slag mortars: Part II. The retarding effect of additions of sodium chloride or malic acid. Cem. Concr. Res, 2000, 30 (9): 1375-1379.

[15] Sakulich A. R., Anderson E., Schauer C., Barsoum M. W.. Mechanical and microstructural characterization of an alkali-activated slag/limestone fine aggregate concrete. Constr. Build. Mater, 2009, 23: 2951-2959.

[16] Holloway M., Sykes J. M.. Studies of the corrosion of mild steel in alkali-activated slag cement mortars with sodium chloride admixtures by a galvanostatic pulse meth-

od. Corros. Sci, 2005, 47 (12): 3097-3110.

[17] Lee W. K. W., Van Deventer J. S. J.. The effect of ionic contaminants on the early-age properties of alkali-activated fly ash-based cements. Cem. Concr. Res, 2002, 32(4): 577-584.

[18] Provis J. L., Walls P. A., Van Deventer J. S. J.. Geopolymerisation kinetics. 3. Effects of Cs and Sr salts. Chem. Eng. Sci, 2008, 63(18): 4480-4489.

[19] Douglas E., Brandstetr J.. A preliminary study on the alkali activation of ground granulated blast-furnace slag. Cem. Concr. Res, 1990, 20(5): 746-756.

[20] Bernal S. A., Skibsted J., Herfort D.. Hybrid binders based on alkali sulfate-activated Portland clinker and metakaolin. In: Palomo A., (ed.) 13th International Congress on the Chemistry of Cement, Madrid. CD-ROM Proceedings. 2011.

[21] Zhang Z., Zhou D., Pan Z.. Selection of setting retarding substances for alkali activated slag cement. Concrete, 2008, 8: 63-68.

[22] Zhu X., Shi C.. Research on new type setting retarder for alkali activated slag cement. China Build. Mater, 2008, 9: 84-86.

[23] Pu X., Yang C., Gan C.. Study on retardation of setting of alkali activated slag concrete. Cement, 1992, 10: 32-36.

[24] Rattanasak U., Pankhet K., Chindaprasirt P.. Effect of chemical admixtures on properties of high-calcium fly ash geopolymer. Int. J. Miner. Metall. Mater, 2011, 18(3): 364-369.

[25] Rovnaník P.. Influence of $C_{12}A_7$ admixture on setting properties of fly ash geopolymer. Ceram Silik, 2010, 54(4): 362-367.

[26] Kukko H., Mannonen R.. Chemical and mechanical properties of alkali-activated blast furnace slag (F-concrete). Nord. Concr. Res, 1982, 1: 16.1-16.16.

[27] Byfors K., Klingstedt G., Lehtonen H. P., Romben, L.. Durability of concrete made with alkali-activated slag. In: Malhotra V. M. (ed.) 3rd International Conference on Fly Ash, Silica Fume, Slag and Natural Pozzolans in Concrete, ACI SP114, 1989: 1429-1444. American Concrete Institute, Trondheim, Norway.

[28] Wang S. D., Scrivener K. L., Pratt P. L.. Factors affecting the strength of alkali-activated slag. Cem. Concr. Res, 1994, 24(6): 1033-1043.

[29] Collins F., Sanjayan J. G.. Early age strength and workability of slag pastes activated by NaOH and Na_2CO_3. Cem. Concr. Res, 1988, 28(5): 655-664.

[30] Puertas F., Palomo A., Fernández-Jiménez A., Izquierdo J. D., Granizo M. L.. Effect of superplasticisers on the behaviour and properties of alkaline cements. Adv. Cem. Res, 2003, 15(1): 23-28.

[31] Criado M., Palomo A., Fernández-Jiménez A., Banfill P. F. G.. Alkali activated fly ash: effect of admixtures on paste rheology. Rheol. Acta, 2009, 48: 447-455.

[32] Kashani A., Provis J. L., Xu J., Kilcullen A., Duxson P., Qiao G. G., Van Deventer J. S. J.. Effects of different polycarboxylate ether structures on the rheology

of alkali activated slag binders. In: Tenth International Conference on Superplasticizers and Other Chemical Admixtures in Concrete (American Concrete Institute Special Publication SP-288): Prague, Czech Republic. CD-ROM. ACI, 2012.

[33] Sathonsaowaphak A., Chindaprasirt P., Pimraksa K.. Workability and strength of lignite bottom ash geopolymer mortar. J. Hazard. Mater, 2009, 168(1): 44-50.

[34] Kong D. L. Y., Sanjayan J. G.. Effect of elevated temperatures on geopolymer paste, mortar and concrete. Cem. Concr. Res, 2010, 40(2): 334-339.

[35] Hardjit D., Rangan B. V.. Development and Properties of Low-Calcium Fly Ash-Based Geopolymer Concrete. Curtin University of Technology Research Report, Curtin University, Perth, Australia, 2005.

[36] Palacios M., Houst Y. F., Bowen P., Puertas F.. Adsorption of superplasticizer admixtures on alkali-activated slag pastes. Cem. Concr. Res, 2009, 39: 670-677.

[37] Palacios M., Puertas F.. Effect of superplasticizer and shrinkage-reducing admixtures on alkali-activated slag pastes and mortars. Cem. Concr. Res, 2005, 35(7): 1358-1367.

[38] Palacios M., Banfill P. F. G., Puertas F.. Rheology and setting of alkali-activated slag pastes and mortars: Effect of organic admixture. ACI Mater. J, 2008, 105(2): 140-148.

[39] Palacios M., Puertas F.. Stability of superplasticizer and shrinkage-reducing admixtures in high basic media. Mater. Constr, 2004, 54(276): 65-86.

第7章 碱激发混凝土——配合比设计标准和早期性能

Lesley S. C. ko，Irene Beleña，peter Duxson，
Elena Kavalerova，Pavel V. Krivenko，Luis-Miguel Ordoñez，
Arezki Tagnit-Hamou and Frank Winnefeld

7.1 简介

2007 年的 RILEM TC AAM 聚集了来自学术界、国家重点实验室、国际企业中的研究碱激发的人员，为今后起草适用于碱激发胶凝材料的标准奠定基础。

众所周知，创新性的非常规技术因现有标准的阻碍和欠缺新标准的支持较难转向实践。对 AAM 来说，它不符合大多数水泥国家标准和国际标准。这些标准主要是基于普通硅酸盐水泥（OPC）或 OPC 复合水泥的组成、化学成分和水化产物。现有的水泥标准大多是关于水泥组成的规定，支持 OPC 和 OPC 基产品，而不是非传统的碱激发胶凝材料。这些标准是基于 OPC 在建筑市场普遍应用的基础上提出的。

一般来说，标准有助于管理已在市场上被接受的产品，确保购买产品的消费者能够确认该产品具有的性质，从而有助于采购和满足监管要求。同时，标准为产品制造商提供市场最低规范，促进竞争，提高效率和最低性能标准的创新，但标准的制定对最优性能并未提供强大的激励[2]。水泥和混凝土标准，包括规范、测试方法、推荐的实验方法、术语、一致性比较等，是与时俱进的文件，通常每隔数年都会进行审查和修订（例如在加拿大 CSA 体系中是隔 5 年修订一次）。标准的内容来自于推荐制造和使用的材料性质，而不是来自先前的技术要求。因此，要改变一个已有的标准或者起草一个新标准需要经过很长的时间。虽然对硅酸盐水泥和混凝土有许多性能测试方法，但是大多数物理化学性能与水泥含量以及后期强度有关。随着水泥化学的标准化，有限的性能测试（例如养护时间、强度和耐久性）不能根据强度等级判断。现在，小范围的水泥测试和标准的需要被广泛接受，以用来描述水泥和混凝土的性能。除非对当前水泥和混凝土的许多性能和通用强度等级及类别之间的关系的理解有根本的改变，水泥标准只会保持现有的大部分规定，使用现有的几个关键性能检测数据。

生产者如果想将 AAM 商业化并引入市场，新标准是必不可少的。新标准的诞生可以为市场中价格低廉、容易生产的商品提供足够的信心。在这个阶段，AAM 开发的工作团体中许多人的成果都已经清楚地证明了这一点。例如，可以使用不同的原料（例如炉渣、粉煤灰、火山灰），碱激发剂（例如碱金属硅酸盐、碱金属氢氧化物、碳酸盐和硫酸盐），通过在混凝土中研磨、混合或直接混合进行生产，以及应用于修补砂浆、预拌混凝土或预制混凝土。在过去 150 年里，大量的方法使得 OPC 市场中难以用同样狭义的组成和方式定义 AAM 这种新材料。

制造工艺与原材料的灵活性要求导致了标准制定过程中很难简化和严谨。AAM 的单一

方法和组成很容易提供标准，但这将消除技术创新的固有灵活性和潜力。随着时间的推移，通过追踪记录和其他因素可以很好地看到规范性路径变得有利于 AAM 标准的开发，如在乌克兰市场的情况，将在下面详细讨论。然而，在更广泛的基于市场的技术利用的这个早期阶段，TC 的意见是，应尽可能采用更灵活开放的方法。

在已有的 EN 和 ASTM 的标准中，只有一种基于性能的水泥标准，就是 ASTM C 1157[3]，其对水泥的组分及结构没有限制。因此，AAM 混凝土可以基于其具体性能被认定为 ASTM C 1157 水泥，例如通用（GU）、高早强（HE）等。然而，市场或当局对 ASTM C 1157 水泥仍然不是普遍接受，在美国 50 个州中只有少数接受这个标准。除了这两种主要的水泥标准之外，还有其他国家的标准可用于 AAM 的使用，例如，加拿大标准（CSA A3004-E1，具有高度重要的基于性能的规定[4]）和苏联及乌克兰标准（包括广泛的规范框架[5-13]）。

例如澳大利亚标准 AS 3972 附录 A 中所述[14]：尽管基于配方的规范是容易制定的，但是这种便利是在创新的基础上实现的，并且能相对容易地并入新的或修订的标准中。用于开发基于性能标准的三个最基本要素是：

（1）性能参数：通常与所需性能最相关的性能。
（2）要求的标准质量水平。
（3）测试方法：符合标准的清晰、可靠、易于使用、价格低廉的测试方法。

7.2 TC 224-AAM 标准编制规划——基于性能实施方案

基于对 AAM 现有和相关标准的评估，TC 决定同意使用现有测试方法获得主要性能来制定新标准。推出新的测试方法耗时又富有挑战，所以未来的 AAM 标准尽可能从现有的标准（特别是测试方法）中获得。正如该领域的研究人员指出的，制备碱激发混凝土有两种可能的途径——首先制备碱激发用胶凝材料，然后与水、砂和骨料混合以形成混凝土；或直接制备碱激发混凝土，通过将所有的胶凝材料组分与水、砂和骨料混合，而不是首先生成胶凝材料。因此，起草了两项标准的框架：碱激发"水泥"的标准和碱激发"混凝土"的标准。

碱激发水泥标准的框架包括以下几条：
（1）碱激发水泥的定义。
（2）合适的矿物组分和激发剂的实例。
（3）基于硬化前后的性能进行分类。
（4）对应提供给客户的信息的要求。
（5）测试方法。

为了将未来发展的范围纳入 AAM 标准，建议不要规定使用的材料，而是确定材料必须满足的基本性能要求。关于耐久性的各个方面，在碱激发"水泥"标准中不需要调节这些，但是这些将在碱激发混凝土基于性能的标准中直接讨论。

类似的，对碱激发混凝土标准的框架包括以下几条：
（1）碱激发混凝土的定义。
（2）有效组分（例如碱激发水泥，和/或矿物组分、激发剂、水、细骨料和粗骨料和/或化学外加剂）和配合比。

(3) 满足性能要求的新拌合硬化的混凝土特性。
(4) 耐久性测试（包括适用的现有测试方法）和每个必须达到的性能要求。
(5) 健康和安全问题。

7.3 碱激发胶凝材料的标准

7.3.1 范围

本文涉及用于砂浆或混凝土中的碱激发胶凝材料的标准。它不仅仅是建立一个标准，而是通过建立一个标准为这类材料创建一个标准系列。

由于一般类别的碱激发胶凝材料涵盖的材料范围很大，并且合适材料的范围远未被完全研究，因此开发基于性能的标准体系是必不可少的（例如类似于 ASTM C 1157[3] 或 CSA A3004-E1[4]），而不是一系列规范标准（如 EN 197[15] 或 ASTM C 150[16]）。这意味着未来的发展可以更容易地纳入标准中去。

7.3.2 碱激发胶凝材料的定义

碱激发胶凝材料是由一种或多种含有铝和硅的氧化物组成的矿物组分与一种或多种激发剂组成。激发剂含有碱金属离子并能够产生较高的 pH 值环境（例如碱金属硅酸盐、氢氧化物、硫酸盐和碳酸盐）。

矿物组分和激发剂可以作为干胶凝材料预混合，然后将预混的胶凝材料与水、砂、骨料和其他组分混合以获得砂浆或混凝土。或者激发剂可以作为水溶液单独加入到矿物组分中，然后可将该双组分胶凝材料与外加水（在碱激发剂是浓缩物的情况下）、砂、骨料和其他组分混合以获得砂浆或混凝土。当前，尚未考虑自激发过程的新方法随时间的推移将逐步被人所接受和推广。最终，所有工艺路线产生的碱激发胶凝材料可以单独销售，例如销售到砂浆或混凝土生产商，因此应该标准化。

7.3.3 矿物组分和碱激发胶凝材料实例

不应指定要使用的材料的具体性质，但需规定具有良好的反应性时要求的最低活性二氧化硅和氧化铝含量（要求对"反应性"的定义和测试要一致），和避免闪凝的最大游离石灰含量。然而，必须确保不能使用有害材料，ASTM C 1157（文献[3]，第 1.5 节）中包含的免责声明可以类推使用，还可以参考各自国家的法律和标准。标准应提供适用材料的清单：

合适的矿物组分，例如：
(1) 磨细粒状高炉矿渣（如根据 EN 15167-1[17] 或 ASTM C989[18]）。
(2) 含二氧化硅的粉煤灰（如根据 EN 450[19]，或根据 ASTM C618[20] 的 C 型或 F 型粉煤灰）。
(3) 煅烧黏土。
(4) 非粒状高炉矿渣，其他工艺的粒状矿渣（如有色冶金、锰铁合金、人造和天然铝硅酸盐玻璃等）。

(5) 其他含铝硅酸盐的材料，包括天然火山灰、底灰、流化床燃烧灰等。

合适的激发剂，例如：

(1) 碱金属硅酸盐。

(2) 碱金属氢氧化物。

(3) 碱硫酸盐。

(4) 碱碳酸盐。

使用其他替代激发剂在技术上肯定是可能的，但是在技术上不太成熟，不值得在标准化的背景下讨论。不同固体-激发剂组合的适用性已在第1章总结，并且在第2、3、4、5章有更详细的讨论，这里不再赘述。对于其他可能存在的材料组合，可以参考关于碱激发胶凝材料的国家标准（特别是来自苏联的标准，包括文献[5-13]）。

7.3.4 分类和要求

以其性能指标为基准的碱激发胶凝材料分类方法利于混凝土生产商的产品设计和对市场中商品特性的理解。此外，不应要求客户购买"黑匣子"材料体系。还应该指出的是，材料应当具有足够的耐久性，我们会根据具体的性能测试结果定义这一术语，而不是在一般意义上的定义。

以下数据和信息应提供给客户（其中一些仅在要求、需要注明时提供）：

(1) 原材料（根据要求）

① 材料来源。

② 元素组成、烧失量。

③ 矿物组成、XRD分析。

④ 密度、比表面积、筛余物（例如 $45\mu m$）。

(2) 硬化前的性能（分类）

① 工作性、流变性（例如流变曲线、扩展度）（根据要求）。

② 在给定温度下的凝结时间（一些材料凝结非常缓慢并需要热处理，例如渗透试验）。

③ 水化热。

(3) 硬化后的性能（分类）

① 力学性能（早期、后期，根据混合料组成和所需养护条件）。

② 抗压强度（早期、后期）。

③ 抗折强度（根据要求）。

④ 拉伸强度（根据要求）。

⑤ 在指定条件下的收缩和/或徐变。

客户可以方便地了解强度等级，例如，类似于 ASTM C 1157[3]。但是，如果定义强度等级或其他规定限值，则在标准中需要有关一致性的规定。强度等级还应包括养护温度，因为基于一些粉煤灰的 AAM 反应缓慢，常常需要热处理以达到合适的强度发展速率。如果在标准中使用强度等级，还应该有"未分类"强度等级，以便适用于某些低强度材料或具有稍高性能变化的材料。这些材料仍然可在某些条件下适用（如液压道路胶凝材料、非典型的混凝土等）。然而，一般来说，为了确保最终混凝土产品的质量以及对性能变化的限制，应给出最低要求。

碱激发胶凝材料的耐久性不一定在激发剂标准中调节，而是直接在碱激发混凝土基于性能的标准中规定；需要在测试方法的发展和验证领域进行更多的工作，然后才能实现这一点，这属于 2012 年成立的 RILEM TC 247-DTA 的职能范围。

7.3.5 测试

合适的测试方法在 RILEM TC 224-AAM 的范围内指定，并将在第 8、9、10 和 11 章中详细给出。然而，当创建 AAM 的标准时，有必要决定是否为材料和测试方法建立单独的标准，或者创建一个包含两者的标准；这里的偏向可能因司法辖区而异。另外，建议对砂浆进行试验，例如：对于强度，固定稠度改变混合物成分（尽管稠度的定义适用于 AAM 是另一个需要详细关注的问题），而不是 EN 196-1[21] 中的固定水灰比的基于质量的配合比设计和其他一些标准。

7.4 碱激发混凝土的标准

7.4.1 范围

本节讨论碱激发混凝土（AAMs）的标准。该标准提出了指导方针，以及为新一代和下一代 AAM 创建标准的一些建议。碱激发混凝土没有像传统混凝土一样被广泛研究，因此，基于性能指标的碱激发混凝土标准的制定利于扩大其未来发展空间，同时可避免使用者局限于碱激发混凝土组成的标准而忽略其材料体系的误区。

7.4.2 碱激发混凝土的定义

碱激发混凝土是由铝硅酸盐或硅铝酸钙、一种或多种混合的碱激发剂（例如碱金属硅酸盐、氢氧化物、硫酸盐或碳酸盐）、水、细骨料和粗骨料以及具有或不具有化学外加剂的一种或多种的矿物组分组成，可以根据材料的预期用途和它们与激发剂的相容性来添加。根据这个定义，每个组分应基于混凝土的一般性能或该成分对混凝土的整体性能的给定效应来选择。

7.4.3 有效胶凝材料配比

由矿物组分、激发剂和水的有效三元组合形成的胶凝材料应根据激发条件（在环境温度或外部加热条件下）配制，并且还要考虑净浆和砂浆中的胶凝材料。

7.4.4 胶砂比

最佳胶砂比应根据具有不同胶砂比的混合物在给定龄期（根据反应进程而不是时间顺序的龄期）下的最高强度（根据选择的标准测试方法，例如 ASTM C109/C109M[22] 测定）来确定。最佳胶砂比可用于计算原料混合物组成。重要的是指定所有测试混合物的流动度，而不是基于水与胶凝材料的比例，尽管由于流变性的基本差异，该值可能最终与硅酸盐水泥的期望值并不相同。工作性应根据流动度测试来定义，并且可以选择参考流动度的值（例如 ASTM C230 /C230M 和 C1437[23,24] 或 EN 12350-5[25]）。可以选择满足目标性能的任何比率，但应该提出细骨料和粗骨料的体系优化方案。

7.4.5 原料混合物组成的计算

绝对体积混凝土配合比设计——ACI 方法（ACI Recommended Practice 211.1[26]）适用于碱激发混凝土，其已在北美洲得到了广泛使用。ACI 配合比设计中需要测定固体激发剂的密度、溶解度以及混合溶液的密度，从而计算相应的体积。

7.4.6 外加水质量的影响

如果外加水的量对激发剂的溶解度有显著影响，则外加水的质量是重要影响因素（ASTM C1602/C1602M 和 D1193[27,28]和 EN 1008：2002[29]）。一般来说，外加水的质量对混凝土的整体质量有着重要的影响，应在 AAM 中明确这一参数。

7.4.7 混合过程对坍落度和抗压强度的影响

混料方法对坍落度及其保持和最终抗压强度有显著的影响[30,31]。应采用能使矿物组分有效激发的混料方式。因此，有两种建议的方法：

第一种方法，把激发剂加入外加水中。在开始时，粗细骨料与少量激发剂溶液预混合，设为时间 t_1。然后将矿物原料加入上述混合物中，并逐渐加入剩余的激发剂溶液，这段时间设为 t_2。静置一段时间 t_3 后，继续混合搅拌一段时间 t_4。通过激发剂溶入外加水的溶解性能（完全溶解和溶解速率）以及初凝时间来设定 $t_1 \sim t_4$ 的时间值。这种方法在北美广泛使用。

在第二种方法中，唯一的区别是将矿物原料加入到预混好的水溶液（充分溶解激发剂的水）中，用一段时间充分激发矿物原料。其他步骤与第一种方法相同。

在确定（或推荐在不严格规定的时候）AAM 混凝土的实验室测试方法的混合标准时，重要的是尽可能模拟混凝土生产的实际过程。这意味着包含激发剂、骨料和水的固体胶凝材料以与商业混凝土生产中类似的方式使用。大多数液体激发剂是碱性的、危险的，因此建议在大规模生产中应避免或尽量少用。此外，从一些固体激发剂中释放的溶解热可有助于加速一些矿物组分的反应性。但这可能是，也可能不是所期望的，这取决于配合比设计、样品几何形状和预期应用。在一些具有低溶解度的激发剂的情况下，需要一些热输入来提高水温或任何其他实用的方法以在合理的时间段内实现溶解。

7.4.8 化学外加剂及其在碱激发混凝土中的稳定性

为了调节碱激发混凝土的工作性，可以使用各种化学外加剂来改善混凝土的工作性和其他性能（ASTM C494/C494M 和 C260[32,33]以及 EN 934 和 EN480[34,35]）。然而，一些化学外加剂的稳定性尚未被广泛研究，如传统混凝土的情况，并且在第 6 章中详细讨论。应评估激发剂的类型和性质、热输入量（如果需要）、化学外加剂的类型和性质，以及活性矿物组分的类型和性质的影响。

7.4.9 骨料的性质和对碱-硅反应的敏感性

骨料的惰性在混凝土的性能中起重要作用。因此，应评估骨料对碱-二氧化硅反应（ASR）膨胀的敏感性（例如，类似于 ASTM C1260 和 1293[36,37]，CSA A23.2-14A 和 25A[38,39]，CR 1901：1995[40]和 AFNOR P18-588[41]）。应对现有的混凝土性能试验如

AFNOR P18-454[42]进行评估,以确定它们是否适用于 AAM 混凝土;对这个问题的进一步讨论将在第 8 章中提出。应采取一些预防措施,因为大多数矿物原料在激发期间对放热敏感。放热的加速效应(根据 ASTM 和 CSA 规范)可能导致在实际激发和养护下矿物原料对 ASR 的真实影响产生误导。所以也有人建议使用不同矿物原料(例如硅灰、F 类粉煤灰、偏高岭土)的二元或三元组合来抑制由碱-骨料反应引起的膨胀。

7.4.10 新拌混凝土性能

应采用满足基于性能标准的新拌混凝土性能的要求。如在普通硅酸盐混凝土中,密度、含气量、坍落度、温度和凝结时间等要求(例如,ASTM C143/143M、C231、C403[43-45]和 EN12350 系列[46])。虽然养护环境(密封、开放、浸没或蒸汽)的问题需要注意,但是实验室中混凝土试样的养护可以根据 ASTM C192/C192M[47]或 EN12390[48]标准进行,并对于硅酸盐水泥产品而言,非常期望的一些固化方法可能由于包括碱浸出/冲洗的问题而证明不适合于 AAM 样品。碱激发混凝土所需的凝结时间可根据温度、凝结时间和激发条件而变化。因此,养护时间应根据碱激发混凝土的性质和激发条件来定。在该领域的进一步研究和开发工作是必要的,以确定是否应强制将养护时间固定至 24h,或者是否优选规定使用的给定时间,从而获得可接受的抗压强度。水化热是另一个需要测量的重要性质。对于 AAM 的具体情况,应考虑 ASTM C186 等测试方法中固有假设的有效性[49]。在这种情况下,预期使用预溶解(液体)或固体激发剂将产生明显不同的结果,因此,在 7.4.7 节中提出的混合方案的问题再次变得重要。

7.4.11 混凝土硬化性能

硬化性能通常在不同的固化龄期测量。用于测试普通混凝土的不同指标也可用于碱激发混凝土中,用于测量抗压强度、干燥收缩率、抗折和劈裂抗拉强度以及其他参数(例如,出现在 ASTM C39/C39M、C426、C78、C496[50-53]以及 EN 12390[48]的各个章节)。还应进行氯离子渗透性、风化、浸出和其他耐久性试验,这些都将在第 8、9、10 章中详细讨论。

7.4.12 健康预防

在与碱激发水泥(AAC)和碱激发混凝土接触之前,必须做好预防保护措施,以确保施工过程能避免潜在的健康风险。跟普通硅酸盐水泥及其混合材一样,碱激发水泥也是有一定潜在危害的,它包含:

(1)碱性化合物(碱氢氧化物和硅酸盐)。

(2)微量的可浸出元素,这可能是危险的(应该使用浸出测试来评估,并对于含有粉煤灰的混合物是至关重要的)。

在混凝土中掺入工业固废(特别是粉煤灰)的辐射排放问题一直是讨论的话题,特别是放射性元素氡。这在一些辖区和特定应用领域中可能是非常重要的,但在这一领域仍有许多开放的科学问题,因此目前对碱激发材料提出具体建议似乎为时尚早。

1. 皮肤接触

应避免皮肤与新拌碱激发浆体或混凝土直接接触;急救程序应在产品随附的材料安全数据表中说明。

2. 过敏性皮肤反应，眼睛接触，吸入

这三个问题应由相关专家根据大多数激发剂的性质以及根据矿物组分与激发剂的相互作用的性质（协同或拮抗反应）分开描述。

7.5 与 AAM 有关的现行标准调研

RILEM TC 224-AAM 已经研究了与碱激发材料（AAM）相关的现有水泥和混凝土标准。以下部分总结了这些标准在 AAM 方面的相关性和缺点。

7.5.1 乌克兰和苏联的 AAM 标准

在 1961 年至 2007 年期间，苏联和乌克兰制定并实施了超过 60 种不同的标准。这些标准涵盖了各个领域，如原材料成分、水泥成分、混凝土配合比设计、制造工艺、结构和设计以及在特殊领域使用的建议。这些标准大多数是规定适用于 AAM 应用的材料，调节用 AAM 制造的各种类型的混凝土或指导制造和应用过程。每当一种新型原料可用时，它们会被提交给标准化委员会进行评估，并进一步规定它们在 AAM 中的用途。这些规范和法规的广泛接受帮助行业成功地推动 AAM 的开发和利用。苏联开发的 AAM 成功应用的重要因素之一，是当时建筑领域 OPC 的匮乏导致可替代建筑材料迫切的需要。

这些文件与 AAM 标准未来的开发高度相关，但是它们是规定性的。因此不符合 7.1 章节中为 AAM 标准的初步开发提出的大纲。初始标准不应该被定为覆盖所有规定材料、配方和如何生产及利用 AAM 的方法的可能性。乌克兰标准为改进特定 AAM 产品提供了长期目标的良好展示，为选择合适原材料和指导工艺参数提供了可选择的参考和实例。众所周知，在各个管辖区域，起草和完成新的或具体的水泥标准是一个耗时的过程。因为大多数利益相关者（消费者、制造商、政府官员等）在标准化委员会中必须达成多数同意接受。各利益相关者的利益常常不仅关注产品质量、健康和安全问题，而且还关注商业和技术优势。这与乌克兰的情况不同。在这种情况下，有权力的授权组织能够发布规范和指南，但这种情况不是全球范围都适用的。AAM 领域的未来国家标准和国际标准必须基于性能标准，并开放整合或纳入潜在的新技术。

7.5.2 澳大利亚标准

目前，澳大利亚标准框架（AS 3972[14] 和 AS 1379[55]）在不偏离 OPC 标准的基础上可做出一些优化。AS3972 的附录 A，如 7.2 节所述，为今后实施基于性能的标准框架提供了基础支持。通过展示对性能的实际经验来开辟对非特定材料（如更宽范围的原料，如非鼓风炉炉渣、底灰和碱激发剂）的使用限制。复合硅酸盐/AAM 胶凝材料在澳大利亚标准制度的使用范围也显得相对宽泛；GB（通用共混）水泥需要含有一定量的硅酸盐水泥，但是没有规定最小含量。存在最大允许的硫酸盐含量为 3.5% 的 SO_3，这似乎限制了硫酸盐激发剂的使用，但是除此之外，标准中规定的限制 AAM 使用的约束条件很少。

7.5.3 ASTM C1157——水泥性能测试标准

ASTM C 1157 涵盖了用于一般和特殊应用的水硬性水泥。这是一个给出性能要求规格的标准，它将水泥分为六种类型：通用（GU）、高早强（HE）、中等耐硫酸盐（MS）、耐高硫酸盐（HS）、中等水化热（MH）、低水化热（LH）。

该标准对水泥或其组分的组成没有限制。基本上，AAM只要符合分类的水泥类型之一，就能满足性能要求。然而，ASTM C 1157 在美国的接受程度目前非常有限（50个州中只有5个州）。

7.5.4 加拿大标准 CSA A3004-E：可选择的辅助胶凝材料（SCMs）

CSA A3004-E[4]涵盖了除常见SCM（FA、炉渣、硅灰）以外的可能适用于混凝土但不完全满足CSA A3001[56]第5条要求的材料。它规定了ASCMs（替代辅助胶凝材料）的水化动力学或火山灰性质的测定以及关于强度和耐久性的混凝土试验的要求。材料的评估包括四个阶段：

（1）材料的表征。材料的完整化学和矿物学分析，包括环境评估。不允许将危险废弃物归类为ASCM。

（2）最佳细度的确定。当ASCM的生产包括破碎和研磨时，可以从砂浆的抗压强度测试获得最佳细度。

（3）混凝土性能测试。新拌合硬化混凝土中ASCM的性能应在广泛的混凝土混合物中进行评估，以反映材料预期用途的范围。在使用混凝土项目之前，混凝土配料应该与当地材料结合进行资格预审。

（4）现场试验和长期性能及耐久性测试（3年）。ASCM在新拌合硬化混凝土中的性能应在广泛的混凝土混合物中进行评估，以反映材料预期用途的范围。

这种标准实践可能打开新型材料（ASCM）的大门，可能包括提供碱激发胶凝材料用于混凝土中的依据。

7.5.5 EN 206-1：混凝土——第1部分：规格、性能、生产和一致性

由于EN 206-1[57]中的定义和措辞，在混凝土中使用AAM胶凝材料是符合本标准的；似乎没有明确和严格的水泥要求符合EN 197-1[15]，因为使用了"一般适用性"和"应该"。然而，这种缺乏明确的要求没有经过合法测试。此外，允许欧洲技术批准（产品不符合任何其他现有标准）和国家标准规定扩大胶凝材料的范围，如瑞士国家标准的附录中提供了可用于更宽范围的胶凝材料，包括碱激发材料。

7.6 原材料分析

在选择用于AAM胶凝材料和混凝土的原料时，重要的是分析原料的潜在的火山灰活性（和反应产物）；在碱激发条件下的反应性可能不同于当将相同的材料作为SCM加入到硅酸盐水泥中时观察到的反应性。有许多现有的标准描述和指定SCM作为水泥组分，但是仍然要确定哪些测试在碱激发条件下是有效的。SCM的分析，包括其表征方法，属于RILEM TC 238-SCM[58]（2011年使用）的范围，因此超出了TC 224-AAM的范围。然而，通过TC 224-AAM工作计划提出的一些需要注意的关注点，作为未来标准制定的一个组成部分，包括以下内容：

(1) 原料的哪些物理化学参数很重要？

(2) 使用可用的激发剂范围，可以给出在与 AAM 合成相关的粉煤灰中的"反应性"氧化铝和二氧化硅含量或"玻璃含量"的定义。如果是这样，如何实现这一点？用这样的方式与硅酸盐水泥混合物中有用的（无疑是不同的）定义相比，没有混淆。在这一领域的工作中利用 HF 侵蚀选择性溶解玻璃相[59]，但由于健康和安全的原因，使用 HF（氢氟酸）作为浸出剂越来越受到限制；它在一些管辖区被管制为有毒化学品，因此不能被广泛使用。那么，是否有另一种测试（优选涉及碱，在碱激发中的反应性）可以提供可靠的结果？

(3) 对于矿渣的情况，是否可以选择矿渣的质量系数（或相关的氧化物比率）来给出可靠的反应性预测？目前有许多公式[60]，它们彼此不一致，或者与一般的材料性能不一致。

(4) 粉煤灰中未燃烧碳的含量应该受到限制，还是仅仅作为潜在的复杂因素？

(5) 如何监测潜在有害元素从粉煤灰中的释放？这取决于激发剂的选择和在其下发生浸出的氧化还原环境[61,62]。这些在大多数标准浸出试验中没有很好地提出（见第 8.2 节）。

(6) 什么技术可以测量不同原料的粒径？Blaine 细度[63,64]被广泛使用，Wagner 浊度计也由 ASTM[65]标准化，但这些方法很少使用。另一个问题是，这些方法对于高长径比的颗粒是否敏感（如偏高岭土在一些标准中也使用筛余物），高级仪器可以在非常广泛的粒径范围内提供可靠的粒径分布信息（在每种粒径范围内使用适当的仪器），但是非常困难，如果有可能，将一个特定的仪器纳入标准体系，对于复杂颗粒形状或宽尺寸分布原料，不同的仪器或测量技术不会总是给出精确的结果。测量期间的颗粒团聚也是潜在的问题。该领域目前正在 TC 238-SCM 中在混合硅酸盐水泥的背景下进行讨论，并且 TC 的这一结果在 AAM 的应用中也将具有很大的价值。

7.7 养护的重要性

在科学界有关于最佳养护形式应用于碱激发剂以获得最佳强度发展和耐久性存在广泛争论。这一领域有许多科学出版物[66,77]一致认为最佳养护温度取决于配合比设计的具体细节。普遍认为延长密封养护时间对于致密度和耐久性是重要的。众所周知，对于硅酸盐水泥与 SCM 的混合物[78,79]，这些非熟料材料的反应速度比硅酸盐水泥的反应慢，并且水的结合通常在这些低钙体系中较弱，弱于在通过硅酸盐水泥水化形成的富 Ca 的 C-S-H 相和/或硫铝酸钙水化物相（AFm 或 AFt）。这意味着控制用于碱激发混凝土的养护条件和环境是必要的；否则，会观察到严重的碳化、开裂和/或表面起尘。这可以通过提供充分的养护环境来防止；"充分"的定义强烈依赖于配合比设计和其他参数（固体原料和激发剂性质，以及水胶比和温度）。

养护条件在标准化方面提出了挑战，对于一种产品理想的养护条件可能完全不适用于另一种产品。这是另一个领域，其中提出了基于性能的规范作为惯例的替代；在测试之前养护条件的选择可能需要变得更灵活，而不是在测试方法中明确定义，尽管会让养护方案中的这种变化带来更多的问题。这仍然是 TC 和 TC 247-DTA 以及更广泛的科学界内开放的讨论领域。

7.8 小结

RILEM TC 224-AAM 建议制定基于性能的标准体系，为碱激发胶凝材料和混凝土提供说明和规范。胶凝材料和混凝土的标准最好单独起草，以便独立生产不同的胶凝材料（例如作为干拌水泥），或者使胶凝材料直接与骨料结合形成混凝土，而不用先生产碱激发"水泥"，再生产混凝土。一些现有的标准比其他标准更适合在现有标准框架内提供碱激发胶凝材料；特别是，ASTM C1157 是一项基于性能的标准，如果该标准被更广泛的接受，它就可以用于 AAM 混凝土。通过本章的讨论充分体现了标准的重要性，许多领域在制定标准体系时需要做大量工作，特别是在材料养护方面。

致谢：作者感谢 Galal Fares 在本章撰写过程中的支持和帮助。

参考文献

[1] Hooton R. D.. Bridging the gap between research and standards. Cem. Concr. Res, 2008, 38(2): 247-258.

[2] European Committee for Standardization (CEN): Hands on Standardization-A Starter Guide to Standardization for Experts in CEN Technical Bodies, 2010.

[3] ASTM International: Standard Performance Specification for Hydraulic Cement (ASTM C1157/C1157M). West Conshohocken, PA, 2010.

[4] Canadian Standards Association: Standard Practice for the Evaluation of Alternative Supplementary Cementing Materials (ASCMs) for Use in Concrete (CSA A3004-E1). Mississauga, ON, 2008.

[5] The Ministry for Construction of Enterprises of Heavy Industry of the USSR: Industry Standard "Slag Alkaline Binders. Technical Specifications", (OST 67-11-84). Moscow, 1984.

[6] The Ministry for Ferrous Metallurgy of the USSR: Technical Specifications "A slag alkaline binder made from silicomanganese granulated slag" (TU 14-11-228-87). Kiev, 1987.

[7] The State Committee of Belarus Republic of the USSR for Construction (Gosstroy BelSSR): Technical Specifications "A slag alkaline cement from cupola/iron/granulated slag" (TU 7 BelSSR 5-90). Minsk, 1990.

[8] The Ukrainian Cooperative State Enterprise of Agroindustrial Construction: Technical Specifications "A slag alkaline binder from ferronickel granulated slag" (TU 559-10.20-001- 90). Zhitomir, 1990.

[9] The Ministry of Metallurgy of the USSR: Technical Specifications "A slag alkaline binder from silicomanganese granulated slag" (TU 14-11-228-90). Kiev, 1990.

[10] The Ukrainian Cooperative State Corporation on Agricultural Construction: Technical Specifications "A slag alkaline binder on sulfate-containing compounds of alkali metals" (TU 10. 20 UkrSSR 169-91). Kiev, 1991.

[11] The State Coo perative Enterprise "Uzagrostroy"-The Tashkent Architectural-Construction Institute: A slag alkaline cement from electrothermo phosphorous slags (TU 10. 15 UzSSR 04-91). Tashkent, 1991.

[12] The State Committee of Ukrainian Republic of the USSR for Urban Planning and Architecture: A slag alkaline binder. Technical Specifications (DSTU BV 2. 7-24-95 supersedes RST UkrSSR 5024-89). Kiev, 1995.

[13] The State Committee for Construction of Ukraine: Binders, Alkaline, for Special Uses-Geocements. Technical Specifications(TU U V 2. 7-16403272. 000-98). Kiev, 1998.

[14] Standards Australia: General Purpose and Blended Cements (AS3972-2010). Sydney, 2010.

[15] European Committee for Standardization (CEN): Cement-Part 1: Composition, Specifications and Conformity Criteria for Common Cements (EN 197-1). Brussels, Belgium, 2000.

[16] ASTM International: Standard Specification for Portland Cement (ASTM C150/C150M-11). West Conshohocken, PA, 2011.

[17] European Committee for Standardization (CEN): Ground Granulated Blast Furnace Slag for Use in Concrete, Mortar and Grout. Definitions, Specifications and Conformity Criteria (EN 15167-1). Brussels, Belgium, 2006.

[18] ASTM International: Standard Specification for Slag Cement for Use in Concrete and Mortars (ASTM C989-10). West Conshohocken, PA, 2010.

[19] European Committee for Standardization(CEN): Fly Ash for Concrete-Part 1: Defi nitions, Specifications and Conformity Criteria(EN 450-1). Brussels, Belgium, 2007.

[20] ASTM International: Standard Specification for Coal Fly Ash and Raw or Calcined Natural Pozzolan for Use in Concrete(ASTM C618-08a). West Conshohocken, 2008.

[21] European Committee for Standardization (CEN): Methods of Testing Cement-Part 1: Determination of Strength (EN 196-1). Brussels, Belgium, 2005.

[22] ASTM International: Standard Test Method for Compressive Strength of Hydraulic Cement Mortars (Using 2-in. or [50-mm] Cube Specimens) (ASTM C109 /C109M-11). West Conshohocken, PA, 2011.

[23] ASTM International: Standard Specification for Flow Table for Use in Tests of Hydraulic Cement (ASTM C230/C230M-08). West Conshohocken, PA, 2008.

[24] ASTM International: Standard Test Method for Flow of Hydraulic Cement Mortar (ASTM C1437-07). West Conshohocken, PA, 2007.

[25] European Committee for Standardization (CEN): Testing Fresh Concrete. Flow Table Test (EN 12350-5). Brussels, Belgium, 2009.

[26] American Concrete Institute: Standard Practice for Selecting Proportions for Normal, Heavyweight, and Mass Concrete (ACI 211. 1-91). Farmington Hills, MI, 1991.

[27] ASTM International: Standard Specification for Mixing Water Used in the Production of Hydraulic Cement Concrete (ASTM C1602/C1602M-06). West Conshohocken,

[28] ASTM International: Standard Specification for Reagent Water (ASTM D1193-06). West Conshohocken, PA, 2011.

[29] European Committee for Standardization (CEN): Mixing Water for Concrete-Specification for Sampling, Testing and Assessing the Suitability of Water, Including Water Recovered from Processes in the Concrete Industry, as Mixing Water for Concrete (EN 1008). Brussels, Belgium, 2002.

[30] Hardjito D., Rangan B. V.. Development and Properties of Low-Calcium Fly Ash-Based Geopolymer concrete, Curtin University of Technology Research Report GC1. Curtin University, Perth, 2005.

[31] Palacios M., Puertas F.. Effectiveness of mixing time on hardened properties of waterglassactivated slag pastes and mortars. ACI Mater. J, 2001, 108(1): 73-78.

[32] ASTM International: Standard Specification for Chemical Admixtures for Concrete (ASTM C494/C494M-10a). West Conshohocken, PA, 2010.

[33] ASTM International: Standard Specification for Air-Entraining Admixtures for Concrete (ASTM C260/C260M-10a). West Conshohocken, PA, 2010.

[34] European Committee for Standardization (CEN): Admixtures for Concrete, Mortar and Grout (EN 934). Brussels, Belgium, 2009.

[35] European Committee for Standardization (CEN): Admixtures for Concrete, Mortar and Grout-Test Methods (EN 430). Brussels, Belgium, 2005.

[36] ASTM International: Standard Test Method for Potential Alkali Reactivity of Aggregates (Mortar-Bar Method) (ASTM C1260-07). West Conshohocken, PA, 2007.

[37] ASTM International: Standard Test Method for Determination of Length Change of Concrete Due to Alkali-Silica Reaction (ASTM C1293-08b). West Conshohocken, PA, 2008.

[38] Canadian Standards Association: Potential Expansivity of Aggregates, Procedure for Length Change Due to Alkali-Aggregate Reaction in Concrete Prisms (CSA 23. 2-14A). Mississauga, ON, 2000.

[39] Canadian Standards Association: Test Method for Detection of Alkali-Silica Reactive Aggregate by Accelerated Expansion of Mortar Bars (CSA 23.2-25A). Mississauga, ON, 2000.

[40] European Committee for Standardization (CEN): Regional Specifications and Recommendations for the Avoidance of Damaging Alkali Silica Reactions in Concrete (CR1901: 1995). Brussels, Belgium, 1995.

[41] Association Française de Normalisation: Granulats-Stabilité dimensionnelle en milieu alcalin (essai accéléré sur mortier MICROBAR) (AFNOR NF P18-588). Saint-Denis, 1991.

[42] Association Française de Normalisation: Réactivité d'une formule de béton vis-à-vis de l'alcali-réaction (essaie de performance) (AFNOR NF P18-454). Saint-Denis, 2004.

[43] ASTM International: Standard Test Method for Slump of Hydraulic-Cement Concrete (ASTM C143/C143M-10a). West Conshohocken, PA, 2010.

[44] ASTM International: Standard Test Method for Air Content of Freshly Mixed Concrete by the Pressure Method (ASTM C231/C231M-10). West Conshohocken, PA, 2010.

[45] ASTM International: Standard Test Method for Time of Setting of Concrete Mixtures by Penetration Resistance(ASTM C403/C403M-08). West Conshohocken, PA, 2008.

[46] European Committee for Standardization (CEN): Testing Fresh Concrete (EN 12350). Brussels, Belgium, 2009.

[47] ASTM International: Standard Practice for Making and Curing Concrete Test Specimens in the Laboratory (ASTM C192/C192M-07). West Conshohocken, PA, 2007.

[48] European Committee for Standardization (CEN): Testing Hardened Concrete (EN 12390). Brussels, Belgium, 2008.

[49] ASTM International: Standard Test Method for Heat of Hydration of Hydraulic Cement (ASTM C186-05). West Conshohocken, PA, 2005.

[50] ASTM International: Standard Test Method for Compressive Strength of Cylindrical Concrete Specimens (ASTM C39/C39M-10). West Conshohocken, PA, 2010.

[51] ASTM International: Standard Test Method for Linear Drying Shrinkage of Concrete Masonry Units (ASTM C426-10). West Conshohocken, PA, 2010.

[52] ASTM International: Standard Test Method for Flexural Strength of Concrete (Using Simple Beam with Third-Point Loading) (ASTM C78/C78M-10). West Conshohocken, PA, 2010.

[53] ASTM International: Standard Test Method for Splitting Tensile Strength of Cylindrical Concrete Specimens (ASTM C496/C496M-04e1). West Conshohocken, PA, 2004.

[54] Kovler K.. Does the utilization of coal fly ash in concrete construction present a radiation hazard? Constr. Build. Mater, 2012, 29: 158-166.

[55] Standards Australia: Specification and Supply of Concrete (AS 1379-2007). Sydney, 2007.

[56] Canadian Standards Association: Cementitious Materials for Use in Concrete (CSA A3001-08). Mississauga, ON, 2008.

[57] European Committee for Standardization (CEN): Concrete-Part 1: Specification, Performance, Production and Conformity (EN 206-1). Brussels, Belgium, 2010.

[58] Juenger M., Provis J. L., Elsen J., Matthes W., Hooton R. D., Duchesne J., Courard L., He H., Michel F., Snellings R., de Belie N.. Supplementary cementitious materials for concrete: Characterization needs. Mater. Res. Soc. Symp. Proc. 2012: 1488, doi: 10.1557/opl.2012.1536.

[59] Fernández-Jiménez A., de la Torre A. G., Palomo A., López-Olmo G., Alonso M. M., Aranda M. A. G.. Quantitative determination of phases in the alkali activation

of fly ash. Part I. Potential ash reactivity. Fuel, 2006, 85 (5-6): 625-634.

[60] Shi C., Krivenko P. V., Roy D. M.. Alkali-Activated Cements and Concretes. Taylor & Francis, Abingdon, 2006.

[61] Provis J. L., Rose V., Bernal S. A., Van Deventer J. S. J.. High resolution nanoprobe X-ray fluorescence characterization of heterogeneous calcium and heavy metal distributions in alkali activated fly ash. Langmuir, 2009, 25 (19): 11897-11904.

[62] Zhang J., Provis J. L., Feng D., Van Deventer J. S. J.. The role of sulfide in the immobilization of Cr(VI) in fly ash geopolymers. Cem. Concr. Res, 2008, 38 (5): 681-688.

[63] ASTM International: Standard Test Methods for Fineness of Hydraulic Cement by Air- Permeability Apparatus (ASTM C204-07). West Conshohocken, PA, 2007.

[64] European Committee for Standardization (CEN): Methods of Testing Cement-Part 6: Determination of Fineness (EN 196-6). Brussels, Belgium, 2010.

[65] ASTM International: Standard Test Method for Fineness of Portland Cement by the Turbidimeter (ASTM C115-10). West Conshohocken, PA, 2010.

[66] Bakharev T., Sanjayan J. G., Cheng Y. B.. Effect of elevated temperature curing on properties of alkali-activated slag concrete. Cem. Concr. Res, 1999, 29(10): 1619-1625.

[67] Criado M., Palomo A., Fernández-Jiménez A.. Alkali activation of fly ashes. Part 1: Effect of curing conditions on the carbonation of the reaction products. Fuel, 2005, 84(16): 2048-2054.

[68] Małolepszy J., Deja J.. The influence of curing conditions on the mechanical properties of alkali-activated slag binders. Silic. Ind, 1988, 53(11-12): 179-186.

[69] Bondar D., Lynsdale C. J., Milestone N. B., Hassani N., Ramezanianpour A. A.. Effect of heat treatment on reactivity-strength of alkali-activated natural pozzolans. Constr. Build. Mater, 2011, 25(10): 4065-4071.

[70] Criado M., Fernández-Jiménez A., Palomo A.. Alkali activation of fly ash. Part III: Effect of curing conditions on reaction and its graphical description. Fuel, 2010, 89 (11): 3185-3192.

[71] Izquierdo M., Querol X., Phillipart C., Antenucci D., Towler M.. The role of open and closed curing conditions on the leaching properties of fly ash-slag-based geopolymers. J. Hazard. Mater, 2010, 176(1-3): 623-628.

[72] Kovalchuk G., Fernández-Jiménez A., Palomo A.. Alkali-activated fly ash: Effect of thermal curing conditions on mechanical and microstructural development-Part II. Fuel, 2007, 86(3): 315-322.

[73] Najafi Kani E., Allahverdi A.. Effects of curing time and temperature on strength development of inorganic polymeric binder based on natural pozzolan. J. Mater. Sci, 2009, 44: 3088-3097.

[74] Rovnaník P.. Effect of curing temperature on the development of hard structure of metakaolinbased geopolymer. Constr. Build. Mater, 2010, 24 (7): 1176-1183.

[75] Sindhunata, Van Deventer J. S. J., Lukey G. C., Xu H.. Effect of curing temperature and silicate concentration on fly-ash-based geopolymerization. Ind. Eng. Chem. Res, 2006, 45(10): 3559-3568.

[76] Somaratna J., Ravikumar D., Neithalath N.. Response of alkali activated fly ash mortars to microwave curing. Cem. Concr. Res, 2010, 40, 12): 1688-1696.

[77] Wang S. D., Scrivener K. L., Pratt P. L.. Factors affecting the strength of alkali-activated slag. Cem. Concr. Res, 1994, 24(6): 1033-1043.

[78] Lothenbach B., Scrivener K., Hooton R. D.. Supplementary cementitious materials. Cem. Concr. Res, 2011, 41(12): 1244-1256.

[79] Hewlett P. C.. Lea's Chemistry of Cement and Concrete, 4th edn. Elsevier, Oxford, 1998.

第8章 耐久性和测试——化学基体裂化过程

Kofi Abora，Irene Beleña，Susan A. Bernal，Andrew Dunster，
Philip A. Nixon，John L. Provis，Arezki Tagnit-Hamou
and Frank Winnefeld

8.1 引言

本章和第9、10两章以综述的形式，评估了在各种类型的侵蚀模式下，建筑材料性能的测试方法。这些方法被划分为"化学"法（第8章）、"传输"法（第9章）和"物理"法（第10章）。值得注意的是，这种分类在某种程度上是随机的，以目前的排版方式很难清晰地考虑到这三种方法之间有明显交叉。在一些地方进行了更为详细的讨论，要么是因为它们是碱激发技术相关领域的关键点，或者仅仅是因为关于碱激发材料（AAMs）的一些形式的侵蚀数据有限；生物腐蚀就是一个特殊的例子，在公开的文献中数据很少。因为对AAM劣化机制目前了解有限，本章主要对未来提出一些问题，而不提供详细的解答，但是会尽可能地吸收相关的建议。

8.2 抗硫酸盐测试

8.2.1 引言

在文献和现有的相关标准中有大量的有关矿物胶凝材料、砂浆和混凝土抗硫酸盐侵蚀性的试验方法、试验结果和裂化机理。然而，在大多数建筑材料的常见测试方法中，测试过程一般是基于普通硅酸盐水泥（OPC）和复合OPC胶凝材料体系的固有设计。相反，没有提供对碱激发材料的抗硫酸盐侵蚀性能方面有关改进的测试报告、案例研究或劣化机理。虽然在OPC耐硫酸盐测试过程中考虑了化学和热力学的影响因素，并通过有效的加速测试方法对耐硫酸盐侵蚀的机理有了很好的理解[1, 2]，但许多国家（包括欧盟在内）仍需要通过行政命令而不是性能测试来评价耐硫酸盐性能。在此问题上，学者们需要考虑的关键点是在硫酸盐侵蚀下OPC和AAM环境下劣化机理的不同。其主要与在OPC中C-S-H型凝胶相形成脱钙作用以及含硫钙（通常是膨胀的）裂化产物的形成有关。AAMs低的钙含量导致了两者的差异。因此需对测试硫酸盐侵蚀下的OPC性能测试参数进行修订。

RILEM TC 224-AAM的任务之一是为AAM制定适宜的测试方法。基于OPC体系的测试方法，对其进行改性以满足AAM的特殊要求（如样品制备、养护和其他AAM的具体问题）。本节介绍：（1）对一般OPC体系抗硫酸盐测试简短的文献综述；（2）对AAM的抗硫酸盐侵蚀性能的一些案例研究；（3）对一般测试抗硫酸盐侵蚀的可能途径的一些考虑。硫

酸（矿物或生物）侵蚀在这一章不予考虑，将在本章后面讨论。

RILEM 技术委员会 TC 211-PAE（水泥基材料在腐蚀水中的性能）已经建立了最先进的 OPC 基体系的抗硫酸盐的耐久性测试报告。本报告的初始版本[3]在本文件的创建过程中进行了协商，这里提出的讨论将不会是一个详尽的总结，读者可参考初始版本，里面有更为详细的论述。这部分的其他主要信息来源于 CEN 报告[4]和瑞士的报告[5]。

8.2.2　OPC 体系的抗硫酸盐测试

一般来说，抗硫酸盐侵蚀测试方法可分类如下：

（1）外部硫酸盐侵蚀。通过将试样浸泡在一个测试溶液中，或通过润湿和干燥循环来模拟外部环境。在这两种情况下，使用含硫酸盐的水溶液（最常见的 Na_2SO_4 和 $MgSO_4$）需要及时更换。外部侵蚀测试了胶凝材料本身由于含硫酸盐膨胀相的形成而带来膨胀的"化学"潜能。还有这种方式的侵蚀下整个体系（净浆、砂浆、混凝土）的性能，以及外部硫酸盐侵蚀下体系的孔隙率影响行为。

（2）内部硫酸盐侵蚀。添加额外的石膏胶凝材料使它"过硫酸盐化"。这里测试了胶凝材料的抗硫酸盐性，净浆或砂浆中石膏的直接加入使水和离子运输的影响被排除在外。

世界各地的各种研究人员实施的特定应用的测试方案和不同方法将会影响所获得的结果。测试环境：真实的环境条件，加速的条件，如非常高的硫酸盐浓度或较高的温度；使用各种评价方法（膨胀和抗折强度）和评价标准。

考虑到浓度和温度作用下硫酸盐溶液本身（和它们的相互作用的结晶产物与水泥基胶凝材料）的相化学，加速的测试方法有效性已成为当前的研究重点。不仅仅是硫酸盐侵蚀，多种类型的化学侵蚀的核心内容将在本章和下面几章中分别讨论。

以下介绍一些常见的测试方法以及特性。

1. 内部侵蚀

（1）ASTM C 452[6]

设计：OPC（不是复合水泥）。

样品：砂浆，OPC 中石膏加至 7% SO_3，$w/c=0.485$（0.460 引气砂浆），（水泥+添加石膏）/砂=2.75。

样品尺寸：25mm×25mm×285mm。

养护：在模具中 23℃养护 1d，在 23℃水中养护，用水不时更新。

评价：第 1d 和第 14d 之间的膨胀性能。

（2）Duggan 测试[7]。

设计：OPC 混凝土。

样品：混凝土，水/水泥=0.40。

样品尺寸：76mm×76mm×356mm。

养护：热循环（最高温度为 85℃），然后在水里养护，没有添加硫酸（延迟钙矾石生成测试）。

评估：90d 膨胀性能。

（3）Le Chatelier-Anstett 试验[8,9]

设计：OPC。

样品：含50%石膏和6%额外补充水分的硬化水泥浆体。

样品尺寸：圆柱体直径80mm，高度30mm，20kgf/cm^2（1.96MPa）压制成形。

养护：水。

评估：90d膨胀性能。

2. 外部侵蚀

(1) ASTM C 1012[10]

设计：OPC，与火山灰或BFS复合的OPC，复合水硬性水泥。

样品：OPC砂浆，$w/c=0.485$（0.460引气砂浆，或调整至流动的复合/水硬水泥），水泥/砂=2.75。

样品尺寸：50mm立方体，用于测试强度，25mm×25mm×285mm棒状测试膨胀。

养护和浸置：在模具中35℃养护1d，然后在23℃的石灰水中养护，直到抗压强度>20MPa，然后浸入23℃的Na_2SO_4（50 g/L）溶液中，或其他可能的溶液（如$MgSO_4$）。

评价：在硫酸溶液中12~18个月后的膨胀性能。

(2) CEN测试[11]

设计：OPC，复合OPC。

样品：$w/c=0.50$砂浆，水泥/砂=1:3。

样品尺寸：20mm×20mm×160mm。

养护和浸置：20℃相对湿度>90%，养护1d，在20℃水中养护27d，然后浸置于Na_2SO_4溶液中（16g/L SO_4=24g/L Na_2SO_4），每个月更换一次溶液，对照空白样品放在水中。

评价：1年后的长度变化。

(3) Koch and Steinegger[12]

设计：OPC，BFS复合OPC。

样品：$w/c=0.60$砂浆，水泥/砂=1:3。

样品尺寸：10mm×10mm×60mm。

养护和浸置：在模具中养护1d，在去离子水中养护20d，然后浸置于10% $Na_2SO_4 \cdot 10H_2O$溶液（44 g/L的Na_2SO_4），对照空白样品放置在去离子水中。

评价：用酚酞滴定硫酸根来更新硫酸盐，抗折强度，目测观察，持续77d。

(4) Mehta and Gjørv[13]

设计：OPC，复合OPC（BFS、火山灰）。

样品：水泥浆，$w/c=0.50$。

样品尺寸：12.5mm立方体。

养护和浸置：50℃湿养护7d，然后浸置于$CaSO_4$溶液（0.12% SO_3=2 g/L $CaSO_4$）或Na_2SO_4溶液（2.1% SO_3=37 g/L Na_2SO_4），用H_2SO_4滴定溴百里酚蓝来更新硫酸盐。

评价：28d后抗压强度。

(5) NMS测试[11]

设计：砂浆和混凝土，混凝土不养护。

样品：用$w/c=0.50$的砂浆或混凝土，样品不养护。

样品尺寸：40mm×40mm×160mm（砂浆/混凝土），芯的直径为50mm，长度为

150mm（混凝土）。

养护和浸置：20℃，相对湿度＞90%，在模具中养护2d，然后在水中养护5d；20℃，65%相对湿度养护21d；之后浸置于真空度为150mbar、饱和度为5%的Na_2SO_4溶液（50g/L的Na_2SO_4）中，再浸置于8℃、50g/L的Na_2SO_4溶液中，对照空白样品浸置于水中。

评价：56～180d后抗拉强度，取决于测试设置。

(6) SVA 测试[14]

设计：砂浆和混凝土，混凝土不养护。

样品：砂浆 w/c=0.50，水泥/砂=1∶3。

样品尺寸：10mm×40mm×160mm（砂浆），芯的直径为50mm，长度为150mm。

养护和浸置：20℃，相对湿度＞90%，在模具中养护2d，然后在饱和石灰溶液中养护12d；分别在20℃和6℃时，浸置于饱和硫酸钠溶液中（44 g/L Na_2SO_4）。对照空白样品分别在20℃和6℃的饱和石灰溶液中养护，每14d换一次溶液，最初的程序只考虑在20℃下测试。

评价：91d后的长度变化。

(7) Swiss Standard SIA 262/1 Appendix D[15]

设计：混凝土。

样品：一般混凝土样品。

样品尺寸：芯的直径为28mm，长度为150mm。

养护和浸置：根据EN 12390-2标准28d湿养护，然后4次硫酸盐侵蚀循环。实施步骤如下：50℃干燥48h，20℃冷却1h，50g/L Na_2SO_4 溶液中浸置120h。

评价：长度变化，质量增加。

（注：本标准目前正在修订，在修订后的版本中一些细节将略有不同。）

(8) wittekindt[8]

设计：OPC，BFS 复合 OPC。

样品：砂浆 w/c=0.60，水泥/砂=1∶3。

样品尺寸：10mm×40mm×160mm。

养护和浸置：模具中养护2d，水中养护5d，然后在0.15M的Na_2SO_4溶液（21g/L）浸置，溶液在不同的时间间隔更换。

评价：膨胀性能（持续时间变量，数月至数年）

8.2.3 AAM 的抗硫酸盐性能

目前，关于AAM的抗硫酸盐侵蚀性能的研究报道较少。在一般情况下，外部硫酸盐侵蚀使用砂浆或混凝土样品作为检测手段。

芬兰[16]和加拿大[17]的早期工作通过测试共振频率分析抗压或抗拉强度、尺寸稳定性、超声波脉冲速度和弹性模量等方法测量暴露在硫酸盐中（在$MgSO_4$和Na_2SO_4溶液中）4～6个月碱金属硅酸盐激发矿渣混凝土的性能，结果表明其抗硫酸盐侵蚀良好。

然而，Kukko 和 Mannonen[16]的研究表明，在10%$MgSO_4$条件下持续12个月以上会使样品（碱激发BFS，OPC和大体积复合BFS-OPC）裂变。然而，在1%$MgSO_4$或在10%

Na_2SO_4 中，样品保持完整并保持（或增加）抗压强度长达 25 个月。

Shi 等人[18]对 AAM 的抗硫酸盐侵蚀性能做了综述，但其测试方法不是分开的。通常 AAMs 性能与 OPC 相当甚至优于 OPC。但 AAMs 的性能主要取决于原材料的化学组成（矿渣、粉煤灰或其他）、激发剂的种类以及测试所用的硫酸溶液的组成和浓度。

Bakharev 等人[19]在 ASTM C 1012 的基础上为 AAM 开发了一种可应用于碱激发矿渣混凝土的抗硫酸盐性能的测试方法。其具体方法如下：直径为 100mm、长度为 200mm 的混凝土圆柱体在蒸汽室养护 28d 后，浸置于 50g/L 的 Na_2SO_4 和 50g/L 的 $MgSO_4$ 溶液中，并定期更换。经过 12 个月后进行了抗压强度测试，并与存储在饮用水的参考样品进行比较。在 Na_2SO_4 溶液中 AAM 混凝土比 OPC 混凝土的性能好，然而在 $MgSO_4$ 溶液中两者相似。在硫酸钠溶液中，未有劣化的迹象，而硫酸镁的侵蚀造成了石膏的形成和 C-S-H 的分解。在同一研究者的第二个论文[20]中，碱激发粉煤灰的抗硫酸盐性能用类似方法进行了测试。在这种情况下，热养护的试块浸置于上述相同的溶液中，使用抗压强度作为评价标准。发现不同种类激发剂及浸置条件下 AAMs 的耐久性有显著的变化（NaOH 激发剂+热养护性能最好）。

Puertas 等人[21]以 Koch-Steinegger 和 ASTM C 1012 方法研究碱激发矿渣和粉煤灰砂浆的抗硫酸盐，结果表明其具有良好的抗硫酸盐性，而 NaOH 激发的高炉矿渣对硫酸盐侵蚀较敏感。同时有石膏和钙矾石形成。然而，Ismail 等人[22]的研究表明，碱激发矿渣粉煤灰混合物对硫酸盐侵蚀的反应是基本取决于硫酸盐阳离子的性质；Na_2SO_4 浸渍导致净浆样品的轻微损伤，而 $MgSO_4$ 引起胶凝材料的严重脱钙、石膏的形成、结构和尺寸完整性的损失。

根据 Swiss Standard SIA 262/1 Appendix D[23]，测试一个特定的非标准养护制度下（在 80℃下热养护，然后储存在空气中）四种粉煤灰基 AAM 混凝土的抗硫酸盐侵蚀性，表明其具有良好的抗侵蚀性。Škvara[24]观察到碱激发粉煤灰砂浆在实验室环境中养护 28d 后浸置于含 44 g/L 的 Na_2SO_4 或 5g/L $MgSO_4$ 溶液中，并没有劣化。其他使用类似测试方法的报告[17,24-29]也报道了 AAM 对硫酸盐溶液具有较好的抵抗性。

8.2.4 关于 AAM 测试方法的评价

在一般情况下，要使用的测试方法应接近于 OPC 和混合 OPC 体系使用的测试方法，有助于接受 AAM 耐久性能测试的有效性。因此，必要时只需轻微变动现有标准测试，在某种程度上将适用于 AAM。改进应主要限于样品的制备和养护，而不是测试方法本身。

1. 外部或内部的抗硫酸盐侵蚀测试

使用硫酸盐含水溶液对砂浆或混凝土进行外部侵蚀测试是最为必要的。整个体系进行测试需考虑其孔隙率和透气性。

2. 测试胶凝材料或混凝土

这个测试需要的样品是混凝土而非胶凝材料，AAM 混凝土可以由没有形成胶凝材料的原料直接制备，从某种意义上来说，OPC 是从筒仓或袋中获得的。

最终产品的耐久性应是决定性的参数，孔隙率和渗透性能会对耐久性产生影响。这是从实验室测试净浆和砂浆取得的重要成果，混凝土测试标准也有所提及。

3. 固定配合比组成或混凝土

AAM 基混凝土有很宽范围的水胶比、砂胶比和外加剂。因此，应该测试"真实"应用于施工现场的混凝土，而不是测试在已经存在的测试方法中定义的特殊混合组成。

4. 与硫酸盐作用前的养护

根据 AAM 的组成，经常会应用特殊养护制度，包括热养护。在水或石灰水中养护并不总是适合 AAMs，因其可能会导致碱浸出。可以用湿养护来代替。一般来说，测试方法中不应指定养护制度（除了样品龄期，可能是 28d），但特定的 AAM 必须使用适宜的养护制度。

5. 硫酸盐反应测试

除了 AAM 样品制备与养护需要调整，建议不修改基本的测试方法，AAM 的测试环境和条件要与 OPC 相同，再与 OPC 性能进行对比。

6. 应该使用哪种测试方法

对样品制备和养护进行改进后出现了几种合适的测试方法，包括 ASTM C 1012、CEN、NMS 和 SVA 测试。然而在这些测试中的硫酸盐剂量、评价方法和标准有很大的不同。从这些差异中透漏出重要的化学和物理意义。目前不可能得出它将是最合适的方法的结论（值得注意的是，这在 OPC 中并没有被实现）。

8.3 碱-骨料反应

胶凝材料的碱含量是非常关键的，关系到硅酸盐水泥混凝土的碱-骨料反应（AAR）[30]。碱-骨料反应可以分为两类：碱-硅反应（ASR）和碱-碳酸盐反应（ACR）。前者发生在水泥中出现的潜在活性骨料和碱之间，Na_2O、K_2O 和 $Ca(OH)_2$，反应产物是碱硅酸凝胶，受潮后易发生膨胀，造成混凝土的膨胀和开裂[31]。碱-碳酸盐反应是混凝土中碳化骨料的白云石与碱反应引起的一种膨胀去白云石化过程，ACR 较 ASR 并不常见。

在 OPC 混凝土中，限制 ASR 反应的破坏性通常是通过减少混凝土的碱含量来实现。OPC 中，Na_2O 含量不应超过水泥质量的 0.6%（其中 Na_2O 含量是包含在一个摩尔当量的基础上）。当 OPC 中有 50% 或更多的 BFS 矿渣时，加入的碱含量超过 1% 是可以接受的[32,33]。然而，在 AAMs 中碱激发剂的使用使其碱含量远高于 PC。例如，碱激发矿渣水泥（AAS）中的碱含量可达到 2%~5%[34]。因此，在碱激发材料（AAMs）中的 ASR 反应的研究已经吸引了很多的关注，有人担心 AAMs 可能会受到 ASR 的影响。

另一方面，由于初始组成材料如 BFS 或 F 类粉煤灰的低钙组成，基于这些材料的碱激发胶凝材料相对于 OPC 的碱-骨料反应可能会表现出不同的行为，如钙在 ASR 过程中在确定的速率和潜在破坏程度方面起决定性作用[35]。

接下来将概述 AAM 胶凝材料中的 ASR 反应，并讨论现有的预测潜在有害膨胀的测试和标准。

8.3.1 测试方法

RILEM TC 219-ACS 评价了 AAM 体系中 ASR 膨胀的常用测试方法的适用性[36]。许多 AAM 研究者已经开发 OPC 体系的 ASTM 或加拿大标准（CSA）。在北美洲出台了用于骨料潜在碱活性的评价测试标准和一个特定的水泥骨料的结合反应性，如快速砂浆棒法

第8章 耐久性和测试——化学基体裂化过程

（ASTM C227）和混凝土棱柱体法（ASTM C1293，CSA 23.2-14），也可以使用化学分析法和岩相（光学显微镜）法（ASTM C289 and C295）。

混凝土棱柱体试验在实际混凝土材料性能中被认为是一个更合适和准确的方法。因为相同的混合物可以用作实验室样品，也可用于所提出的预筑混凝土产品或结构中[37]。然而，这两个具体的测试方法至少需要1年的时间才能完成。

化学分析法（ASTM C289）是一种更快速的检测方法，但并未对骨料的膨胀趋势进行评价。岩相法（ASTM C295）评价骨料可以确定骨料中潜在的活性材料，但未能提供特定的水泥骨料组合的膨胀性数据。

快速砂浆棒法（ASTM C1260）是一种快速的测试方法并且结果是保守的，是评价碱激发胶凝材料中 ASR 最常用的测试方法[38-42]。然而，这种测试方法需要砂浆浸泡在 80℃ 1M NaOH 中，专为骨料的评价设计，不是为骨料/水泥体系设计，用在一个固有的高碱体系如 AAMs 中似乎会被怀疑。Xie[42]等认为，即使用来评价骨料，这种方法往往会给一个高估的结果；即用此方法去评价骨料，一种具有良好性能的骨料可能会被列为是有害膨胀的。

当水泥暴露在正常环境中时，自收缩和 ASR 膨胀的时间尺度不同。硬化期间以自收缩为主，而 ASR 主要作用在后期，造成长期的膨胀。采用快速砂浆棒法提供的加速度（ASTM C1260）使这两种机制部分重叠，这显然影响了 ASR 膨胀的评价[42]。当测试 AAM 的 ASR 潜力时这是一个必须考虑的因素，对于 OPC 砂浆也一样。图 8-1 显示，碱激发粉煤灰的 ASR 膨胀比 OPC 要低，这是因为早期自收缩的补偿作用（图 8-1b，c 里特别展示了测试过程中的收缩变化）。

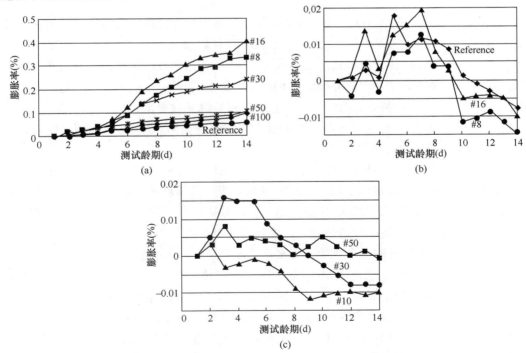

图 8-1 用 10% 的玻璃取代骨料中的石英砂后 ASR 的 ASTM C 1260 测试期间观察到的膨胀
（a）OPC 和（b，c）硅酸钠激发粉煤灰。砂浆中的骨料根据 ASTM C 1260 的建议分级，然后指定尺寸的骨料（标记）部分（或全部，♯8 大小）用活性玻璃取代（内容改编自 Xie 等人[42]）

此外，在 ASTM C1260 的实验方法中，铝硅酸盐前驱体的激发还没有达到一个真正完全的（或满意的）程度就有 ASR 反应可能发生了。Yang[43]研究表明 ASR 主要发生在前 30~60d，随后到达平稳期。一些作者[44]也指出，在含有 BFS 的砂浆中，最初的膨胀速度缓慢，意味着 ASTM C1260 加速试验方法可能不适合测试在一个更长使用寿命中砂浆的 ASR 潜在膨胀的最合适测试方法。

这种测试方法的目的是检测 16d 内使用 OPC 砂浆棒特殊骨料有害的 ASR。然而，在 AAS 砂浆中当测试时间是有限的 16d 时，有害 ASR 可能并没有被观察到，给人一种砂浆"安全"的误导性结果。有人建议，对 AAS 砂浆的试验时间应该持续至少 6 个月[38]，虽然对具有相同测试时间的 OPC 和混合 OPC 样品的长期测试结果的解释可能还需要进一步验证。

除了 ASTM 和 CSA 的方法，Ding[45]用 RILEM 加速混凝土棱柱体法（RILEM TC-106[46]）与养护温度 60℃，Al-Otaibi[47]在 1995 草案中 BSI DD218 指定的方法（注意，这个标准已经由 BS 812-123：1999[48]代替），用了持续 12 个月的试验时间。

测试硅酸盐水泥（OPC）为基础的体系中的 ASR 时，加碱会加速反应是常见的。在 ASTM 砂浆棒法（C227）中将测试水泥的碱含量调至 1.25% Na_2O_{eq}。然而，在 AAMs 中，由于在最初的混合成分中碱含量较高，同时在测试过程并未添加多余的碱，其含碱量一般为 3.5%~6% Na_2O_{eq}[34,37,38,41,47,49]。

8.3.2 砂浆棒测试结果

ASTM 砂浆棒法（C227）（或这种方法的改变）是研究者的热门选择。这种测试方法中使用了 AAMs 中最早的 ASR 研究[16]，活性蛋白石骨料替代部分石英砂，并发现它们的"F-混凝土"样品膨胀比控制硅酸盐水泥膨胀小。Yang 和同事评估了碱激发矿渣水泥砂浆使用活性玻璃的碱-骨料反应[43,50]。硅酸钠（$Na_2O \cdot nSiO_2$）、Na_2CO_3 和 NaOH 碱激发矿渣水泥中，在给定 Na_2O 用量和测试条件下，$Na_2O \cdot nSiO_2$ 碱激发矿渣水泥的膨胀最大，NaOH 碱激发矿渣水泥的膨胀最小（图 8-2）。无论激发剂的选择如何，Yang 声称膨胀随着碱用量和矿渣碱度的增加而增加。

Chen 等[34]用含碱量 3.5% 的 Na_2O 激发含有石英玻璃骨料的矿渣砂浆棒，并证实由于 ASR 硅酸钠激发 BFS 膨胀得到了发展。同时研究表明，高碱度矿渣的使用会导致砂浆膨胀增大（图 8-3）。硅酸钠模数（SiO_2/Na_2O 比；Ms）对硅酸盐激发矿渣水泥混凝土的膨胀也有显著效果，用模数 1.8 的硅酸钠激发时，观测到的膨胀最大[50]。Al-Otaibi[47]解释说，当 Ms 更高时，由于碱激发硅酸盐形成的部分水化产物会导致膨胀的下降。

García-Lodeiro 等人[39]用 ASTM C1260 加速砂浆棒测试，棒浸在 80℃ 的 1 M NaOH 中，并与 8M NaOH 激发的低碱（0.46% Na_2O）OPC 和粉煤灰的性能作比较。他们发现，样品虽然在粉煤灰体系中有小的膨胀，但远小于在 OPC 体系中的极端膨胀和开裂。使用相同的测试方法，Fernández-Jiménez 等人[40]发现，用硅酸钠溶液激发的粉煤灰比用 NaOH 激发的粉煤灰的膨胀更严重，但仍低于低碱 OPC。Xie 等[42]也使用 ASTM C1260 进行砂浆试验，包含玻璃骨料。由硅酸钠激发的粉煤灰砂浆膨胀很小，而 OPC 砂浆表现出相当大的膨胀。

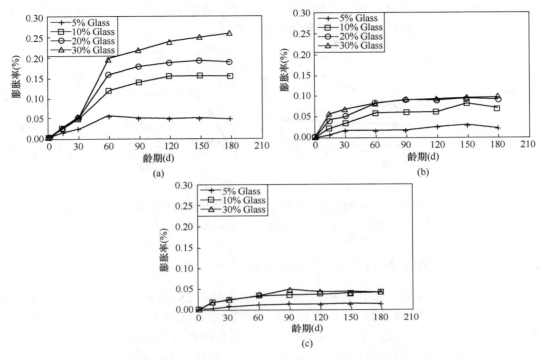

图 8-2 激发剂和石英玻璃含量对砂浆棒膨胀的影响：(a) $Na_2O \cdot nSiO_2$ 激发，
(b) Na_2CO_3 激发和 (c) NaOH 激发矿渣水泥（数据来于 Yang[43]）

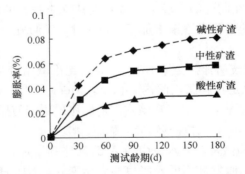

图 8-3 矿渣类型对砂浆棒膨胀的影响（改编自 Chen 等人[34]），碱激发粉煤灰用于
测量 ASR 的激发剂是 8M NaOH[39] 或 $Ms=1.64 \sim 3.3$ 的硅酸钠[40,42]

8.3.3 混凝土样品的测试结果

Bakharev[37]用 ASTM 混凝土棱柱体法（ASTM C1293）评价了用 0.75 模数的硅酸钠激发的矿渣混凝土的反应（图 8-4），并报道 AAS 混凝土比类似等级的 OPC 混凝土更容易发生 ASR 劣化。研究表明，AAS 混凝土中的 ASR 膨胀可以通过强度的快速发展来降低。因此，使用至少两年的时间观察 BFS 混凝土样品是必不可少的。

Al-Otaibi[47]使用在 OPC 体系中有活性的英国砂岩骨料按照 BS 812-123 制备棱柱体法混凝土，于 38℃和 100%相对湿度条件下养护，测试了一系列硅酸钠激发的 BFS 混合物的膨胀性，发现碱激发 BFS 体系产生很小的膨胀，远低于有害的标准。其膨胀随混合物中 Na_2O 的增加而增加，但随着硅酸盐激发剂模数的增加而降低。

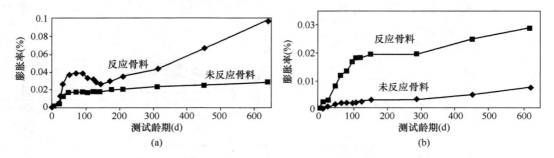

图 8-4 （a）OPC 和（b）硅酸钠激发矿渣混凝土中的膨胀测量 ASR 混凝土棱柱体试验
（ASTM C 1293-95）（改编自 Bakharev 等[37]）

8.3.4 AAM 体系中骨料对 ASR 膨胀的影响

研究表明，骨料掺量、大小和活性骨料性质对 AAMs 的 ASR 膨胀性能产生影响。众所周知，影响普通 OPC 砂浆的 ASR 膨胀量的重要因素是活性骨料的大小；最高的 ASR 膨胀是由骨料的性质和组成决定的。这是因为大量潜在的膨胀凝胶形成骨料颗粒，减小了尺寸，另外高活性和/或极细骨料进行了火山灰反应，替代 ASR 过程，二者的竞争效应可能减小了膨胀[51]。例如，在一个特定的 OPC 体系研究中，最大膨胀的蛋白石粒径为 20～50mm，打火石粒径为 1～3mm，废玻璃骨料粒径为 300～600μm[42]。然而，在硅酸钠激发粉煤灰砂浆的 ASR 膨胀在同样的实验中，被发现其反应骨料的尺寸比总含量的影响小；当玻璃骨料在惰性骨料中的替代量逐渐增加至 100％时，碱激发粉煤灰砂浆的 ASR 膨胀会增加[42]。

Yang[43]表明，当活性玻璃骨料含量小于 5％时，无论碱用量和激发剂的性质如何，碱激发矿渣水泥体系的膨胀总是在膨胀极限内。相反，Metso[52]用含蛋白石活性骨料的碱激发矿渣水泥砂浆棒测 ASR 膨胀，碱激发矿渣水泥的膨胀取决于使用的 BFS 的性质以及蛋白石含量，并在蛋白石含量约为 5％时观察到最大膨胀。

Pu 和 Yang[53]研究了碱激发矿渣水泥中的 ASR 膨胀和微观结构。他们发现骨料含有活性成分时发生 ASR，但指出激发剂不同膨胀也会发生变化。NaOH 作为激发剂以及活性骨料含量小于 15％时没有破坏性的膨胀发生。Na_2CO_3 或 $Na_2O \cdot nSiO_2$ 作为激发剂时，允许的最大活性骨料含量可高达 50％。

Chen 等人的结果[34]表明，ASR 被认为"有害"只发生在当测试 18d 活性骨料用量为 15％的 AAS 体系中，甚至粒径为 80～150μm 的级配骨料也会导致其出现最高的 ASR 膨胀，随着活性骨料含量增至 50％，膨胀程度逐渐增加。

这些结果表明，已公布的研究数据至今都不能明确或统一说法，显然在这方面还需要进一步的科研工作和测试方法验证。

8.3.5 OPC 和 AAM 胶凝材料中的 ASR 比较

虽然碱激发矿渣（AAS）水泥比 OPC 钙含量低，无 $Ca(OH)_2$（认为这是有益的）也是一个特性。但其碱浓度高，通常超过 3%，而 OPC 小于 0.8% 的 Na_2Oeq。这导致用户和/或相关人事认为，当使用碱活性骨料时碱含量高可以促进 ASR。然而，当碱金属被化学结合在反应产物中而不是在孔溶液中时[38,54]，高浓度的活性氧化铝是可用的（直接加入分子筛前驱体或富铝原料如偏高岭土）[55,56]，膨胀危险也相应减少。

在前面的章节中讨论过，有研究报道采用 ASTM C1260 方法（加速砂浆棒法），AAM 体系比 OPC 的 ASR 膨胀发展得少。Puertas 等人[41]用三种骨料[硅质、非活性石灰和活性石灰（白云）]比较 OPC 以及硅酸钠活性矿渣（AAS）的膨胀性。持续 4 个月的测试时间内，在 ASTM C1260 试验条件下硅质骨料 OPC 样品比相应的 AAS 显示的膨胀大四倍。所有 AAS-钙质骨料样品的膨胀比相应的 AAS 稍高，但这些样品在 80℃ 1N NaOH 溶液中时，4 个月后最糟糕的膨胀是在 0.05% 左右[41]，所以事实上它高于 OPC 砂浆也不算特别成问题。

相反，正如在 8.3.2 提到的 Bakharev 等人[37]报道的，当采用混凝土棱柱体法（ASTM C 1293）时，AAS 混凝土受 ASR 劣化影响比同类级 OPC 混凝土更容易。他们认为，在碱活性矿渣混凝土中早期的 ASR 膨胀可以通过快速增加强度来减轻，因此，对 BFS 样品的长期性能测试是很重要的。

8.3.6 碱-骨料反应测试结果的讨论

如上所述，研究表明 AAM 体系可产生破坏性的 ASR 膨胀。大多数人采用小规模的人工试验方法或不常用的活性骨料的砂浆棒加速试验方法。当骨料跟碱反应时，在碱激发水泥体系中有潜在的破坏碱激发反应存在。从这些试验可以得出的结论是膨胀性由骨料种类和胶凝材料组成以及每种组成如激发剂、BFS 或粉煤灰的性质决定。然而，结果表明，AAM 体系中潜在的破坏性碱性反应比 OPC 体系中少。使用有限数量的混凝土样品进行的加速试验，给出了相反的迹象，无论 AAMs 受 ASR 影响比 OPC 混凝土是多还是少。

一些学者指出，试验方法对预测 ASR 膨胀有明显的影响。高温和高湿度（或浸渍）用来加速 ASR 反应特别适合于促进 AAM 体系的硬化反应。这样的测试可能不会提供确切的有效结果。在正常使用条件下展开长期的观察，一个详细科学的对控制 AAM 体系中 ASR 机理的理解被推荐。这很重要，由于较低的早期反应速率和特殊环境下 AAM 的高收缩，可能导致对 ASR 性能的误解。

但是，暴露在自然环境中天然骨料的 AAM 混凝土样品的 ASR 的具体研究尚无报道。在本报告中的第 2 章和第 11 章有所总结，表明已被分析的样品上没有任何有害膨胀的反应证据。此情况与 20 年前辅助胶凝材料如 OPC 体系中复合的粉煤灰、矿渣相似。使用快速测试和人工活性骨料的测试实验结果相矛盾会导致行业的混乱。只有关注包含一系列的天然骨料的混凝土样品，暴露在实验室和自然环境中时，才能达成共识。如果按照严格的标准，这些材料将是减轻 ASR 膨胀危害的宝贵资源[57]。

8.3.7 文献中测试方法的总结

不同的研究人员使用的各种 AAM 体系适用的 ASR 评价测试方法在表 8-1 中作了总结。

表 8-1 适用于碱激发材料的碱-骨料反应分析的测试方法总结

来源	地域	测试方法	样品组成	骨料	样品尺寸	碱含量	养护和曝光
Kukko 和 Mannonen[16]	芬兰	ASTM C 227-71	"F-混凝土"（有专用激发剂的矿渣）$w/b=0.27$	标准砂+活性蛋白石 3%～15%部分取代最细的部分	—	OPC: 1.54%; F-混凝土中含矿渣 1.89%，激发剂升至 2%	24h 后脱模，然后（40±2）℃，RH98%养护 75d 后测试时间
Gifford 和 Gillott 1996[49]	加拿大	CSA A23.2-14A-94	胶凝材料：420kg/m³，$w/b=0.43$，粗骨料/砂=1.5	含白云石的石灰石（ACR）；活性硅质石灰石	75mm×75mm×305mm	OPC: 1.25% Na₂O_eq（水泥质量的），AAS: 含 BFS 质量 6%Na₂O	70℃下热养护或正常湿养护 6h，再存放在 38℃和 RH100%的密封、隔热、持续加热的柜中
Bakharev 等 2001[37]	澳大利亚	ASTM C 1293	硅酸钠激发 BFS，$M_s=0.75$，$w/b=0.5$	惰性砂+活性粗骨料	75mm×75mm×285mm	含 BFS 质量 4% Na₂O	24h 后脱模，再储存在（38±2）℃，RH≥95%下养护
Chen et al. 2002[34]	中国	"砂浆棒法"	Na₂CO₃/NaOH/Na₂SO₄/水玻璃激发的 AAS，骨料/胶粘剂=2.25，$w/b=0.4$	级配的石英玻璃	10mm×10mm×60mm	—	1d 后脱模，在（38±2）℃，RH≥95%下养护
Fernández-Jiménez 和 Puertas 2002[38]	西班牙	ASTM C 1260-94	NaOH 激发 BFS，溶液/BFS=0.57，骨料/矿渣=2.25	活性蛋白石骨料（21%活性二氧化硅）	25mm×25mm×230mm	含 BFS 质量 4% Na₂O	25℃，RH99%养护 24h，脱模，浸置在 80℃、1M NaOH 溶液中
Xie 等 2003[42]	美国	ASTM C 1260	硅酸钠激发粉煤灰，$M_s=1.64$，$w/b=0.47$，骨料/胶凝材料=2.25	10%废玻璃骨料替代普通河砂 筛余 2.375mm 10%； 1.18mm 25%； 600μm 25%； 300μm 25%； 150μm 15%	—	—	OPC: 24℃，RH100%养护 24h; AAFA: 60℃，24h 脱模后（24h），浸在水浴中 80℃，24h; 测量长度为初始长度，然后放置在 1M NaOH 中

续表

来源	地域	测试方法	样品组成	骨料	样品尺寸	碱含量	养护和曝光
Li 等. 2005 2006[58, 59]	中国	ASTM C 441-97	w/b=0.35, 骨料/胶凝材料=2.25, 碱激发偏高岭土, w/b=0.36	活性石英玻璃细骨料,每个级配20%: 2.5~5mm; 1.25~2.5mm; 0.63~1.25mm; 0.315~0.63mm; 0.16~0.315mm	—	对于地质聚合物: 12.1%, 其他混合物: 0.94%, 0.57%, 0.47%	20℃下养护24h, 脱模, 测量初始长度, 然后在38℃下浸置
Garcia-Lodeiro 等 2007[39]	西班牙	ASTM C 1260	OPC w/b=0.47 AAFA, 8 M NaOH, 溶液/FA=0.47 骨料/胶粘剂=2.25	活性蛋白石骨料: 2~4mm, 10%; 1~2mm, 25%; 0.5~1mm, 25%; 0.25~0.5mm, 25%; 0.125~0.25mm, 15%	25mm×25mm ×285mm	—	OPC: 混合后, 在21℃和RH99%下养护1d, 脱模, 在85℃的水中养护1d。AAFA: 混合后, 放置在85℃和高湿度下20 h后脱模。对于OPC和AAFA砂浆棒实验: 浸置在85℃, 1 M NaOH中, 测量90d后长度
Fernandez-Jimenez 等 2007[40]	西班牙	ASTM C 1260-94	8 M NaOH (NH), 85% 12.5M NH+15% Ms 3.3 硅酸钠 (SS), 骨料/FA=2.25, 溶液/FA=0.47 (NH) 或 0.64 (SS) OPC, w/b=0.47	非活性骨料	25mm×25mm ×285mm	—	85℃, RH99%下养护20h, 脱模, 85℃浸置在1M NaOH中
Al-Otaibi 2008[47]	科威特	BS DD218: 1995	固体或液体硅酸钠, Ms=1 or 1.65, w/b=0.48	活性好的骨料	75mm×75mm ×280mm	含BFS质量的4%或6% Na₂O, 对于OPC, 1% Na₂O	—
Puertas 等 2009[41]	西班牙	ASTM C 1260-94 用EN-UNE 196-1准备的砂浆	水玻璃-AAS Ms=1.08 溶液/胶凝材料=0.52 控制(OPC), w/b=0.47 骨料/胶凝材料=2.25	硅质/钙质骨料 2.36~4.75mm, 10%; 1.18~2.36mm, 25%; 0.6~1.18mm, 25%; 0.3~0.6mm, 25%; 0.15~0.6mm, 15%	砂浆棒, 25mm×25mm ×287mm	4% Na₂O/矿渣质量	在水中: 99%RH, T(21±2)℃ (1st d), 脱模, 80℃浸置在水中, 测量第一个收缩, 然后80℃浸渍在1M NaOH中4个月
Kupwade-Patil 和 Allouche 2011[60]	美国	ASTM C 1260	F类或C类FA, 14 M NaOH+硅酸钠	活性骨料: 石英, 砂岩和石灰石	51mm×51mm ×254mm	—	样品浸置于1M NaOH, 80℃的养护箱中

8.3.8 关于 AAMs 试验方法的评价

有很多 AAM 体系可能产生破坏性 ASR 的相关研究。许多研究都是在砂浆棒的基础上结合了快速测试和非常规骨料测试。结果如下所示：

当骨料是碱反应性，且在水化过程中与碱金属结合得较少时，AAM 体系中存在潜在的有破坏性的碱-骨料反应。早期收缩及温度对 AAM 体系的强度发展的影响意味着测试应该在较低的温度下进行，也需要比 OPC 基混凝土更长的测试时间。

在这些测试中，膨胀已被证明与激发剂的类型和掺量有关，也与胶凝材料和/或激发剂的组成有关。结果表明，膨胀随混合物中 Na_2O 含量的增加而增加，并随水玻璃模数 (SiO_2/Na_2O) 的增加而减少。总的来说，使用类似的骨料时，AAM 体系中的破坏性碱反应比 OPC 体系低。

使用有限数量的混凝土样品进行的加速试验，出现相反的现象，无论 AAMs 受 ASR 影响是高还是低，更适合回到基本原理并研究处于自然环境中的混凝土样品中的天然活性骨料。

8.4 浸出试验

8.4.1 浸出试验简介

人们普遍认识到，施工中使用工业副产品材料最重要的环境风险之一是污染物从材料中浸出，并随后迁移到环境中。在原材料放置、使用中、回收再利用之后以及最终处置之后，污染物都可能会释放污染。排放的污染物与水接触可能对地下水、地表水以及土壤造成危害。

用于 AAM 胶凝材料和 AAM 混凝土的原材料通常是工业生产中的副产品及废料，不仅包括有用及惰性组分，也有许多有害成分（如重金属、芳香族多环烃的有机质、天然存在的放射性物质等）。当这些材料暴露在环境中时，与雨水、地表水或地下水发生作用就会浸出有害物质。如果某种有害成分浓度非常高的话，会威胁环境及人类身体健康。为确保材料对环境及人类健康无害，浸出试验是非常重要的试验。

在一些国家，基于浸出测定的重要性以及环境因素，某些建筑材料受到规范限制。1995年在欧盟成立浸出/萃取试验协调网络，旨在统一不同的浸出/萃取试验和评估标准。同时建立和发布相应的规定和欧洲标准，其中包含了不少有用的观点和论述[61-65]。

本节将简要介绍不同的浸出试验过程和适用于建筑及废料的标准。将根据以前的工作经验分析其在 AAM 中的应用情况，并提供一些建议或修改意见。

8.4.2 废弃物处置和材料使用中的浸出试验

按照定义，浸出是固相在矿物的溶解、吸附/解吸平衡、络合作用、pH 值、氧化还原环境、溶解的有机质和生物活化共同作用下向周围水中释放无机质、有机质和放射性物质的过程。这个过程是非常普遍的，暴露在水中的任何物质都将会从它表面或者内部依靠材料的多孔性结构浸出组分，在从地球化学到萃取冶金、文化遗产保护、核废料处理等方面，浸出试验已被深入研究。

在开发和使用建筑材料过程中,元素的"可浸出总量"是在极端条件下长时间浸出的元素总量。然而,设计测试条件来重现真实环境条件以获得该参数的实际值是非常重要的,因为元素的可浸出总量不一定等于其元素的总浓度。相反,通常可浸出总量低于总浓度,这取决于外部情况和条件。

可能影响浸出率的一些固有和外在参数包括(图 8-5):

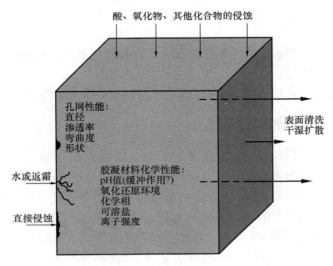

图 8-5 影响材料中有害元素释放的因素(摘自 Van der Sloot and Dijkstra[62])

(1)固-液平衡作为 pH 值的函数:溶解度、吸附作用、释放、氧化还原电位、酸缓冲能力。

(2)固-液平衡作为固/液比的函数:孔隙水组成、离子强度、可溶性盐浓度。

(3)质量传递和释放机制:扩散、溶解、扩散溶解、表面现象、洗脱(与外部环境中水的流失相关)、湿/干循环以及两种或更多这些方法的组合。

(4)物理性质:形状(粉末、颗粒状、整体)、湿度、孔隙率、密度、渗透性。值得注意的是,由于质量传输速率的差异,颗粒状和整体块状样品通常会具有非常不同的浸出行为。

因此,对碱激发材料的浸出评估方法推荐如下:

(1)根据材料应用的需要定义潜在释放方案和浸出参数。

为了评估方案,应该了解不同的参数,如:

① 土工技术规范;

② 原材料来源及加工方法;

③ 规划应用项目的水文条件;

④ 材料的内在特性。

(2)选择测试方法来衡量最相关的浸出参数。

(3)材料测试确定浸出参数。

(4)在真实环境暴露下对实验室测试的验证。

根据真实环境和现场条件计算有毒元素的释放(默认环境、现场特定条件、现有数据库）。

（5）基于默认环境、特定条件和现有经验评估环境影响。

由于预测中的浸出行为的复杂性与难度，在（几乎按定义）使用条件下，提出了一大批不同的测试方法，这可大致分为三个类别：

① 基本特性：对材料浸出行为有关的特定性能的测定。不同因素各不相同，包括液固比、浸出介质组成、pH 值、氧化还原电位、络合能力、材料老化、材料的物理参数和其他因素。

② 验证实验：用于比较材料的浸出参数（在前期的基本测试中确定）与"可接受"的已有浸出行为的基准参数。

③ 验证和质量控制测试：快速测试以验证材料是否符合规范的测试。

8.4.3 筛选测试：有效性或浸出潜力

这些测试为了测出材料最大的潜在浸出能力，通常在极端条件下使用高液固比（≈100mL/g）以及特定 pH 值以使其释放：pH<4 时释放阳离子，pH 值为 7 时释放阳、阴离子。通常采用 EDTA 络合剂进行滴定的方式测量。测试结果要给出待测物质的有效性，或浸出溶液浓度（这些数据不能在不同测试方案之间照搬，所以只能用单一测试来比较不同材料），或在持续的测试时间中物质释放的总量，或每种被释放的元素的初始百分比含量（如果固体材料中的总浓度是已知的）。测试方法见表8-2，标准粒度分布的分类信息见表8-3，在本章其他小节也出现过。

上述测试作为普遍的浸出测试方法，并不仅只应用于水泥或胶凝性材料中。因此考虑到碱是从胶凝性材料中释放到浸出液中，在此测试方法中需要添加酸来维持一个较低的 pH 值环境。相反，如果用于测试的原料（特别是粉煤灰和矿渣等工业副产品）显酸性，则需要添加碱来维持酸碱平衡。氧化还原反应环境在测试方法中并未给定，但氧化还原条件的控制在确定反应速率和过渡金属如铬的释放程度尤为重要，这是测试中要考虑的重要问题。TCLP 测试是相对传统的方法，表8-2 中列出的几种其他测试方法，已在很大程度上解决了 TCLP 的某些不足，但它仍然是一个相对被广泛接受的测试方法。EDTA 测试（AV002.1）要比直接浸出试验（AV001.1）更具说服性。

表8-2 标准筛选试验总结

测试方法	限制条件	测试条件
US EPA TCLP Method 1311 浸出试验[66]	粒径<9.5mm，或比表面积在1~3.1cm^2/g	浸出时间为18h； 液固比为20L/kg； 样品容器以30r/min 转速旋转搅拌； 在 pH=2.9 的醋酸溶液中； 用硝酸溶液使浸出液酸化到 pH<2（为了避免沉淀产生）
AV001.1 在 pH=4 和 pH=8 时的可浸出量[67]	接触时间取决于粒径（表8-3）	两者同时浸出； 液固比为100mL/g； 加入硝酸或者氢氧化钠溶液使 pH 值达到 4 或者 8

续表

测试方法	限制条件	测试条件
EA NEN 7371:2004[68]-建筑物和废弃材料的浸出特性。浸出过程中无机组分最大可浸出量的测定。"最有用的浸出试验"	物料粒径降低（<125μm）	两者依次浸出； 累计液固比为 50mL/g 干物料； 硝酸溶液提供 pH 值为 7 和 4； 每种 pH 值条件下浸出时间为 3h； 早期版本的测试[69]也使用较高的液固比（100mL/g）
AV002.1，pH=7.5 时的可浸出量，用 EDTA 表征[70]	接触时间取决于粒径，如 RU-AV001.1 规定（表8-3）	单一浸出； 液固比为 100mL/g； 添加硝酸或氢氧化钠溶液使 pH 值最终达到 7.5

表 8-3　几种浸出测试中不同粒径所用的接触时间

粒径（mm）	接触时间（h）
<0.3	18
<2	48
<5	168

测试方法应用时很少固定参数，如样品制备方法、养护和调整实验条件（如果有的话）。这些参数在确定 AAM 胶凝材料或 AAM 水泥固化废弃物时非常重要，因此应给出完整的测试结果。

8.4.4　平衡浸出测试

表 8-4 中所列的测试方法是把材料磨细，在特定的化学条件下（通常为 pH 值和液固比）或在静态或动态条件下检测污染物释放。

表 8-4　一些基于平衡的浸出测试方法总结

测试方法	限制条件	试验条件	试验结果
SR002.1，pH 值作用下的溶解与释放[70]	物料粒径减小；时间与粒径有关（表8-3）	11 个平行溶出度提取； 3≤pH≤12； 去离子水，加入 HNO_3 或 KOH； 液固比：10mL/g 干物料； 以（28±2）r/min 的转速旋转搅拌	数据结果为滴定曲线和成分溶解释放曲线
PrEN 14429，pH 值滴定测试[71]	—	批次试验（初始添加硝酸或氢氧化钠），最终 pH 值为 4～12； 通过自动酸/碱添加进行 pH 控制	数据结果为 mg/L 或以 mg/kg 计的浓度
SR003.1，以不同的 L/S 比率进行批量提取—溶解与释放作为 L/S 比的函数[67]	物料粒径减小；接触时间与粒径有关（表8-3）	5 个并行提取； 去离子水； 液固比：0.5mL/g，1mL/g，2mL/g，5mL/g 和 10mL/g 干物料	给出孔隙水中活性元素的浓度

8.4.5 传质浸出测试

这些测试（表8-5）是在整块材料或压实材料上进行的，目的是根据材料的物理和化学（渗透性）性质来确定污染物的释放速率。

表8-5 一些基于传质的浸出试验的总结

测试方法	适用性	试验条件	测试结果
槽浸出试验（MT00x.1）[67]	可应用于以下两种样品之一：单片（MT001.1）或压实颗粒（MT002.1）	最初去离子水（即自身pH值条件）；液面比10cm³/m²；刷新间隔：2h、3h、16h、1d、2d、3d	累积释放量作关于时间的函数—样品表面积每平方米释放多少毫克
槽浸出试验（NEN 7347，NEN 7375）[72,73]	整块（NEN 7375）或细小颗粒（NEN 7347）样品	蒸馏水（自身pH值）或控制pH值（pH值为4或者7~8）；液体与样品体积比为2~5；6h到64d之间分八个时间间隔更换水	累积释放量作关于时间的函数—样品表面积每平方米释放多少毫克
CEN/TS 14405，（柱）渗滤浸出实验[71]	平衡或质量传递速率；液固比取决于渗透率、密度和应用的高度	超过24h达到饱和并并达到预平衡；通过纵列向上流动；液固比在0.1~10L/kg之间	累积释放量作关于液固比的函数，以mg/L或mg/kg表示
ANSI/ANS 16.1[74]	用于核废料玻璃的单片样品	静态浸没试验；5d或90d以上的持续测试时间；指定的时间间隔更换矿物质水；液面比10cm³/m²	根据形成网络的元素的溶出率计算可浸性指数

8.4.6 OPC和AAM体系中影响浸出的因素

1. 浸出pH值

通过溶解含钙物质来进行酸性pH值加速浸出。白云石和石灰石骨料也被酸侵蚀裂化，在水中浸出形成可溶性盐。根据酸的类型，劣化机理不同：一些酸（草酸、磷酸）形成在物质表面的不溶性沉淀；具有可溶性钙盐的酸破坏性更强；弱酸因缓冲效应在相同pH值下比强酸更具侵蚀性。如8.2节所述，特别是硫酸和富含钙的胶凝材料，易于和硫酸根离子发生反应，形成钙矾石或石膏。

如果碳酸盐含量很高的话，碱性pH值起保护作用，不会导致脱钙现象，但往往会造成二氧化硅和氧化铝的溶解。

2. 浸出用水的组成（阳离子或阴离子浓度和性质）

离子强度低（如蒸馏水或去离子水），当软水和胶凝材料相接触时由于较高的浓度梯度，从而使浸出增强。如前一节所述，硫酸盐与水泥浆中的钙和铝酸盐反应。镁或铵盐溶解在富含钙的胶凝材料产品中，形成可溶性盐类（如水镁石）或膨胀的化合物（如碳硫硅酸钙），导致脱钙。在硅酸盐水泥中氯化物可以形成弗里德尔盐，被认为是预埋钢筋的破坏因素而非胶凝材料本身的影响。

3. 水的流态

水的流态是确定浸出率的关键。流动水要比静止水侵蚀性更强，因为扩散是质量传递的唯一机制，而流动水更容易增强质量传递。由于流动而引起的物理侵蚀或由搅拌引起的颗粒破碎，都会加快材料的劣化。

4. 养护和老化

在测试之前，废物材料足够的养护时间是提高其化学稳定性和微观结构的关键，否则会导致结果不准确。然而，样品的过度老化或风化（包括干燥、碳化和其他造成粘结性降低的形式）导致样品的劣化（强度降低、孔隙率的变化、开裂、成盐、水化产物的变化、可溶性元素的氧化形式的变化）将影响浸出速率。

8.4.7 AAMs 中的浸出测试

Comrie、Davidovits 等人的早期研究表明，碱激发胶凝材料可应用于废物固化和控制重金属浸出[75,76]。在这类应用中碱激发胶凝材料的性能（以及用途）将在本报告第 12 章展开详细讨论；在这一点上，足以说明一些胶凝材料的性能从极好到极差的变化取决于胶凝材料配合比、养护条件、会浸出的元素和具体的测试应用。

碱激发胶凝材料的分析测试包括上面列出的这些标准测试和许多非标准测试方法。一般主要使用传统的 TCLP 测试方法和微调的测试方法，而 ANSI/ANS 16.1 测试在碱激发核废料方面普遍应用。然而，应用于碱激发胶凝材料的大多数浸出试验在文献中没有具体的、能够遵循的一套标准化的测试方法，导致对不同研究中的性能进行对比时相对困难。然而，在一般情况下，阳离子比阴离子更能有效地固化，大的阳离子比小的阳离子更能有效地固化。在碱激发胶凝材料中废料固化的详细讨论，本质上和浸出阻力有关，参考第 13 章以及相关文献[77-79]，这里不再重复。然而，和浸出相关的不同于废物固化的一个重要问题是碱浸出问题；其与胶凝材料的微观结构有很大关系，通过孔隙网络限制扩散的成分参数也会影响碱浸出[80]。对于硅酸盐水泥和复合水泥，已知基体本身的酸中和能力在确定浸出性能方面是重要的[81]，AAM 的情况也是如此，但这仍有待详细测试研究。

8.4.8 关于 AAM 测试方法的评价

AAM 相关的文献总结表明，有许多不同的浸出测试方法。一些人采用传统现有的标准或方法，而另一些人则采用非标准的和改进后的方法。这取决于研究的目的，应该尽可能地选择标准的或者改自标准的浸出实验。测试方法的选择取决于测试流程能有效评估质量控制、特定的产物和应用的无害化、迁移机制的研究或者其他参数。

首要问题是明确要获取什么信息，然后再去选择最合适的测试方法，或者（必要时）适当地在特定需要或者特定情况下进行测试。这种方法对所有材料（包括传统材料和非传统材料）都是适用的。特别是待测试样品也应与研究目的相一致，因此应以合适的方式选择组合物、固化条件以及净浆、砂浆或混凝土作为目标测试样品。然而，许多已发表的研究成果并没有遵循这种方法。

为了确保 AAM 的特定产物符合环境和健康规范，则应基于不同条件和参数分析实际环境下产品的浸出性以及老化特性。在这方面，应该注意到碱激发胶凝材料与传统的 OPC 体系中的老化受不同的化学性质和显微结构的影响。这是因为 N-A-S-H 体系中的胶凝材料组成中没有足够量的钙[82]。

正如上面讨论的，固体材料或废弃物作为特殊建筑材料（通常 OPC 体系）需制定标准。一些给定条件，比如液固比、pH 值和颗粒大小都会对具体实验方法和浸出元素产生影响，因此需要对材料进行独立分析。然而，样品规格或浸出的时间长短等因素直接关系到样品的性质。因此，样品取样要具有代表性。所以在这种情况下，讨论磨细的样品和块状样品净浆、砂浆和混凝土是不同的。

如上所述，即使在不同的体系中（C-A-S-H 或 N-A-S-H）有非常大的差异。浸出时间对于在 OPC 体系一样条件下老化的碱激发胶凝材料是没有影响的；在碱激发胶凝材料的样品中，pH 值影响元素的浸出程度，但是在一些情况下和在硅酸盐水泥中观察到的现象完全不同。特别是，OPC 体系在酸性介质所发生的浸出行为影响碱激发胶凝材料，这将在下面第 8.5 节进行讨论。在一些情况中，不同配合比设计的碱激发胶凝材料暴露在碱性介质中也可能对碱激发胶凝材料有益[83]。

材料的浸出性与其耐久性紧密相关，因此这是关键系数（无论是通过破碎基体还是通过孔隙扩散），能够决定相应的浸出时间，这在不同的环境和浸出方法中都是适用的。如果为某种材料和环境选定的浸出时间比需要的短，得到的结果将不可靠，将会导致错误的结论。在非均质相（即活性物质更容易附着在试样中心而不是表面）中的关键因素是非常重要的，否则浸出速率很可能受到氧化还原平衡条件的影响，当氧化还原反应电位达到某个特定值时也可能导致某种特定物质突然释放。

最后，可以得出结论（在广泛意义上）：

（1）根据研究的目的选择实验方法和设计实验方案是非常重要的；几乎没有一种方法是普遍适用的。

（2）通过耐久性（劣化条件）和所需的浸出时间来获得可靠的结果需要做更多的工作，以确保材料在特定时间、特定环境下无害化。

8.5 耐酸性

8.5.1 简介

虽然大多数混凝土没有受到高酸性条件的影响，但在某些应用中，这确实成为一个问题，在这种情况下，混凝土的寿命可能会严重缩短。酸雨[84]、酸性硫酸盐土壤[85,86]、畜牧业[87,88]和工业过程[89]都可能发生酸化而降低混凝土的寿命。然而，生物硫酸侵蚀是经济和工业方面对基础设施所面临的最严重的酸侵蚀。其经常发生在下水管道[90-93]。随着各种工艺方案（要么与管道本身混凝土处理有关，要么与表面的涂料有关）的发展和实施[94-96]，下水管道的酸侵蚀已成为世界研究项目中的一个主要研究点。

一般意义上，混凝土的许多酸侵蚀测试方法与第 8.4 节所述的几个涉及暴露在酸性条件下的浸出实验是一样的。在强酸的侵蚀下，有一些非常重要的机理将会在这一节简单介绍。据悉，RILEM TC 211－PAE[3]最近的"技术报告"详细地解释了硅酸盐水泥的这些问题，并且读者认为这篇文献充分分析了不同的酸侵蚀进程、影响和测试。

无论是基于 OPC 还是基于 AAM，酸侵蚀的一个最重要的模式就是通过离子交换反应而使混凝土发生劣化。这将会导致基体的纳米结构和微米结构发生破坏而使混凝土的强度降低。在某些条件下这种情况发生得极其快速和严重，并且酸性条件可以由工业或生物过程诱导。

在实验室测试中，调整不同的参数以尽可能地模拟现实环境或加速劣化，更快地获得结果，并且加速程度和方式将影响检测结果。这些参数包括 pH 值和酸性溶液的浓度、样品的物理性质（块状或粉末状；砂浆或混凝土）、温度、酸补加率、是否存在机械运动或者机械流、干湿交替、冷热交替以及载荷。应该认真选取这些参数并且和测试结果一同呈报。劣化度的选取（强度损失、质量损失和侵入深度）可能对水泥的相关性能带来不同的结果，特别是当不同样品之间胶凝材料的化学性质完全不同的时候[97]。通常情况下，将需要多个相关的指标结合。此外，样品制备、养护条件、侵蚀环境以及测试时的成熟度都是至关重要的。

8.5.2 测试方法的分类

在实践中，由于暴露在酸性环境中，在这些条件下大量的性能参数对确定一种材料的成功与失败是非常重要的，绝大多数的酸侵蚀实验都是采用非标准化实验方法。如下所示，现有的测试方法将会以不同的方式进行分类。

1. 侵蚀物质的类型

化学劣化和微生物劣化机制是有区别的。化学劣化过程包括有机酸（如乳酸和醋酸）以及无机/矿物酸（如硫酸或盐酸）等的攻击。分析这些劣化机制的试验方法是将胶凝材料样品直接浸没在一种或者多种浓度的酸溶液中。

微生物劣化的机制是通过微生物产生侵蚀性物质；事实上在单纯的化学劣化过程中可能存在同一种酸，但是在微生物诱导的物理化学条件下很可能更复杂，这可能导致其他的劣化效果。一个特别重要的例子是下水道的氧化还原循环导致的硫化物氧化并且引起对混凝土管道生物质硫酸盐侵蚀[90,91]。因此，这可能需要使用特殊的测试方法。蒙特尼等人[92]提供了对硫酸侵蚀的混凝土进行化学、微生物和现场试验方法的综述，并描述了在试验期间通过提供硫杆菌来实现 H_2S 氧化为 H_2SO_4 的试验程序，样品在合适的营养环境中，以表面的胶凝材料的损失率作为评价性能的关键指标。

2. 测试方法的适用范围

测试方法的适用范围对测试结果有明显的影响，因为测试方法可能会影响到如下参数，如比表面积与液体的比值、补加物中是否存在侵蚀性物质、界面过渡区（混凝土与砂浆试样）、工业上使用的或者是模拟实验使用的侵蚀性液体，以及与服役条件相当的加速程度。因为界面过渡区引起的显微结构和渗透性差异，砂浆或水泥浆样品的研究不能总是外推到混凝土中，骨料（硅质或碳质）的选择差异也很重要。

在低液固比和无补加酸的情况下，实验侵蚀液的 pH 值变化很大。由于 AAM 胶凝材料的强碱性，在某些情况下，浸出溶液实际上可能会变成碱性的。如大量文献中报道的那样，在 AAM 胶凝材料中的"酸侵蚀"测试会形成其他类型的沸石；我们并不希望在实际测试中真正保持酸性。

3. 是否加入机械作用

当只有化学作用时,材料劣化层的生长相对缓慢,这就会减缓进一步的反应。当机械作用同酸侵蚀结合,磨损会去除劣化层并且形成新鲜的表面,通过物质层间无障碍的扩散作用而受到化学侵蚀;因此,这会加速劣化过程。增加磨损的常用方法是手动冲刷或自动冲刷(使用洛杉矶仪器已被证明在 AAM 的研究中是有效的[98]),或浸在水中摇晃或搅拌。在某些情况下,还可以通过施加干湿循环来加速酸的进入,从而允许通过对流过程吸收侵蚀性试剂,这比纯扩散快得多。

4. 模拟试验中加速劣化的参数

正如 8.2 节所述的硫酸盐的侵蚀一样,侵蚀性溶液的浓度和温度都可以加速进程。如果所选的浓度或温度太高的话,热力学/相平衡混杂将会变得非常明显[92]。使用表面积和体积之比大的试样也可能提高活性表面,但如果溶解二氧化硅的浓度增大到允许的范围,将会出现不可预期的结果[82]。

5. 所用酸的性质

侵蚀性溶液中酸的强度在确定酸侵蚀率和酸侵蚀程度时很重要;对于给定浓度的强酸更具有腐蚀性,因为它们产生更低的 pH 值环境。然而,一些弱酸也可能由于缓冲作用表现出强腐蚀性,弱酸与易溶性钙盐作用比强酸与难溶性钙盐作用可能对 BFS 基胶凝材料更具危害性[99]。在同 pH 值条件下,硫酸比硝酸更具有腐蚀性[100-102],因为硫酸的二次性质也可以提供低 pH 值缓冲作用和额外的劣化[97]。

干的、不吸湿的固体酸不侵蚀干混凝土,但有些会侵蚀潮湿混凝土。如果发生侵蚀的话,干燥的气体在混凝土内部接触到足够的水分,二氧化碳的侵蚀导致胶凝材料碳化,这种特殊的情况将在第 9 章展开讨论。

8.5.3 测试劣化样品的方法

测定劣化的参数包括尺寸的改变(由于硫酸钙盐的沉积导致的膨胀,由于胶凝材料的溶解导致截面损失)、试样的质量、块状样品的强度或者弹性模量的损失,酸渗透/侵蚀变的深度,浸出液或者胶凝孔溶液 pH 值的改变,钙离子或释放到溶液中的网络骨架中元素的浓度等。通过 SEM、XRD 或红外光谱分析侵蚀凝胶区域的微观结构和纳米结构的改变。

关于混凝土类型的相关表现,劣化方式的选择可能导致不同的结果[97]。因此,一个简单的方法可能不足以充分描述劣化,需要使用多个相关指标,认真选择测试条件、胶凝材料的性质,以及寻求的信息。例如,由于受部分胶凝材料溶解(造成质量损失)和硫酸钙沉积的竞争影响,在硫酸作用下的 AAMs 的劣化程度并不是完全相关的;然而和硝酸作用的样品则表现出完全不同的效果和趋势[97]。

8.5.4 AAMs 耐酸测试的应用

AAMs 具有高度耐酸性;近年来这已经成为此领域学术和商业发展的主要驱动力[103-107]。然而,许多声明中表示对暴露在酸中的 AAMs 的使用率并没有足够详细的测试,这对研究它的长期性能是至关重要的。此外,硅酸盐水泥胶凝材料设计的实验通常已经开始应用,然而对 AAMs 目前尚未被验证,所以对于真正性能有可能不会提供预想的信息[108,109]。

由于在 AAMs 的耐酸性研究中使用不同原材料、固化时间、配合比设计、样品形态、酸暴露环境和性能参数，因此很难对实验结果进行比较。同时很多测试方法和预期进行情况差别很大，例如室温下 70% 的硝酸[104] 或 100℃时 70% 的硫酸[110]。使测试条件更接近预期是为了获得更具有代表性的结果，但是可能需要更长的测试时间。这在本章所讨论的加速测试方法的开发中或多或少是普遍的。

绝大多数用作测定酸侵蚀劣化的水泥和混凝土质量损失的方法，几乎没有关于强度损失和深层侵蚀的。抗压强度损失是研究材料在多种情况下的性能中最有意义的指标。在几周或几个月时间快速试验方法中，劣化方法的使用可能由于试验中完好的胶凝材料变得复杂，在一定程度上抵消了劣化的胶凝材料的强度损失。在抗压强度测试中重复试样的差异，可能产生样品被破坏或者已经劣化这种问题，如样品几何形状的改变（当样品部分劣化很难使样品端部平整），样品的不均匀性（内部核心强，边缘部分弱）也可能在加压情况下使样品断裂。然而，作为衡量酸侵蚀劣化的残留强度的最大缺陷是，在给定的侵蚀程度的质量损失百分比很大程度上取决于样品的几何性状；在相同侵蚀程度下，较大的样品比较小的样品损失强度小，导致在比较调查结果的时候比较困难。

相反，腐蚀的深度[97,99-102] 提供了更直接测定抵抗酸侵蚀裂化的能力，针对组成或者是强度发展差别很大的物料组不会产生体系的或随意的错误。侵蚀程度可以通过高精度和假设整个测试过程中样品完整部分尺寸稳定来测量，使样品几何、理论上更具重复性，进行比较的不同实验是相对独立的。

8.6 耐碱性

AAMs 耐碱侵蚀的研究只在有限的测试环境中进行了测试。一般在氢氧化物及碳酸盐的强碱环境下，碱硅酸盐激发和氢氧化物激发粉煤灰都观察到性能良好[111,112]，然而当暴露在碳捕收溶剂中时，BFS 基胶凝材料并没有表现出良好的性能[83]，可能是由于钙离子的离子交换作用。中温煅烧也能增强低钙铝硅酸盐胶凝材料在碱性浸出过程中的耐蚀性[112]，但是当胶凝材料产物是水化硅（铝）酸钙时，该方法可能不太管用。现有的标准化测试方法无法模拟测试环境，其碱性溶液环境相对复杂，因此为每种应用设计特定的测试条件是非常有必要的。

8.7 海水侵蚀

模拟海水中浸泡碱激发 BFS 胶凝材料实验表明其耐腐蚀性能良好[16]。Puertas 等人[21] 的研究发现，在海水浸泡过程中胶凝材料的孔隙结构的改变程度比只用氢氧化钠激发剂要低，证明所有样品在 180d 后很难劣化。Zhang 等人[113-115] 开发了碱硅酸盐激发纤维增强 BFS/偏高岭土胶凝材料作为海洋混凝土的结构涂层（图 8-6），在间歇浸入海水的条件时表现出很好的耐腐蚀性和耐久性，而硅酸盐水泥在此条件下很容易破坏（轻微的表面开裂但却与基层混凝土牢牢结合）。

图 8-6　碱激发胶凝材料涂料在我国海洋混凝土构件中的应用
（图片由 Z. Zhang 提供，南昆士兰大学）

8.8　软水侵蚀

根据相关浸出机理，学者认为流动的软水[116,117]对硅酸盐水泥具有攻击性。但是软水攻击下的 AAMs 的性能差异并没有任何研究。关于静态条件下 AAM 试样在蒸馏水中的性能研究较多，结果表明样品的抗酸或者其他侵蚀性溶液的能力对其各项性能起着决定性作用。然而，最初蒸馏水环境中离子强度会由于溶液孔隙组成的淋洗而迅速增大，很快就到达胶凝材料成分浸出变慢的临界点。因此，这仍然是需要研究的领域。

8.9　生物诱导侵蚀

正如在第 8.5 节关于酸侵蚀 AAM 胶凝材料的讨论，生物诱导对建筑材料有腐蚀性。在实验室测试条件下关于生物诱导侵蚀 AAMs 的相关研究至今还没有。然而，碱激发 BFS 混凝土经过长时间在储物沟槽里衬中使用（物质的衰变导致了有机酸的生成）之后的劣化程度非常低[118]。这将在第 11 章中进行讨论。

8.10　小结

化学侵蚀引起碱激发裂化的方式很多，其中最主要的可能是关于硫酸盐的侵蚀，以及碱-骨料反应将基体组分或固化物质浸出到中性或酸性环境中。在这些领域中，有大量现有的测试方法和程序，过去已经有几个应用于 AAM 的分析。在每种情况下，都有比其他方法优于 AAM 分析的测试方法，因为它们更接近地重现了材料可能暴露于其中的预期浸置条件（从而导致可用的劣化机制）。然而，不可能在这些情况下制定某个特定测试作为测试AAMs 的最优选择，因为在实验室中必须模拟一系列反应条件，因此需要不同的测试来模拟不同的反应条件和相应的劣化过程。以下是对现有测试方法的一些主要建议：

（1）AAMs 对硫酸盐侵蚀的抵抗能力是非常好的，尽管在验证实验室测试对插入数据的结果方面需要做更多的工作。因为在侵蚀的结果样品中观察到很小的变化，所以存在着潜在的问题。也就是说，不容易确定测试方法是否能够表示真实的化学进程。

（2）认为浸泡在氢氧化钠中的碱-骨料反应的分析不具有代表性，是由于在这种条件下胶凝结构自身的变化。有必要进行长期测试分析 AAM 中的 ASR 问题。在硅酸盐水泥中导致膨胀反应的反应机制是否相同，这个问题依然存在。

（3）由于形成部分完整的腐蚀层，腐蚀深度也就是 AAMs 的耐酸性成为比质量或者抗压强度损失更有用的研究方法。

（4）浸出实验必须有特定的侵蚀条件，否则结果肯定会出错。

（5）需要做更多的工作才能提出有关通过浸没在碱溶液、软水、生物诱导的侵蚀性环境或海洋条件中进行劣化试验的建议。

参考文献

[1] Monteiro P. J. M.. Scaling and saturation laws for the expansion of concrete exposed to sulfate attack. Proc. Natl. Acad. Sci. U. S. A, 2006, 103(31): 11467-11472.

[2] Glasser F. P.. The thermodynamics of attack on Portland cement with special reference to sulfate. In: Alexander M. G., Bertron A. (eds.) RILEM TC 211-PAE Final Conference, Concrete in Aggressive Aqueous Environments, Toulouse, France, 2009, 1: 3-17. RILEM.

[3] Alexander M., Bertron A., De Belie N. (eds.). Performance of Cement-Based Materials in Aggressive Aqueous Environments: State of the Art Report of RILEM TC 211-PAE. Springer/RILEM, Dordrecht, 2013.

[4] CEN/TC 51: Technical Report CEN/TR 15697: Cement-Performance Testing for Sulfate Resistance, 2008.

[5] Leemann A., Loser R., Lothenbach B.. Stand des Wissens: Sulfatbeständigkeit von Beton und sulfatbeständige Zemente. Cemsuisse Report 200806, 2009.

[6] ASTM International: Standard Test Method for Potential Expansion of Portland-Cement Mortars Exposed to Sulfate(ASTM C452-10 / C452M-10). West Conshohocken, PA, 2010.

[7] Grabovski E., Czarnecki B., Gillot, J. E., Duggan C. R., Scott J. F.. Rapid test of concrete expansivity due to internal sulfate attack. ACI Mater. J, 1992, 89: 469-480.

[8] Wittekindt W.. Sulphate-resistant cements and their testing. Zement-Kalk-Gips, 1960, 13: 565-572.

[9] Talero R.. Kinetochemical and morphological differentiation of ettringites by the Le Chatelier-Anstett test. Cem. Concr. Res, 2002, 32: 707-717.

[10] ASTM International: Standard Test Method for Length Change of Hydraulic-Cement Mortars Exposed to a Sulfate Solution(ASTM C1012-10 / C1012M-10). West Conshohocken, PA, 2010.

[11] Stark J., Wicht B.. Dauerhaftigkeit von Beton. Birkhäuser Verlag, Basel, 2001.

[12] KochA., Steinegger H.. A rapid method for testing the resistance of cements to sulphate attack. Zement-Kalk-Gips, 1960, 13: 317-324.

[13] Mehta P. K., Gjørv O. E., A new test for sulfate resistance of cements. J. Test. Eval, 1974, 2: 510-514.

[14] Mielich O., Öttl C.. Practical investigation of the sulfate resistance of concrete from construction units. Otto-Graf-J. 2004, 15: 135-152.

[15] Schweizerisches Ingenieur and Architektenverein, SIA): Determination of the resistance to sulfates of core test specimens, fast test(SIA 262/1 Appendix D). Zürich, Switzerland, 2003.

[16] Kukko H., Mannonen R.. Chemical and mechanical properties of alkali-activated blast furnace slag(F-concrete). Nord. Concr. Res, 1982, 1: 16. 1-16. 16.

[17] Douglas E., Bilodeau A., Malhotra V. M.. Properties and durability of alkali-activated slag concrete. ACI Mater. J, 1992, 89(5): 509-516.

[18] Shi C., Krivenko P. V., Roy D. M.. Alkali-Activated Cements and Concretes. Taylor &Francis, Abingdon, 2006.

[19] Bakharev T., Sanjayan J. G., Cheng Y. B.. Sulfate attack on alkali-activated slag concrete. Cem. Concr. Res, 2002, 32(2): 211-216.

[20] BakharevT.. Durability of geopolymer materials in sodium and magnesium sulfate solutions. Cem. Concr. Res, 2005, 35(6): 1233-1246.

[21] Puertas F., Mejía de Gutierrez R., Fernández-Jiménez A., Delvasto S., Maldonado J.: Alkaline cement mortars. Chemical resistance to sulfate and seawater attack. Mater. Constr, 2002, 52: 55-71.

[22] Ismail I., Bernal S. A., Provis J. L., Hamdan S., Van Deventer J. S. J.. Microstructural changes in alkali activated fly ash/slag geopolymers with sulfate exposure. Mater. Struct, 2013, 46(3): 361-373.

[23] Lucuk M., Winnefeld F., Leemann A.. Unpublished results of Empa project No. 841186, 2007.

[24] Škvára F., Jílek T., Kopecký L.. Geopolymer materials based on fly ash. Ceram.-Silik, 2005, 49(3): 195-204.

[25] Rodríguez E., Bernal S., Mejía de Gutierrez R., Puertas F.. Alternative concrete based on alkali-activated slag. Mater. Constr, 2008, 58(291): 53-67.

[26] Mauri J., Dias D. P., Cordeiro G. C., Dias A. A.. Geopolymeric mortar: study of degradation by sodium sulfate and sulfuric acid. Riv. Mater, 2009, 14: 1039-1046.

[27] Hu M., Zhu X., Long F.. Alkali-activated fly ash-based geopolymers with zeolite or bentonite as additives. Cem. Concr. Compos, 2009, 31(10): 762-768.

[28] Zhang J., Provis J. L., Feng D., Van Deventer J. S. J.. Geopolymers for immobilization of C^{6+}, Cd^{2+}, and Pb^{2+}. J. Hazard. Mater, 2008, 157(2-3): 587-598.

[29] Zhang J., Provis J. L., Feng D., Van Deventer J. S. J.. The role of sulfide in the

immobilization of Cr(VI) in fl y ash geopolymers. Cem. Concr. Res, 2008, 38(5): 681-688.

[30] Helmuth R., Stark D., Diamond S., Moranville-Regourd M.. Alkali-Silica Reactivity: An Overview of Research. Strategic Highway Research Program, SHRP-C-342, Washington, DC, 1993.

[31] Cong X.-D., Kirkpatrick R. J., Diamond S.. 29 Si MAS NMR spectroscopic investigation of alkali silica reaction product gels. Cem. Concr. Res, 1993, 23(4): 811-823.

[32] Smolczyk H. G.. Slag structure and identifcation of slags. In: 7th International Congress on the Chemistry of Cement, Paris, France, 1980, 1: Ⅲ-Ⅰ/4-16.

[33] Thomas M. D. A., Innis F. A.. Effect of slag on expansion due to alkali-aggregate reaction in concrete. ACI Mater. J, 1998, 95(6): 716-724.

[34] Chen Y. Z., Pu X. C., Yang C. H., Ding Q. J.. Alkali aggregate reaction in alkali slag cement mortars. J. Wuhan Univ. Technol, 2002, 17(3): 60-62.

[35] Leemann A., Le Saout G., Winnefeld F., Rentsch D., Lothenbach B.. Alkali-silica reaction: the influence of calcium on silica dissolution and the formation of reaction products. J. Am. Ceram. Soc, 2011, 94(4): 1243-1249.

[36] Lindgård J., Andiç-Çakır Ö., Fernandes I., Rønning T. F., Thomas M. D. A.. Alkali-silica reactions(ASR): literature review on parameters influencing laboratory performance testing. Cem. Concr. Res, 2012, 42(2): 223-243.

[37] Bakharev T., Sanjayan J. G., Cheng Y. B.. Resistance of alkali-activated slag concrete to alkali-aggregate reaction. Cem. Concr. Res, 2001, 31(2): 331-334.

[38] Fernández-Jiménez A., Puertas F.. The alkali-silica reaction in alkali-activated granulated slag mortars with reactive aggregate. Cem. Concr. Res, 2002, 32(7): 1019-1024.

[39] García-Lodeiro I., Palomo A., Fernández-Jiménez A.. Alkali-aggregate reaction in activated fly ash systems. Cem. Concr. Res, 2007, 37(2): 175-183.

[40] Fernández-Jiménez A., García-Lodeiro I., Palomo A.. Durability of alkali-activated fly ash cementitious materials. J. Mater. Sci, 2007, 42(9): 3055-3065.

[41] Puertas F., Palacios M., Gil-Maroto A., Vázquez T.. Alkali-aggregate behaviour of alkaliactivated slag mortars: effect of aggregate type. Cem. Concr. Compos, 2009, 31(5): 277-284.

[42] Xie Z., Xiang W., Xi Y.. ASR potentials of glass aggregates in water-glass activated fly ash and Portland cement mortars. J. Mater. Civil Eng, 2003, 15: 67-74.

[43] Yang C.. Alkali-aggregate reaction of alkaline cement systems. Ph. D. Thesis, Chongqing Jiangzhu University, 1997.

[44] Davies G., Oberholster R.. Use of the NBRI accelerated test to evaluate the effectiveness of mineral admixtures in preventing the alkali-silica reaction. Cem. Concr. Res, 1987, 17(1): 97-107.

[45] Ding J.-T., Bai Y., Cai Y.-B., Chen B.. Physical-chemical index of fly ash quality

in view of its effectiveness against alkali-silica reaction. Jianzhu Cailiao Xuebao/ J. Build Mater, 2010, 13: 424-429.

[46] RILEM TC 106-AAR: recommendations of RILEM TC 106-AAR: alkali aggregate reaction. A. TC 106-2-detection of potential alkali-reactivity of aggregates-the ultraaccelerated mortar-bar test. B. TC 106-3-detection of potential alkali-reactivity of aggregates-method for aggregate combinations using concrete prisms. Mater. Struct, 2000, 33(229): 283-293.

[47] Al-Otaibi S.. Durability of concrete incorporating GGBS activated by waterglass. Constr. Build. Mater, 2008, 22(10): 2059-2067.

[48] BritishStandards Institution: Method for determination of alkali-silica reactivity: concrete prism method(BS 812: 123). London, UK, 1999.

[49] Gifford P. M., Gillott J. E.. Alkali-silica reaction(ASR) and alkali-carbonate reaction (ACR) in activated blast furnace slag cement(ABFSC) concrete. Cem. Concr. Res, 1996, 26(1): 21-26.

[50] Yang C., Pu X., Wu F.. Studies on alkali-silica reaction, ASR) expansions of alkali activated slag cement mortars. In: Krivenko P. V. (ed.) 2nd International Conference on Alkaline Cements and Concretes, Kiev, Ukraine, 1999: 101-108.

[51] Tsuneki I.. Alkali-silica reaction, pessimum effects and pozzolanic effect. Cem. Concr. Res., 2009, 39(8): 716-726.

[52] Metso J.. The alkali reaction of alkali-activated Finnish blast furnace slag. Silic. Ind, 1982, 47(4-5): 123-127.

[53] Pu X., Yang C.. Study on alkali-silica reaction of alkali-slag concrete. In: Krivenko P. V. (ed.) 1st International Conference on Alkaline Cements and Concretes, Kiev, Ukraine, 1992, 2: 897-906.

[54] Gruskovnjak A., Lothenbach B., Holzer L., Figi R., Winnefeld F.. Hydration of alkali- activated slag: comparison with ordinary Portland cement. Adv. Cem. Res, 2006, 18(3): 119-128.

[55] Krivenko P. V., Gelevera A. G., Petropavlovsky O. N., Kavalerova E. S.. Role of metakaolin additive on structure formation in the interfacial transition zone "cement-alkali-susceptible aggregate". In: Bilek V., Keršner Z. (eds.) Proceedings of the 2nd International Conference on Non-Traditional Cement and Concrete, Brno, Czech Republic, 2005: 93-95. Brno University of Technology & ZPSV Uhersky Ostroh, a. s.

[56] Krivenko P. V., Petropavlovsky O., Gelevera A., Kavalerova E.. Alkali-aggregate reaction in the alkali-activated cement concretes. In: Bilek V., Keršner Z. (eds.) Proceedings of the 4th International Conference on Non-Traditional Cement & Concrete, Brno, Czech Republic. ZPSV, a. s. 2011.

[57] ThomasM.. The effect of supplementary cementing materials on alkali-silica reaction: a review. Cem. Concr. Res, 2011, 41(12): 1224-1231.

[58] Li K.-L., Huang G.-H., Chen J., Wang D., Tang X.-S.. Early mechanical prop-

erty and durability of geopolymer. In: Davidovits J. (ed.) Proceedings of the World Congress Geopolymer 2005-Geopolymer, Green Chemistry and Sustainable Development Solutions, Saint-Quentin, France, 2005: 117-120. Institut Géopolymère.

[59] Li K.-L., Huang G.-H., Jiang L.-H., Cai Y.-B., Chen J., Ding J.-T.. Study on abilities of mineral admixtures and geopolymer to restrain ASR. Key Eng. Mater, 2006, 302-303: 248-254.

[60] Kupwade-Patil K., Allouche E.. Effect of alkali silica reaction, ASR) in geopolymer concrete. In: World of Coal Ash 2011, Denver, CO, 2011.

[61] Kosson D. S., Van der Sloot H. A., Sanchez F., Garrabrants A. C.. An integrated framework for evaluating leaching in waste management and utilization of secondary materials. Environ. Eng. Sci, 2002, 19(3): 159-179.

[62] Van der Sloot H. A., Dijkstra J. J.. Development of horizontally standardized leaching tests for construction materials: a material based or release based approach. Report ECN-C-04- 060, 2004.

[63] Van der Sloot H. A., Kosson D. S.. A unified approach for the judgement of environmental properties of construction materials (cement-based, asphastic, unbound aggregates, soil) in different stages of their life cycle. In: Ortiz de Urbina, G., Goumans H. (eds.) Environmental and Technical Implications of Construction with Alternative Materials: Wascon 2003: 503-515.

[64] Van der Sloot H. A., Van der Seignette P., Comans R. N. J., Van Zomeren A., Dijkstra J. J., Meeussen H., Kosson D. S., Hjelmar O.. Evaluation of environmental aspects of alternative materials using an integrated approach assisted by a database/expert system. In: Proceedings of the Conference on Advances in Waste Management and Recycling, Dundee, Scotland, 2003.

[65] Van der Sloot H. A., Van Zomeren A., Dijkstra J. J., Hoede D.. Prediction of long term leachate quality and chemical speciation for a predominantly inorganic waste landfill. In: Proceedings of the 9th International Waste Management and Landfill Symposium, Sardinia, Italy, 2003.

[66] US EPA: Test Method 1311-TCLP, Toxicity Characteristic Leaching Procedure, 1992.

[67] Hulet G. A., Kosson D. S., Conley T. B., Morris M. I.. Evaluation of waste-form analysis protocols that may replace TCLP. In: Proceedings of WM '00, Tucson, AZ. Paper 13-1, 2000.

[68] EA NEN 7371: 2004: Leaching characteristics of moulded or monolithic building and waste materials. The determination of the availability of inorganic components for leaching. "The maximum availability leaching test", 2004.

[69] NEN 7341: Leaching characteristics of solid earthy and stony building and waste materials leaching tests: determination of the availability of inorganic components for leaching, 1995.

[70] Van der Sloot H. A., Meeussen H., Kosson D. S., Sanchez F.. An overview of leaching assessment for waste disposal and materials use-Laboratory leaching test methods. In: Wascon 2003, San Sebastian, Spain.

[71] European Committee for Standardization, CEN): Characterization of waste. Leaching behaviour tests. Influence of pH on leaching with initial acid/base addition, CEN/TS14429: 2005). Brussels, Belgium.

[72] NEN 7347: Leaching characteristics of solids earthy and stony building and waste materials -Leaching tests-Determination of the leaching of inorganic components from compacted granular materials.

[73] NEN 7375: Leaching characteristics of moulded or monolithic building and waste materials: Determination of leaching of inorganic components with the diffusion test. 'The tank test', 2004.

[74] American Nuclear Society: Measurement of the leachability of solidifed low-level radioactive wastes by a short-term method (ANSI/ANS-16. 1-2003). ANS, La Grange Park, IL, 2003.

[75] Comrie D. C., Paterson J. H., Ritcey D. J.. Geopolymer technologies in toxic waste management. In: Davidovits J., Orlinski J. (eds.) Proceedings of Geopolymer '88-First European Conference on Soft Mineralurgy, Compeigne, France, 1988, 1: 107-123. Universite de Technologie de Compeigne.

[76] Davidovits J.. Geopolymers-inorganic polymeric new materials. J. Therm. Anal, 1991, 37(8): 1633-1656.

[77] Vance E. R., Perera D. S.. Geopolymers for nuclear waste immobilisation. In: Provis J. L., Van Deventer J. S. J. (eds.) Geopolymers: Structure, Processing, Properties and Industrial Applications, 2009: 403-422. Woodhead, Cambridge.

[78] Provis J. L.. Immobilization of toxic waste in geopolymers. In: Provis J. L., Van Deventer J. S. J. (eds.) Geopolymers: Structure, Processing, Properties and Industrial Applications, 2009: 423-442. Woodhead, Cambridge.

[79] Shi C., Fernández-Jiménez A.. Stabilization/solidification of hazardous and radioactive wastes with alkali-activated cements. J. Hazard. Mater, 2006, B137 (3): 1656-1663.

[80] Lloyd R. R., Provis J. L., Van Deventer J. S. J.. Pore solution composition and alkali diffusion in inorganic polymer cement. Cem. Concr. Res, 2010, 40(9): 1386-1392.

[81] Van Eijk R. J., Brouwers H. J. H.. Study of the relation between hydrated Portland cement composition and leaching resistance. Cem. Concr. Res, 1998, 28 (6): 815-828.

[82] Lloyd R. R.. The durability of inorganic polymer cements. Ph. D. Thesis, University of Melbourne, Melbourne, 2008.

[83] Gordon L. E., Provis J. L., Van Deventer J. S. J.. Durability of fly ash/GGBFS based geopolymers exposed to carbon capture solvents. Adv. Appl. Ceram, 2011, 110

(8): 446-452.

[84] Xie S., Li Q., Zhou D.. Investigation of the effects of acid rain on the deterioration of cement concrete using accelerated tests established in laboratory. Atmos. Environ, 204, 38: 4457-4466.

[85] Soroka I.. Portland Cement Paste and Concrete. Macmillan, London, 1979: 338.

[86] Floyd M., Czerewko M. A., Cripps J. C., Spears D. A.. Pyrite oxidation in Lower Lias Clay at concrete highway structures affected by thaumasite, Gloucestershire. UK. Cem. Concr. Compos, 2003, 25: 1015-1024.

[87] De Belie N., Lenehan J. J., Braam C. R., Svennerstedt B., Richardson M., Sonck B.: Durability of building materials and components in the agricultural environment. Part III: concrete structures. J. Agr. Eng. Res, 2000, 76: 3-16.

[88] Bertron A., Duchesne J., Escadeillas G.. Degradation of cement pastes by organic acids. Mater. Struct, 2007, 40: 341-354.

[89] Chaudhary D., Liu H.. Influence of high temperature and high acidic conditions on geopolymeric composite material for steel pickling tanks. J. Mater. Sci, 2009, 44: 4472-4481.

[90] Davis J. L., Nica D., Shields K., Roberts D. J.. Analysis of concrete from corroded sewer pipe. Int. Biodeter. Biodegr. 1998, 42: 75-84.

[91] Gutiérrez-Padilla M. G. D., Bielefeldt A., Ovtchinnikov S., Hernandez M., Silverstein J.: Biogenic sulfuric acid attack on different types of commercially produced concrete sewer pipes. Cem. Concr. Res, 2010, 40(2): 293-301.

[92] Monteny J., Vincke E., Beeldens A., De Belie A., De Belie N., Taerwe L., Van Gemert D., Verstraete W.. Chemical, microbiological, and insitu test methods for biogenic sulfuric acid corrosion of concrete. Cem. Concr. Res, 2000, 30: 623-634.

[93] Parker C. D.. Species of sulphur bacteria associated with the corrosion of concrete. Nature, 1947, 159: 439-440.

[94] Fourie C. W., Alexander M. G.. Acid resistant concrete sewer pipes. In: Alexander M. G. Bertron A. (eds.) RILEM TC 211-PAE Final Conference, Concrete in Aggressive Aqueous Environments, Toulouse, France, 2009, 1: 408-418. RILEM.

[95] Saricimen H., Shameem M., Barry M. S., Ibrahim M., Abbasi T. A.. Durability of proprietary cementitious materials for use in wastewater transport systems. Cem. Concr. Compos, 2003, 25: 421-427.

[96] Scrivener K. L., Cabiron J.-L., Letourneaux R.. High-performance concretes from calcium aluminate cements. Cem. Concr. Res. 1999, 29: 1215-1223.

[97] LloydR. R., Provis J. L., Van Deventer J. S. J.. Acid resistance of inorganic polymer binders. 1. Corrosion rate. Mater. Struct, 2012, 45(1-2): 1-14.

[98] Pacheco-Torgal F., Castro-Gomes J., Jalali S.. Durability and environmental performance of alkali-activated tungsten mine waste mud mortars. J. Mater. Civil Eng, 2010, 22(9): 897-904.

[99] Shi C.. Corrosion resistance of alkali-activated slag cement. Adv. Cem. Res, 2003, 15(2): 77-81.

[100] Allahverdi A., Škvára F.. Nitric acid attack on hardened paste of geopolymeric cements -Part 1. Ceram. -Silik, 2001, 45(3): 81-88.

[101] Allahverdi A., Škvára F.. Nitric acid attack on hardened paste of geopolymeric cements -Part 2. Ceram. -Silik, 2001, 45(4): 143-149.

[102] Allahverdi A., Škvára F.. Sulfuric acid attack on hardened paste of geopolymer cements. Part 1. Mechanism of corrosion at relatively high concentrations. Ceram. -Silik, 2005, 49(4): 225-229.

[103] Wastiels J., Wu X., Faignet S., Patfoort G.. Mineral polymer based on fly ash. In: 9th International Conference on Solid Waste Management Widener University, Philadelphia, PA, 1993.

[104] Rostami H., Brendley W.. Alkali ash material: a novel fly ash based cement. Environ. Sci. Technol, 2003, 37: 3454-3457.

[105] Silverstrim T., Rostami H., Larralde J., Samadi A.. Conpiler: U.S, 5601643 [p]. "Fly ash cementitious material and method of making a product". U.S. Patent Offi ce, 1997.

[106] Johnson G. B.. WO 2005/049522 A1 "Geopolymer concrete and method of preparation and casting" World Intellectual Property Organisation, 2005.

[107] Shi C.. U.S. patent 6749679 B2: Composition of materials for production of acid resistant cement and concrete and methods thereof. U.S. Patent Office, 2004.

[108] Van Deventer J. S. J., Provis J. L., Duxson P., Brice D. G.. Chemical research and climate change as drivers in the commercial adoption of alkali activated materials. Waste Biomass Valoriz, 2010, 1(1): 145-155.

[109] Provis J. L., Lloyd R. R., Van Deventer J. S. J.. Mechanism and implications of acid attack on fly ash and ash/slag inorganic polymers. In: Alexander M. G., Bertron A. (eds.) RILEM TC 211-PAE Final Conference, Concrete in Aggressive Aqueous Environments, Toulouse, France, 2009, 1: 88-95. RILEM.

[110] Buchwald A., Dombrowski K., Weil M.. The influence of calcium content on the performance of geopolymeric binder especially the resistance against acids. In: Davidovits J. (ed.) Geopolymer, Green Chemistry and Sustainable Development Solutions, 2005: 35-39. Institut Géopolymère, Saint-Quentin.

[111] Sindhunata Provis J. L., Lukey G. C., Xu H., Van Deventer J. S. J.. Structural evolution of fly ash-based geopolymers in alkaline environments. Ind. Eng. Chem. Res, 2008, 47(9): 2991-2999.

[112] Temuujin J., Minjigmaa A., Lee M., Chen-Tan N., Van Riessen A.. Characterisation of class F fly ash geopolymer pastes immersed in acid and alkaline solutions. Cem. Concr. Compos, 2011, 33(10): 1086-1091.

[113] Zhang Z., Yao X., Zhu H.. Potential application of geopolymers as protection

coatings for marine concrete: I. Basic properties. Appl. Clay Sci, 2010, 49(1-2)1-6.

[114] Zhang Z., Yao X., Zhu H.. Potential application of geopolymers as protection coatings for marine concrete: II. Microstructure and anticorrosion mechanism. Appl. Clay Sci, 2010, 49(1-2): 7-12.

[115] Zhang Z., Yao X., Wang H.. Potential application of geopolymers as protection coatings for marine concrete III. Field experiment. Appl. Clay Sci, 2012, 67: 57-60. K. Abora et al.

[116] Grattan-Bellew P. E.. Microstructural investigation of deteriorated Portland cement concretes. Constr. Build. Mater. 1996, 10(1): 3-16.

[117] Adenot F., Buil M.. Modelling of the corrosion of the cement paste by deionized water. Cem. Concr. Res, 1992, 22(2-3): 489-496.

[118] Goncharov N.. Corrosion resistance of slag-alkaline binders and concretes in aggressive organic environments. Ph. D. Thesis, Kiev Red Banner of Labor Order Civil Engineering Institute, 1984.

第 9 章 耐久性和测试——劣化与物质传输

Susan A. Bernal, Vlastimil Bílek, Maria Criado, Ana Fernández-Jiménez, Elena Kavalerova, Pavel V. Krivenko, Marta Palacios, Angel Palomo, John L. Provis, Francisca Puertas, Rackel San Nicolas, Caijun Shi and Frank Winnefeld

9.1 简介

在钢筋混凝土的应用中，材料结构破坏主要与嵌入钢筋锈蚀有关，而非胶凝材料本身的劣化造成。任何结构性混凝土的关键作用是提供足够的覆盖深度和碱度以使钢材长时间地保持钝化状态。侵蚀性物质（如氯化物的侵入和/或通过诸如碳化的过程）会造成碱度的降低从而失去钝化保护造成钢筋锈蚀。这意味着硬化胶凝材料的质量传递性质对于确定混凝土的耐久性至关重要，因此对于碱激发材料的运输相关性质的分析和测试将成为本章的重点，包括碱激发胶凝材料内的钢侵蚀化学以及渗透性（这是在过量碱移动的情况下观察到的现象）。

9.2 渗透性和孔隙率

对硅酸盐水泥基混凝土的微观结构和渗透率的相互关系已进行了大量研究[1-4]。尽管不同的孔隙率测试技术已在不同的司法管辖区得到了标准化规范，但当前仍依赖于商业化的或实验室设备分析样品。因此，本节中的讨论将提供在标准化和非标准化分析方案中获得的结果和数据。RILEM TC 116-PCD[5] 和 189-NEC[6] 的 "最新技术报告" 对混凝土进行了原位渗透分析以及对现有技术进行了描述和分析。

9.2.1 实验室中的孔隙率测定法

复合材料的孔隙率分析方法通常是在一定压力下流体被迫进入样品从而测量进入固体材料的孔隙空间的流体体积和流体通过整体样品的速率或程度。气体渗透性测试通常使用压缩空气或氧气进行，提供关于气体通过样品的直接数据。由浓度梯度而不是压差驱动的气体扩散的测试方法也是可用的[7-9]。

不同孔径范围所对应的测试表征方法如图 9-1 所示。很明显，在研究 AAM（或基于 OPC 混凝土）时，没有一种技术跨越全部孔径范围，因此对这些材料的特征进行表征时需要应用一些不同技术。

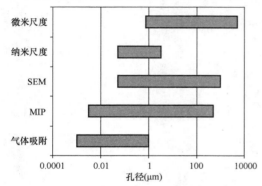

图 9-1 几种可用的分析技术探测到的大致孔径范围

第9章 耐久性和测试——劣化与物质传输

气体吸附分析（通常使用氮气、氩气或氦气）通过各种算法〔如流行的Brunauer-Emmett-Teller（BET）[10]和Barret-Joyner-Halenda（BJH）[11]法〕来计算孔径分布和表面积，压汞法（MIP）广泛用于硬化胶凝材料相的研究。MIP对诸如建筑材料中观察到的复杂孔隙几何形状的适用性尚存争议[12]。目前，多循环MIP和伍德合金侵入等测试方法改善了MIP的缺陷，有效地表征了材料中存在的孔隙几何结构[13-15]。硬化混凝土渗透率分析为其水分渗透性分析也提供了有用的信息。这些测试中的每一个参数都能够提供对于了解AAM混凝土的结构和耐久性以及预测其使用寿命等至关重要的信息。复杂材料如AAM胶凝材料或混凝土的全面多尺度表征需要全面的技术来获取所需的孔径信息。

BJH方法[11]已经在各个管辖区[16,17]中得到标准化认定，但其仅针对多孔材料，而非水泥或建筑材料。与其他更先进的方法相比，BJH方法无明显优势，但其仍广泛应用于碱激发胶凝材料分析计算中[18,19]。Lloyd等[15]和Zheng等人[20]各自使用BJH技术来观察随着激发剂浓度增加的粉煤灰基AAM中的孔隙反应，这符合对第4章所述的这些材料形成的概念性理解。通过BJH分析对碱金属氢氧化物/硅酸盐激发的偏高岭土材料的研究表明，随着激发剂模数的增加，孔半径减小；Sindhunata等人的数据[21]对于类似的粉煤灰基材料通常符合这一观察。然而，在这两种情况下，重要的孔体积似乎都在最大的可测量孔径（刚刚超过100nm）以及低于最小的可测量孔径（在2～3nm之间）之间分布，这为数据的分析带来了复杂性。大孔径主导物质传输对相关性能起着重要作用，但此尺度的孔径并未在BJH数据中表现出来。与孔体积的质量测定相比，BJH技术也可能将总孔隙率低估6～8倍[15]。在BJH和MIP测定计算中均使用圆柱状的[22]标准模型计算孔径分布，造成了同一碱激发粉煤灰胶凝材料中孔径分布结果之间也存在分歧。同时在DIN和ISO文件[23,24]中被标准化的"t-plot"方法仅用于偏高岭土碱激发胶凝材料的结果分析[25]。迄今为止，"t-plot"方法似乎并未被应用于AAM的分析上。

众所周知，MIP在水泥应用具有局限性和不准确性[12]，但因其简单性和仪器的可用性而被广泛使用。MIP对于复杂孔隙几何形状的样品（"墨水瓶"效应）的分析并不准确。因为复合形状孔的全部体积被记录为具有与入口的最窄部分相同的有效孔径，导致平均孔径被低估。压汞法样品中最小孔隙的高压通常会超过其所施加的材料的强度（压缩或拉伸），导致样品出现明显的破碎效应。MIP对土壤和岩石分析的应用[26]和多孔催化剂[27]已经通过ASTM标准化，其在多孔材料中的应用一般通过各种标准[28-31]进行描述。

碱硅酸盐激发BFS的MIP分析[32]显示存在双峰孔径分布，分别为>1nm和<20nm范围内具有显著的孔隙率。但是这些尺寸范围之间孔隙率很小。而硅酸盐水泥分析显示其为单孔孔径分布，大部分孔体积在10～100nm的孔径之间。毛细管孔在物质传递中起着重要的作用。通过AAM的MIP分析可知：由K替代Na导致碱硅酸盐激发的偏高岭土胶凝材料的中值孔径显著降低[33-35]，同样地，将BFS掺入主要是偏高岭土的胶凝材料时也会出现类似现象[36]。

扫描电子显微镜（SEM）从微观方面对AAM的孔结构进行二维分析。SEM孔隙分析的主要挑战是在不含固体物质区域的孔识别。图像分析算法有效地克服了富钙体系中的这些问题（富含钙的体系，如硅酸盐水泥或碱激发的BFS等）。

图9-2显示了随着水化程度的增加，通过氢氧化钠或水玻璃激发的各种矿渣的粗（毛细管）孔（0.05～5μm）的发展过程[39-41]。随着水化程度的增加，粗（毛细管）孔隙率减小，

浆体渗透率降低。从图 9-2 也可以看出，激发剂的性质对于孔隙的发展作用比胶凝材料的化学组成更大。

对于在固体和孔区域之间具有较低相对密度的铝硅酸盐 AAM 体系，伍德合金侵入技术已被证明是有价值的[15]。这种技术把温度低于 100℃ 的熔融合金在中等压力作用下（低于 MIP 中使用的压力，以防止损坏孔隙）以液体形式侵入样品孔隙的部分微结构中；然后冷却样品，合金在孔网络中固化，使高元素量区域与低元素区的胶凝材料进行对比[43-44]。最近将该技术应用于碱激发的粉煤灰和 BFS 胶凝材料[15]中，计算出小于 11nm 的孔能够用熔融合金填充，同时使用时应低于抗压强度以使微结构损伤最小化。

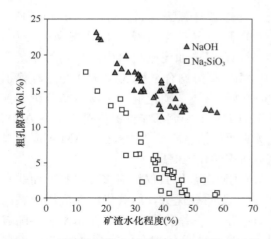

图 9-2 通过使用抛光部分的 SEM 图像分析获得的碱激发矿渣浆料的粗（毛细管）孔隙率（0.05～5μm）（±0.5%）与水化度（±3%）的关系。不同化学成分的 8 种矿渣用 NaOH 或 Na_2SiO_3 以 Na_2O% 激发，水/渣 = 0.40（本哈哈等人的数据[39-41]）

X 射线显微成像是应用于水泥（碱激发和传统）分析的主要三维表征技术。在过去十多年里，已应用此技术对硅酸盐水泥材料进行了大量研究[45-49]。最近首次在 750nm 像素分辨率下提出了碱激发胶凝材料的 X 射线显微成像的体系分析[50-52]。图 9-3 为硅酸钠 AAM 胶凝材料的给定区域的灰度图像和分割图像的实例。对于有限数量的 AAM 样品，还收集了 30nm 像素分辨率的数据[53]，但是这种"纳米断面成像"技术还在试验阶段，尚未广泛使用。

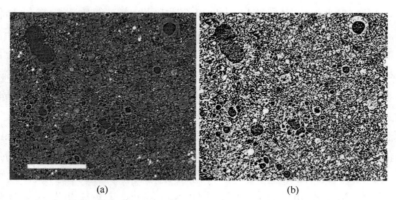

图 9-3 粉煤灰衍生的 AAM 的显微成像图像的二值阈值
(a) 原始灰度图像；(b) 二进制分割图像（比例尺为 200μm）

图 9-3 还显示了与来自粉煤灰的 AAM 样品的显微成像分析的问题[51,52]。图像中大的暗区域表示中空或部分中空（空心或空球）灰分颗粒的内部，其不与凝胶内的孔体积连接，因此不利于材料的输送性质。然而，断层数据集的总孔隙率的计算将包括这些区域，因此它们必须通过使用仅考虑"连接孔隙率"的算法来排除。对基于 BFS（或硅酸盐水泥）的样品来说问题较少，因为前驱体颗粒不包含显著的不可接近的孔体积。

如图 9-4 所示,在重建断层扫描数据中进行的分割和扩散模拟证明[51],含有 50%或更多 BFS 的碱激发胶凝材料的孔隙率随固化持续时间的延长而降低,表明硅酸钠水化物["N-A-S-(H)"]和钙(铝)硅酸盐水化物["C-(A)-S-H"]凝胶共存。正如第 5 章所述,在混合粉煤灰/BFS 的 AAM 胶凝材料体系中发生水的一些化学结合。图 9-4 表明 BFS 代替粉煤灰导致更高的孔隙率。这与一组相似的胶凝材料的氮吸附分析结果一致[15],并且从层析成像获得的孔隙率在该研究中获得的相似混合物设计和固化持续时间的样品的 2%体积内。

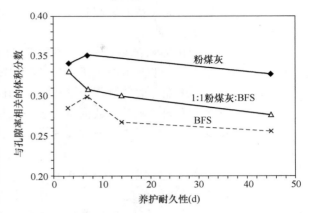

图 9-4 一系列硅酸钠激发的粉煤灰/BFS 体系的分段孔隙率和养护时间之间的关系(数据选自 Purvis 等人[51])

随着凝胶的发展,其孔隙率随时间减少,孔隙网络的弯曲度增加[51]。在早期(<7d)和 45d 龄期之间该参数的增加几乎为 2 倍,表明当与不良固化的相比较时,通过良好固化的富含 BFS 的 AAM 胶凝材料的扩散速率将减半。这进一步突出了在碱激发混凝土中实现高性能和耐久性的充分固化的重要性。然而,从断层扫描结果发现,无论早期强度发展的速率如何,当胶凝材料被使用并暴露于侵蚀性环境时,长时间的密封固化有利于使用寿命延长和总体耐久性能的提高。

在富含灰分的 AAM 体系中,对胶凝材料的理解和控制孔隙率具有重要意义。与硅酸钙凝胶相比较,硅铝酸钠凝胶的孔隙率较高,例如由硅酸盐水泥水化产生的硅酸盐凝胶导致侵蚀性离子通过胶凝材料的高扩散性(因此快速的质量传输)到达嵌入式钢筋,则意味着其耐久性较差。如第 2 章和第 11 章所述,地质聚合物和其他碱激发胶凝材料的使用性能的证据[54-56]以及下文第 9.4 节所述的实验室氯化物渗透测试表明,观察到的性能明显优于原始渗透率数据[54,57-60]。这表明存在孔隙率直接影响渗透性(特别是离子渗透性,也涉及凝胶化学特异性效应和相互作用)以及这些材料的耐久性的附加效应。然而,这些影响尚未得到很好的理解。

9.2.2 透气性

有很多种方法用来分析水泥基胶凝材料和混凝土的气体渗透性,这些方法都是基于达西定律的。测试方法中,一种情况是测试气体以一定速率通过材料进入初始真空腔,另一种情况是气流以一定质量从加压气室通过材料进入较低压力的材料内腔。人们在相同原理下,更喜欢使用浓度差而不是压力差注入气体。常用的扩散气体有氦气、O_2 或 CO_2。根据测试方法的细节、胶凝材料特性、饱和(部分饱和)程度和样品的预处理过程,决定加载的体积压力、表观压力或实际压力,从而得到不同范围的数据[61-65]。RILEM 循环测试[66]显示,商用 Autoclam,Hong-Parrot 和 Torrent 透气性仪器的测试结果之间具有很好的相关性和一致性。相比其他两种测试仪器,Torrent 仪器依据混凝土配合比设计的函数,可以得到更好的渗透性能趋势线[66]。然而,当样品存在裂纹和/或湿度条件有差异时,各种测试方法和应用结果对比就相对困难了。这需要精细地制备样品，细心对比不同水胶比下的变化趋势（如养护程度、孔结构和/或凝胶化学）。

大多数实验室测试体系都是以 CEMBUREAU 氧渗透法为基础进行改进得到的自制或定制的商用测试方法[67]。试验中可以限定不同的样品尺寸，使用两种预处理方法（会得到不同的测试结果）对现浇和取芯混凝土试样进行测试。使用单一测试方法获得的渗透率值会在一定范围内波动，不同实验室的数据对比起来就相对困难。特别是一些样品尺寸和几何形状的差异会引起渗透数据的巨大差异[68]，这就要求在渗透率测试时，需要对所有试验参数详细说明。

透气性的现场测试主要以 Torrent 方法[69]为主，该方法用于标准 SIA 262/1[70,71]，其开发、验证和应用相关的详细信息均可在线获得[72]。Torrent 方法中同心双室装置的设计可减少侧向空气流泄漏的影响。同时便携式测试设备可应用于现场商品混凝土的测试。混凝土的湿度对测试结果有重要影响，而现场电阻率测试时要比实验室的湿度更低。对于新拌混凝土，水从样品蒸发到真空测试室是真空测试中的关键环节[73]。Autoclam 测试方法[74]在湿混凝土中应用的有关问题，已在之前的 RILEM TC 出版物[75]中进行了详细讨论。这就需要在新拌混凝土试验期间仔细监测（或者如果可能控制）温度，以确保结果能被准确地解释。

迄今为止，在已发表的学术文献中，关于碱激发胶凝材料和混凝土的气体渗透性测试报告相对较少。Hakkinen[76]的早期工作表明，当在 70% 相对湿度下测试时，碱激发的 BFS "F-混凝土"的 N_2 渗透率约是同等力学性能和总孔体积的硅酸盐水泥混凝土的十倍。如第 9.2.3 节中所述，干燥"F-混凝土"存在较多的微裂纹（如本报告第 10 章更详细讨论的），导致干燥样品中的气体渗透性较高。但事实是，可用的数据很少，而且似乎没有公开使用任何前边提到的商用设备来对碱激发 BFS 的气体渗透性进行研究，这就无法跟使用相同装置测试的硅酸盐水泥混凝土进行对比。

其实，目前只有一篇关于单一粉煤灰基 AAMs 的气体渗透性研究的报道。Sagoe Crentsil 等人[77]发现，蒸汽养护的粉煤灰基 AAM 混凝土的透气性与类似抗压强度（40MPa）的硅酸盐水泥混凝土的透气性相当，但遗憾的是在测试时没有提供测试方法或样品的龄期。

9.2.3 透水性

有必要分别考虑 AAM 的气体渗透性和水分渗透性，因为混凝土的这些参数之间的关系通常是间接的[78]。大多数透水性测试的缺点是预处理和/或测试期间的持续时间长（几周或几个月），虽然已经开发和出版了预处理样品的加速程序[79]，但这些方法并未得到推广。在长时间预处理期间，样品成熟度的差异在碱激发胶凝材料和混凝土的分析中可能是显著的。

在各个管辖范围内，标准化的最简单的渗水试验涉及通过煮沸样品测量总孔体积，并测量干燥质量和饱和质量的差异来计算水吸收量（以此得到孔体积）。描述该测试方法变体的标准包括 ASTM C642[80]和 AS 1012.21[81]。该测试被用作一些司法管辖区具体耐久性的主要预测手段[82]，并且能够测量以水化物或其他固相填充的硅酸盐水泥基产品中的体积，具有良好的重复性和精度，尽管它与实际渗透性（因此与氯化物渗透和碳化）的关联性较弱[83]。在将这种类型的测试应用于 AAM 时，必须注意铝硅酸盐凝胶相中弱结合水的作用问题，应当具体说明在沸腾试验前的预干燥步骤中是否已除去这种弱结合水[84]。目前弱结合水的作用问题尚不清楚，虽然通过使用本测试对基于 BFS 的 AAM 混凝土获得了有趣且具有潜在指导意义的数据[85,86]，但是该测试结果的全面解释将需详细地检查在干燥、再润湿和沸腾期间，AAM胶凝材料中的化学和微结构变化过程。涉及浸没无沸腾的类似试验，可证明更适用于AAM，例如使用20℃浸渍的BS 1881-122[87]，但样品预处理（包括使用压力或真空饱和度）仍然是一个讨论点。还有人指出，吸水测试方案的可用变体虽然在性质和实施上极为相似，但可以根据测试方法给出不同的结果[78]。

水渗透到混凝土中的直接测量通常是在与第9.2.2节讨论中的透气性试验相似的几何形状下进行，并且一些相同的装置（例如Autoclam体系）可用于水和气体渗透性试验，尽管这通常仅限于使用高压（而不是真空）将测试流体驱动到样品中的体系。因为水通过样品的渗透速度比气体的渗透速度慢，所以通常通过在规定的时间之后分割样品来测量水的渗透深度（例如在EN 12390-8[88]），而不是等待稳定状态的流动。一个显著的例外是CRD-C 163-92方法[89]，它是使用最初真空饱和的样品和相对较小（直径3×最大骨料尺寸）的圆柱形样品进行稳态流动测试。Figg测试方法：将钻孔钻入混凝土样品并填充水，记录水从混凝土中流出的速率，可用于现场混凝土的测试。但其钻孔过程中发生裂纹的可能性以及周围混凝土的湿度状态（例如由于钻井过程中的加热）的不确定性仍会导致本次测试的重复性较差。Figg测试也可用于气体渗透分析，但不管测试中使用的流体是什么，都有一个缺点，即通常不可能用这种方法来直接计算渗透率系数。

EN 12390-8[88]、ISO 7031[91]和DIN 1048-5[92]标准分别涉及施加压力迫使水进入预先干燥的混凝土样品，并且测量分裂渗透深度。在这些测试中施加的压力从150kPa增至700kPa，ISO 7031测试使用的压力随测试的每一天而增加。每个测试还具有指定的一组干燥和样品预处理条件以及指定的持续时间，并且这些差异与施加压力差异的组合意味着测试的结果彼此不能直接相比较。CRD-C 48方法还使用施加的压力来将水压入大样品中，但是渗透程度是根据储层体积的变化而不是测试后的样品自身进行分析计算的。

碱激发胶凝材料和混凝土直接的透水性和透气性测量显示出一系列性能，这主要取决于所测试的混合设计。具有良好固化的胶凝材料和低水胶比的AAM在这些测试中表现良好[77,93,94]，但通常不提供特别突出的结果，这可能是由于低水平的空间结合水与关键结合凝胶相关。AAM和硅酸盐水泥之间的比较显示出非常宽泛的结果。Shi[94]获得的硅酸钠激发的BFS混凝土的水渗透率比硅酸盐水泥低约1000倍，而Wongpa等人[95]发现他们的碱硅酸盐激发的混凝土的渗透性比类似强度级别的硅酸盐水泥混凝土高100～10000倍。这些研究都使用达西定律来计算加压装置中的水渗透性。然而，大多数据落在该范围的中心内，与硅酸盐水泥混凝土相比，碱激发粉煤灰[77,93]、BFS[96]或粉煤灰/BFS掺混三种碱激发胶凝材料[97]的透水率值类似（在一个数量级内）。Zhang等人[98]还研究了碱激发偏高岭土/BFS胶凝材料，并发现通过添加更多的BFS或降低w/b比能够降低透水性，正如第5章提到的凝胶化学的讨论中所预期的一样。

9.2.4 毛细管作用

水渗透到混凝土中的另一个主要方法是通过毛细管虹吸作用，其中将干燥（或部分干燥的）样品的一端与水接触，并且通过毛细管作用将水吸入样品。通常通过规律间隔称重样品来获得数据，然后将获得的数据用作孔网几何和连通性的间接测量。该测试由Fagerlund[99]详细分析，已经以各种形式进行了标准化，包括ASTM C1585[100]、EN 1015-18[101]、SIA

262/1附录A[102]以及在RILEM TC 116-PCD[103]建议中的详细说明。RILEM方法包括对温度和孔隙压力的具体控制，但仅推荐在四个不同时间对样品进行称重，因此不能满足扩散方程（$t^{1/2}$依赖性）的要求，然而在其他协议中，这种依赖性通过更加规范的称重和延长的测试持续时间（根据标准方法长达15d，有时需要更长的持续时间来实现碱激发的BFS混凝土的毛细管系统的完全饱和[60,86]）来实现。与其他长期测试一样，扩展测试持续时间的主要缺点是，样品的水化程度在测试期间增加。

毛细管吸附性测试表明，碱激发的BFS混凝土的孔网络已经足够精细和曲折，产生了相当低程度的毛细管吸附性[32,58-60,104-106]。尽管总孔隙率在大多数情况下类似于或高于硅酸盐水泥的总孔隙率。在碱激发的BFS体系中使用较高模数的激发剂[107]或较低的含水量[108]降低了吸水速率。同时其吸附性随着在潮湿条件下固化时间的增加而降低[108]。尝试了通过描述具有椭圆孔形状模型的毛细管孔[109,110]来模拟经过碱激发的BFS混凝土的孔网络的水流速率的方法，这在一定程度上是成功的。尽管已知碱激发胶凝材料的实际孔网几何结构涉及显著的"墨瓶"效应，其中较大半径的孔体积仅可通过狭窄的颈部[15,51]。

高度多孔的碱激发偏高岭土或基于天然火山灰的胶凝材料具有非常高的毛细管吸力。毛细管吸力加剧了碱的运动，会导致硬化浆体风化，此过程可以通过提供蒸发冷却水源进行热控制[111-112]。

另一种标准化测试方法是初始表面吸收试验（ISAT），如BS1881-208[113]中所述，其使用与干燥（或在使用中）混凝土试样的表面接触的窄毛细管，并通过计算水从毛细管到材料的移动速率来计算混凝土的吸附性。该试验具有相对快速（每个样品约1h）的优点，但是在目前公开的文献中似乎没有应用于碱激发的混凝土中。类似的方法包括EN 772-11[114]试验，该试验规定了测试样品为砂浆，并测量水从外管流入材料孔结构中的流量。

9.2.5 干/湿循环

大约200年来交替的干/湿循环条件可能会损坏建筑材料[115]，但是似乎没有任何具体的标准化测试方法来描述和分析在特定干/湿循环下的混凝土性能。当前氯化物或硫酸盐渗透或冻融过程等的方法涉及干/湿循环，但不太适用于在淡水中的润湿和干燥。这可能是因为硅酸盐水泥基材料的干/湿循环的有害影响主要与物质传输的加速有关，而不是干/湿循环过程本身的直接影响。在与硅酸盐水泥混凝土有关的一个主要的ASTM出版物中指出，"作者不知道交替润湿和干燥本身导致变质的情况"[116]，尽管有人认为这种变质也可能是由于Ca和碱的浸出（包括废气）[117]导致的。尽管如此，在引入一类新的材料如AAMs时，要考虑到广泛使用的材料相对无害的机理，可能对具有不同化学性质但在类似的应用中起作用的材料具有不利影响。

针对这个概念，学者们对碱激发胶凝材料、砂浆和混凝土的干/湿循环进行了系统研究。Puertas等人[118]在70℃干燥条件下18h和21℃浸泡6h，交替循环50次的条件下，研究了硅酸钠激发的BFS和NaOH激发的粉煤灰和1∶1的粉煤灰：BFS砂浆的干/湿循环特性。同时冲击试验方法表明，湿法循环过程对材料的性能没有显著的不利影响。第一次破裂的影响数量普遍增加，但样品完全断裂所需的数量减少。这可能说明，循环处理提供额外的固化持续时间来提高材料的表面硬度，但是由于温度和/或湿度条件的变化可能引起一些微裂纹，在第10章将详细探讨微裂纹问题。Slavik等人[119]和张等人[98]还指出，偏高岭土-煤粉炉粉煤灰和偏高岭土-BFS胶凝材料体系的样品在最初几个月龄期间湿法循环的强度没有显著差异。

Häkkinen[120]还在交替的低/高相对湿度（40%/95%）条件下研究了碱激发的BFS混凝土，每个循环在每个环境中进行1周，进行了10个试验循环，与连续维持在95%相对湿度下的样品相比，发现并没有展现出任何抗压或抗折强度的显著影响。

9.3 界面过渡区

骨料与胶凝材料之间的界面过渡区（ITZ）一直是混凝土技术研究的重点。在胶凝材料与骨料接触的区域中，与"基体"胶凝材料区域相比，往往具有微观结构差异，这意味着这些界面区域对于物质传输、拉伸和流动性会不成比例地影响地质聚合物混凝土的性能（通常是不利的）。人们认为结构的关键差异是由于骨料的存在导致更高的异质性，并且混凝土搅拌期间浆料和骨料的相对运动也会引起ITZ的微观结构发生大的变化。已知该区域的宽度为15～20μm[3,121]，并且通常具有水泥颗粒的限制，这意味着水泥浆中的水灰比更高。因此，ITZ具有不同的化学性和多孔性。当与本体相比时，ITZ通常具有较高浓度的硅酸盐晶体以及较低浓度的水化硅酸钙（C-S-H）[121-123]。由于硅酸盐晶体大于C-S-H晶粒，并且不能紧密堆积，ITZ的孔隙率显著高于水泥浆料的孔隙率[124,125]，因此ITZ通常是混凝土中最薄弱的区域。ITZ较高的孔隙率也可能导致孔隙渗透，从而使得有害物质如氯化物更容易渗透到混凝土中[126,127]。

没有一种标准化的方法可用于量化界面过渡区的存在和结构。因为我们通常在学术研究中对其详细分析，而不是站在混凝土生产者的角度进行实际参数测试。扫描电镜中的背散射电镜与组成分析平行应用，被认为是区分孔隙率、反应颗粒和未反应颗粒的最佳二维技术[37-42]。这种技术的主要缺点是，它提供了关于孔隙率的信息但没有涉及该区域的连通性或渗透性，而这些在研究耐久性问题时恰恰是必需的。

对骨料和碱激发胶凝材料之间的相互作用几乎没有人进行详细的关注，人们对于该区域的性质尚未达成共识。这也与第8章中讨论的碱聚集反应的问题有关，而且区分该区域发生的"理想"和"有害"过程的进一步详细工作是相当必要的。一些作者，包括Brough和Atkinson[128]已经报道，在OPC混凝土中存在界面过渡区，但与OPC体系的情况相比，它与净浆试块的区别较小。其他作者[57,129-136]没有在碱激发的粉煤灰、偏高岭土或BFS砂浆中观察到任何明显的界面过渡区。一般来说，一致认为该区域并不比块状胶凝材料弱得多，如OPC体系的情况。Krivenko等人[137]特别指出，将偏高岭土加入到基于BFS的胶凝材料中有助于进一步致密化这些材料的ITZ，其中胶凝材料和骨料包括铝之间的化学相互作用，被认为增加了这个区域中的材料显微硬度。

因此，AAM胶凝材料与传统硅酸盐水泥相比具有显著的优点。硅酸盐水泥的化学成分容易导致在骨料颗粒周围形成含有大的机械弱化晶体的多孔区域[3,125]，并且这些区域的渗透是混凝土中力学破坏和物质传输的关键途径。碱激发混凝土的碱激发剂不仅与铝硅酸盐前驱体相互作用，而且与骨料表面相互作用，并且这种相互作用容易导致水化产物在接触区宽度上的均匀度增加。这对最终性能可能是有益的，但需要进一步的研究来验证这些观点。

9.4 氯化物

9.4.1 耐氯离子渗透的重要性

氯离子即使在高碱性条件下也具有破坏钢钝化膜的能力,尽管已知 Cl^-/OH^- 比率存在阈值效应,在确定腐蚀发生中具有一定的作用[138,139]。通过胶凝材料基质和/或界面过渡区域将氯化物渗透到增韧钢筋的表面,因此会导致钢筋腐蚀和钢筋混凝土结构的损坏。氯化物引起的腐蚀是沿海海岸以及使用除冰盐的寒冷地区的混凝土劣化的常见原因,对由氯化物引起的腐蚀受损混凝土进行结构修复是非常昂贵的。因此,使用具有低氯化物渗透性的优质混凝土材料在耐久性钢筋混凝土结构的施工中是非常重要的。因此,必须确定和明确碱激发胶凝材料体系的化学和微观结构的渗透性,氯化物渗透速率和钢腐蚀之间的关系。本节将重点介绍氯化物渗透的测量,以下部分(第9.5节)将介绍碱激发材料孔隙溶液环境中钢腐蚀的分析。

9.4.2 氯化物渗透测试

值得注意的是,本报告内不可能全面地描述分析混凝土的每种氯化物渗透试验。氯离子测试方法众多,近年来,Stanish 等[140]和 Tang 等[141]提供了许多可供选择的测试方案的详细评估。RILEM TC 在这一领域也一直很活跃,其中包括 178-TMC 和 235-CTC,读者也参考了他们的重要著作[142,143],以便广泛地分析氯化物检测。表9-1 显示了一些更常见的氯化物测试方法,这些方法大致根据应用于通过样品的氯化物的迁移加速度进行分类。

表9-1 常用的不同氯化物渗透测试方法,列举了从低浓度到高浓度加速氯化物迁移试验

标准试验方法	时间	测量	应用评价
EN 13396-04[144]	6个月	浸入3%的 NaCl,测量三个深度的氯化物含量	6个重复样品;没有计算传输参数; 2002 初步版本规定40℃; 2004 批准版本 23℃
ASTM C1543-10a[145] (基于 AASHTO T259)	90d	用3%NaCl 浸渍,通过钻孔在特定深度取样氯化物含量	提取样品的空间分辨率; 非常粗糙(12.5mm 间距);使用部分干燥的样品可加速初始氯化物侵蚀[146]
Nordtest NT Build 443[147] (&相关标准,如 ASTM C 1556-11[148])	35~40d	浸入 16.5 wt%的 NaCl(或 ASTM 测试中15%),通过成型研磨测量氯化物渗透深度分布,由 Fick 第二定律计算的非稳态扩散系数	一式三份; 非常高的 NaCl 浓度; 测试后样品的分析可能是耗时的
加速氯化物迁移测试[149,150]	5~15d	RCPT 标准中的变量,具有更大的溶液体积和更长的持续时间以降低热效应 主要是通过氯化物通过量决定测试结果,但是在稳态电流,电荷通过和氯化物通过 OPC 材料之间存在相关性[151]	使用高电压(60V)电加速试验的持续时间相对更长

续表

标准试验方法	时间	测量	应用评价
Nordtest NT Build 355[152]	至少 7d	在 12V 直流电下的稳态迁移,测量顺流电池中的氯化物浓度	低渗透混凝土的应用极限范围
Nordtest NT Build 492[153] (基于 CTH 测试[154])	通常 24h (6~96h)	在 10~60V 之间的直流电流下,非稳态迁移的氯化物迁移系数;从色谱分析的渗透深度	近年来作为一种快速测试方法明显流行了;在 Nordtest 的各种方法中具有最优的精度[155]
快速氯化物渗透测试 (ASTM C1202[156]—基于 AASHTO T277[157])	6h	在测试期间施加 60V 直流电压下通过的电流;低电荷通过与低氯化物渗透性相关	高可变性和低重现性[158]; 由氢氧化物和阳离子而不是氯化物的运动控制[159]; 测试期间有放热问题;建议使用更短的测试持续时间[146]; 与 OPC[160] 比,特别是复合水泥[161,162] 的 90d 结果的相关性较差

9.4.3 AAM 的氯化物渗透试验

快速氯离子渗透性测试方法（ASTM C1202 或 AASHTO 227）广泛用于预测混凝土的氯化物渗透性,尽管其基本上是对取决于孔结构和孔溶液化学性质的电导率的测量,而不是直接测量氯离子运动。而且测试条件相当严格,在样品上施加 60V 的电压将引起加热效应和样品结构的其他变化。由于这个和其他一些原因,测试具有很大争议（现已修订标准应用）。当比较具有不同胶凝材料化学性质的混凝土混合物时,电导率可能受孔溶液组成变化的影响。基于电化学的分析和计算表明,用辅助胶凝材料代替硅酸盐水泥会由于孔隙溶液组成的变化而使通过的电荷减少高达 10 倍,而氯离子运输的实际变化是远远小于此[162,163]。尽管如此,这种测试已经被最广泛地使用（并且被世界范围内的规范和监管机构广泛要求并被其接受）来预测在硅酸盐水泥材料和 AAM 中的氯化物渗透率。以下将论讨科研人员通过使用这个测试而得到的结果。

Douglas 等人[164]使用快速氯化物渗透性测试来测量 6 批硅酸钠激发的 BFS 混凝土,在 28d 时获得的值在 1300~2600C 之间,在 91d 时测试的混凝土的值为 650~1850C。低于 2000C 的值由标准分类为"低",低于 1000C 为"非常低"。此研究提供了关于这些材料的耐久性的正面数据。然而,电通量随着激发剂/BFS 比的增加而增加,这表明测量可能与混凝土孔溶液中的碱浓度有关,较高的激发剂含量可以使硅酸盐激发的 BFS 材料具有更致密的结构是众所周知的,而这些测试结果却是相反的。

高炉矿渣掺入硅酸盐水泥可以减少硬化水泥浆、砂浆和混凝土中的氯化物扩散[165]。同时,Roy 等人[166]发现向波特兰-BFS 混合水泥中加入碱能降低氯化物的扩散速率,这可以通过氯化物自然扩散的硬化浆料盘加以测量。

通过 Na_2SiO_3、Na_2CO_3 或 NaOH 激发的 BFS 砂浆和 ASTM III 型硅酸盐水泥砂浆[94]之间的比较发现，碱硅酸盐激发材料具有比其他样品更高的抗压强度、更低的孔隙率和致密的孔结构。但是由 Na_2CO_3 和 NaOH 激发的样品，在每个测试龄期的 3～90d 内的快速氯化物渗透性测试中显示出较低的电通量。这与其透水性测试结果一致。Wimpenny 等人[85]的研究表明，与相同掺量的合成纤维作为增强材料的硅酸盐水泥相比，AAM 具有类似的总孔隙率，但其浸水试验中测量的氯化物渗透性要低得多。

与碱激发材料相关传输性能的所有测试一样，测试时样品养护程度的问题对于测试结果的正确解释至关重要。对于硅酸盐水泥体系，表 9-1 中列出的大多数测试通常在 28d 或 56d 龄期的样品上进行，但是也经常从现场结构中取芯进行测试。对于碱激发胶凝材料，其通过超过 28d 龄期的显微结构和物理化学演化而获得成熟的碱激发胶凝材料，在 30d 或更长的测试持续时间期间，样品的发展（并且考虑到几周的预调节周期在"固化"结束和测试周期开始之间应用）将有可能导致从短时间测试和长时间测试获得的结果之间存在差异。在硅酸盐水泥材料的不同测试方案之间已经进行了许多比较性的实验室研究，但还没有公开这种碱激发胶凝材料的研究。使用 ASTM C1202 RCPT 方法与稳态迁移试验（由 Mejía 等人[167]使用并且类似于 Nordtest NT Build 355 方案的方法）并行的研究碱硅酸盐激发的 BFS 混凝土[86]孔隙溶液化学支配 RCPT 的结果，在静电加速条件下氯离子通过样品的直接测量似乎对胶凝材料的孔提供更大的敏感性。与此一致，Bernal 等[60]也表明，AAM 混凝土中浆体含量对 RCPT 试验过程中电荷的影响最小，但其 300～500kg/m³ 胶凝材料的混凝土的渗透性能显著不同。

Shi[94]和 Bernal 等人的研究[86]表明，碱激发的 BFS 中 RCPT 结果随固化时间的函数并没有明显变化。而 Husbands 等人[168]用商用碱激发的 AAM 产品在 7d 或 28d 通过的电荷显示出类似的趋势。但是 1 年样品的电通量出现了显著的倒缩（通过电荷的范围被归类为"可忽略的"）。Pyrament 样品沉淀试验的结果表明其对氯化物渗透性具有极好的耐久性。Zia 等人[169]发现 Pyrament 的 RCPT 结果受到骨料选择的显著影响，其中泥灰岩骨料（主要是碳酸盐）比硅质骨料具有更高的电荷，表明孔溶液化学（和其与聚合物的相互作用）主导测试结果。

Adam[58]使用 RCPT 方法和该方案的加速试验研究了碱金属硅酸盐激发的粉煤灰和基于 BFS 的混凝土。加热效应困扰了碱激发样品的分析。加速试验中，碱硅酸盐激发的 BFS 混合物显示出比相应的粉煤灰基 AAM 低得多的电荷，激发剂模数增加会降低电通量，这与样品的 56d 或 90d 测试结果不同。然而，尽管 RCPT 方案预测在碱激发粉煤灰样品中的氯离子扩散显著高于对照组硅酸盐水泥的水泥/BFS 共混物或碱激发 BFS。与其他任何样品相比，沉积试验显示氯化物进入碱激发粉煤灰样品的进入量最低。粉煤灰中的碱富集孔溶液的化学反应和收缩造成的表面微裂纹导致了其氯离子通过量较高。Al-Otaibi[170]还发现，电通量随着激发剂模数的增加而减少，养护 182d 后更显著。

9.4.4 AAM 中氯离子渗透机理

弗瑞德盐（Friedel's salt）的形成降低了氯化物通过硅酸盐水泥胶凝材料的孔隙网络的速率。弗瑞德盐是一种由氯化物与硫代铝酸钙的相互作用形成的氯铝酸钙相。在碱激发的胶凝材料体系中并未发现弗瑞德盐（也不像在低钙体系中所预期）以及其他结晶氯化物的存在，这表明使用氯化物的特定化学键合途径难以进入 AAM。与硅酸盐水泥相比，AAM 中存在的任何氯化物吸附差异仍然需要得到解释。因为这些效应与硬化的胶凝材料中的氯化物迁移试验中孔几何形状差异的影响相矛盾。同时目前尚未有有效方法对两组因素分离进行试验。另一方面，胶凝材料化学组分没有转化成弗瑞德盐被认为是碱激发胶凝材料比硅酸盐水泥在高浓度 $CaCl_2$ 溶液[171]环境下具有更高稳定性的原因，因此可以证明这在一些情况下是有利的。

本节的讨论表明了在确定碱激发胶凝材料的耐久性时，进行尽可能多的性能测试的重要性，而不是基于特定的差的结果选择丢弃材料或配方测试。在许多情况下，通过 RCPT 试验方案预测的碱激发胶凝材料的氯化物渗透性出现极好或极差的极端情况，而其他渗透性参数（如第 9.2 节中所述）没有显示相同的极端变异性预测性能。非常希望在 AAM 胶凝材料和混凝土的未来研究中进行替代氯化物渗透试验，以获得在材料使用时氯化物渗透问题的研究；必须得出结论，这种现象尚未得到很好的解释。

9.5 钢筋锈蚀直接分析

埋入混凝土中的钢筋由于混凝土的高碱性环境而形成一层保持在其表面的薄氧化层，从而保护其免受腐蚀。对于硅酸盐水泥，环境 pH 值通常超过 12.5，并且对于未受损的碱激发胶凝材料[173]可能高于 12.5。高 pH 值导致在混凝土中的钢筋表面形成一层非常致密的不可渗透的薄膜钝化层，阻碍钢筋的进一步腐蚀。然而随着时间的推移，浆体的 pH 值降低，同时氯化物迁移到钢筋表面，此时钝化层极不稳定，易造成钢筋的严重腐蚀。如第 9.4 节所述，一旦钝化层被破坏，钢筋会非常快速地被腐蚀掉。诱导腐蚀发生的氯化物浓度通常以 Cl^-/OH^- 比作为阈值[138,139]，因此高碱度可以降低氯化物的破坏作用。

ASTM C876[174]描述了一种通过比较其电化学电位与 $Cu/CuSO_4$ 参比电极的电化学电位来测试钢筋腐蚀的方法。电位差高于 $-200mV$ 时没有被腐蚀的概率为 90%，而电位差低于 $-350mV$ 时对应于 90% 的腐蚀概率。这些标准被认为提供了在含氯化物的体系中腐蚀的可能性的良好指标[175]，但是腐蚀的发生也受到 O_2 浓度（其取决于渗透性）、碳化［其可以产生大的腐蚀变化对于氧化还原电位（Eh）的小偏移[175]］和其他参数的影响。在硅酸盐水泥体系中使用腐蚀抑制剂如亚硝酸钙可以通过控制 Eh 来稳定钝化层[176]，同时强氧化剂或还原剂在合适的使用方法下可能是有效的。诸如 ASTM G109 [177] 或 EN 480-14 [178] 的方法与诸如 ASTM C876 等文献的建议并行地用于测试这些添加剂的功效。然而，当前研究中似乎并未对碱激发胶凝材料进行详细的描述。由于测试所需的时间较长（例如，ASTM G109 将氯化物沉积与电化学测试结合在一起），人们提出了应采用更加快速的测试方法的建议，例如"快速宏单元测试"[179]，并且这可以提供在分析 AAM 方面向前推进的良机。

Gu 和 Beaudoin[175]也总结了影响混凝土腐蚀电位测量的因素，见表 9-2。大多数情况下，这些因素与 AAM 相关但仍需要通过科学分析来详细验证。Poursaee 和 Hansson[180]研究评估混凝土中钢筋腐蚀的各种加速技术的适用性（或缺乏），将该主题一般性地描述为"雷区"，并指出"任何旨在加速腐蚀的过程都应该被怀疑"。在此基础上，实验室测试不可能提供 AAM 混凝土在钢筋腐蚀方面的在役性能的完整表示，但是仔细设计的测试也可能提供一些有用的用于预测不同胶凝材料类型的耐久性信息。

表 9-2　影响硅酸盐水泥混凝土腐蚀电位测量的因素

参数	半电池电位迁移	钢筋锈蚀速率变化
O_2 浓度	O_2 浓度越高，正迁移越多	因环境而定
碳化	负迁移增大	增强
氯化物	负迁移增大更多	增强
氧化物（阳极阻锈剂）	正迁移	降低
还原剂（阴极阻锈剂）	负迁移	降低
环氧涂覆的钢筋	很难构成电池	涂层不破坏的话，降低
镀锌钢筋	负迁移	有效的，但在高碱水泥中略有减弱
涂覆、修补和杂散电流	测试无法确定	影响不同

与常规混凝土相当的碱性下碱激发混凝土将能够在 Pourbaix 图的惰性区域中保护钢筋，从而在钢筋混凝土结构中提供低腐蚀水平。Wheat[182]将 Pyrament 碱激发混凝土浸渍在质量分数为 3.5% 的 NaCl 溶液中，同时钢筋保持在被动状态下以及相同溶液的干/湿循环中，3 年后仍具有优异的性能。这归因于胶凝材料具有高的不可渗透性。钝化能力和钝化状态的持续时间取决于胶凝材料的性质和剂量、激发剂的类型和溶液的条件[183-185]。碱激发的 BFS 混凝土和高体积 BFS-OPC 混合物的孔隙溶液 Eh 因高炉炉渣中硫化物的存在而减少几百毫伏[186]，导致这些材料中的钢筋腐蚀问题。事实上，在许多需要高耐久性的应用中必须使用特定的高容量 BFS-OPC 混合物如 EN 197 CEM Ⅲ 或 CEM Ⅴ，这表明标准电化学论据不能捕获这些材料的化学性质对于嵌入钢保护层的孔溶液的全部细节。

9.5.1　碱激发粉煤灰中的钢筋

Bastidas 等人[184]和 Fernández-Jiménez 等人[187]研究硅酸盐或碳酸盐激发的粉煤灰基地质聚合物钢筋混凝土的 80mm×55mm×20mm 的棱柱样品在质量分数为 2%$CaCl_2$ 环境下的，通过测量 6mm 直径的碳钢钢筋的测试电极与外直径为 50mm 的不锈钢盘，其中心具有孔，以容纳用作参考的饱和甘汞电极（SCE）的辅助电极之间的电极差来测量其腐蚀程度。在此过程中用湿海绵增强电解质（混凝土基体）和辅助电极（不锈钢盘）之间的接触以促进电测量，并且用胶带控制测试电极（10cm^2）的表面积（图 9-5）。

新拌浆体凝固后将样品脱模并在环境温度下储存，测试开始时首先在 95% 相对湿度（RH）下 90d，然后在 30% RH 下 180d，最终恢复到 95%RH 760d 直到实验结束[187]。在这些条件下，碱激发粉煤灰砂浆在相对湿度较高的环境下成功钝化钢筋（图 9-6）。所有样品的干燥增加了电位，降低了腐蚀的可能性。氯化物在所有情况下的存在都是有害的。

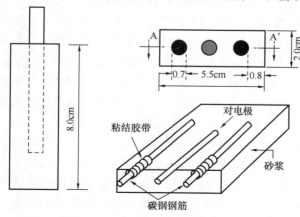

图 9-5　用于 Bastidas 等人[184]和 Fernández-Jiménez 等人[187]的腐蚀试验的棱柱样品

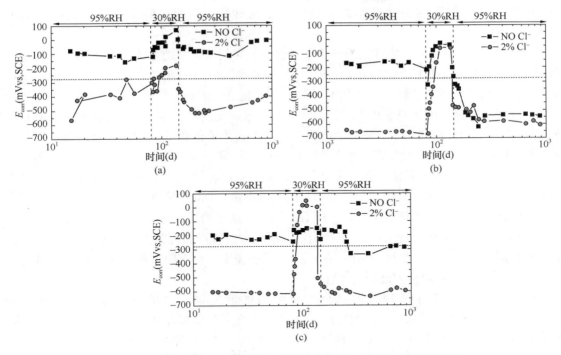

图 9-6 腐蚀电位 E_{corr}，记录为对饱和甘汞电极（SCE；电位相对于 $Cu/CuSO_4$ 电极加入 75mV）的半电池电位

(a) 硅酸盐水泥 CEM I 砂浆；(b) 飞灰＋8M NaOH 砂浆；(c) 硅灰石＋硅酸钠砂浆。灰线和点是具有 2%Cl⁻ 以 $CaCl_2$ 添加的样品，黑线和点不含添加的氯化物（来自 Fernández-Jiménez 等人的数据[187]）。虚线水平线显示 90%的腐蚀可能性的 ASTM C876 规范；在此线下的值表明根据标准测试方法的高腐蚀可能性

在干燥期后，NaOH 激发的粉煤灰样品中的钝化被视为失败（根据测试方法的分类），可能由于在中等相对湿度下富碱且多孔的基质的碳化，而硅酸盐激发的灰分和硅酸盐水泥没有显示出这种效果。腐蚀电流（i_{corr}）的趋势和样品中腐蚀产物的观察结果与腐蚀电位数据相一致[187]。

几十年前，不锈钢（SS）增强元件首次应用于硅酸盐水泥混凝土，已被证明即使在非常恶劣的环境中仍能抵抗腐蚀[188]。然而，与碳钢相比，SS 的高成本阻碍了推广使用。通过其他元素取代镍元素产生新的低镍不锈钢，将可能成为常规碳钢的替代品[189]。当其处于浓度高达 2%氯化物以及 95%RH 环境下的碱激发粉煤灰砂浆中时，180d 的腐蚀行为类似于有无（AISI 304）不锈钢的 AAM 砂浆腐蚀行为，相对于饱和甘汞电极的电极的差值约为 －200～－100mV，表明嵌入在灰泥砂浆中的低镍 SS 的良好耐久性。

9.5.2 碱激发 BFS 和其他矿渣胶凝材料中的钢筋

Kukko 和 Mannonen[190]的研究表明，碱激发 BFS "F 混凝土"材料浸入在模拟海水中 1 年时间内对嵌入钢筋提供良好的保护。在此试验中仅仅从视觉角度判断在样品上尚未形成腐蚀产物，并未进行电化学测量。Deja 等[191] 和 Malolepszy 等人[192]通过测量极化曲线、腐蚀电流和增强质量损失，研究了在水中和 5%$MgSO_4$ 溶液中碱激发矿渣砂浆中的钢筋腐蚀。结果表明硫酸镁浸泡仅对钢筋有轻微的影响，淡水中几乎没有影响。腐蚀电流数据表明，碱

激发矿渣砂浆中钢筋的腐蚀速率高于硅酸盐水泥砂浆,但两种砂浆中的电流随时间而降低。

Holloway 和 Sykes[193]将详细的电化学特性分析应用于硅酸钠激发的 BFS 砂浆中的低碳钢筋分析。其中苹果酸作为缓凝剂,NaCl 作为加速腐蚀剂。试验结果表明,高浓度的 NaCl 减少了初始腐蚀电流。这种趋势从基本化学角度不能清楚解释。然而,Bernal[194]关于胶凝材料碳化对腐蚀速率的影响获得了类似的结果,与未碳化的材料相比,部分碳化的胶凝材料在浸入水中 12 个月期间显示出较高的耐腐蚀性。上述均未能用标准理论直接解释,因此碱激发的 BFS 胶凝材料中钢筋腐蚀所涉及的机理仍需要进一步详细的科学分析研究。Holloway 和 Sykes[193]基于孔隙溶液化学和电化学的观点提出,炉渣中的硫化物造成了此类试样的电化学复杂性,影响了腐蚀动力学和测量的腐蚀电流。同时在他们的研究中发现,与混合水中高的氯化物剂量的预期效果相比,所有腐蚀速率测量的结果都是"低"。

Bernal[194]和 Aperador 等人[195]也指出,其研究中所有样品的腐蚀电位落在 ASTM C876 表明存在高腐蚀可能性的地区。Montoya 等人[196]还提出,在有限元模拟的基础上,碱激发的 BFS 砂浆似乎可以增强阴极的耐久性。

碱性副产物也可用作 BFS 和其他冶金炉渣的激发剂,这是苏联特别感兴趣的领域[54,197]。然而,含有氯化物和硫酸盐的一些副产物会引起混凝土中钢筋的腐蚀。Krivenko 和 Pushkaryeva[198]使用两种混合的激发剂(一种由 90% Na_2CO_3 + 10% NaOH 和另一种 45% Na_2CO_3 + 40% Na_2SO_4 组成)在 15% NaCl 环境下研究了碱激发混凝土中钢筋的腐蚀。表 9-3 的结果表明,钢筋的腐蚀取决于炉渣的性质、碱激发剂的性质和用量以及混凝土的碳化。当使用碳酸盐-氢氧化物激发剂时,质量损失随激发剂浓度的增加而增加。碳酸盐-硫酸盐-氯化物激发剂表现了类似的腐蚀性能。磷渣和最高激发剂浓度制备的样品被快速碳化并且产生高浓度的氯化物,导致了比在任何其他样品中观察到的更严重的腐蚀现象。

表 9-3 在干/湿循环下由不同矿渣和碱激发剂制成的碱激发矿渣水泥混凝土中钢筋的质量损失[198]

矿渣类型	激发溶液的密度 (kg/m³)	增韧钢筋的质量损失(g/m²)(表头最下一行是按照日计时)							
		90%Na_2CO_3+10%NaOH+15% NaCl				45%Na_2CO_3+40%Na_2SO_4+15% NaCl			
		6d	9d	12d	18d	6d	9d	12d	18d
碱性	1100	0.52	0.53	0.52	0.52	0	0	0	0
	150	0.70	0.73	0.71	0.72	0.89	0.91	0.90	0.91
	1200	0.98	0.96	0.98	0.97	1.07	1.09	1.07	1.06
酸性	1100	0	0	0	0	0	0	0	0
	1150	0.41	0.43	0.40	0.42	0.63	0.61	0.62	0.62
	1200	0.71	0.70	0.72	0.71	0.58	0.60	0.59	0.59
中性(电-热-磷)	1100	0	0	0	0	0	0	0.36	0.36
	1500	0.71	0.76	0.73	0.72	46.91	74.73	78.54	84.10
	1200	1.12	1.14	1.11	1.13	59.12	85.73	86.04	87.40

添加剂会改变硬化混凝土的结构致密化并改变孔溶液的化学性质。表 9-4 显示了添加剂对磷渣和含有 45% Na_2CO_3 + 40% Na_2SO_4 + 15% NaCl 的激发剂制备的碱激发混凝土中干/

湿循环过程中的质量损失的速率和程度的影响。添加5%的硅酸盐水泥熟料、10%的铌矿渣（其富含Al）和5% CaF_2 有助于降低腐蚀速率（降至高达100倍）。然而，由于铌铁矿渣的低可用性（以及通常还有放射性特征[199]），以及氟化物可以大大加速钢筋的腐蚀（如果存在于低温下），同时不适用任何浓度仅在较高浓度（~$100×10^{-6}$或更高）下提供腐蚀抑制[200]。

表9-4 在干/湿循环下，激发剂由45%Na_2CO_3+40%Na_2SO_4+15%NaCl
组成的碱激发磷矿渣混凝土中增韧材料的质量损失的影响[198]

外加剂	增韧钢筋的质量损失（g/m^2）		
	6个月	12个月	18个月
空白混凝土（不添加外加剂）	46.9	78.5	84.1
NaOH（5wt%）	18.0	31.1	42.2
OPC熟料（5wt%）	0.41	0.38	0.39
铌铁矿渣（10wt%）+CaF_2（5wt%）	8.81	0.52	0.51

9.5.3 关于AAM试验方法的评价

目前，对AAM胶凝材料中钢筋的腐蚀化学性的理解可能不足以使得能够开发适用于AAM的化学试验方法。对于基于BFS或其他含有硫化物的冶金炉渣的AAM而言，其在胶凝材料体系内产生的还原环境导致电化学中的复杂性。从大量BFS与OPC的混合物的分析中学习（由于该研究课题的成熟度更高，已经达到了比AAM的分析更高级的阶段），对基本理解硫化物以复杂方式影响钢筋腐蚀速率方面是有必要的。高浓度碱的条件下，地质聚合物的碳化、氯化物和碱之间的相互作用以及在钢筋-胶凝材料界面处的传输性能和钢腐蚀化学之间的关系的作用将成为未来研究的重点。因此，AAM的任何测试分析方法至关重要。试验条件和完整的试验报告对读者理解和利用试验结果是至关重要的。这在耐久性测试的实施中是普遍重要的。但是在诸如腐蚀测试的领域中是特别关键的，其中存在许多不完全理解的参数，这些参数可能潜在地影响从每个测试中获得的结果。

9.6 碳化

9.6.1 引言

已知二氧化碳（CO_2）通过碳化的劣化过程而长期地显著影响水泥基材料的耐久性[201-203]。这种现象由气体扩散和化学反应机理控制[204]，主要由基质的结构特征和材料的渗透性决定。碳化通常导致材料碱度的降低，主要反应产物（在OPC基材料中的硅酸盐、C-S-H凝胶和钙矾石）的脱钙以及力学性能的降低和渗透性的增加，造成了氯化物或硫酸盐的透过，增加钢筋的腐蚀程度[201,204-206]。这就是碳化被认为是水泥基结构破坏的主要原因之一。

普通硅酸盐水泥基砂浆和混凝土的碳化已被广泛研究，碳化被认为是相对容易理解的现象。来自大气的CO_2通过材料的孔扩散，溶解在孔溶液中形成HCO_3^-，与存在的富含钙的水化产物反应[202,204,207]造成脱钙。然而，对于 AAM 的混凝土碳化机理和影响因素的研究相对较少。

9.6.2　普通硅酸盐水泥基材料的碳化

众所周知，硅酸盐水泥体系中碳化的程度取决于构成与 CO_2 反应的胶凝材料的反应产物以及调节 CO_2 扩散性的因素，例如孔隙网络和暴露环境（主要是相对湿度和温度）。图 9-7 为调控碳化进程的主要参数，也是控制 CO_2 扩散率和 CO_2 与胶凝材料的反应性的主要参数[204,208]，例如：胶凝材料类型的作用与水化过程中形成的相的量和类型相关，其中包含碱金属或碱土金属阳离子的那些最容易与环境中存在的 CO_2 反应[209]。同时还发现有机物质和阴离子易与硬化浆料中的水化产物反应，增加 CO_2 扩散率造成碳化程度的恶化[204]。另一方面，加速碳化过程中较高的 CO_2 浓度（其最初促进固体密度的增加）导致浆料的超细孔隙率，增加材料中的毛细管吸附性[210]。在更高的 CO_2 浓度下，$CaCO_3$ 相形成也可能有变化，它会形成亚稳相，改变孔径分布。

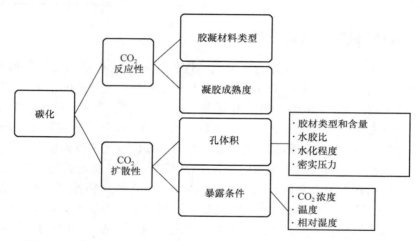

图 9-7　根据 Fernández-Bertos 等人的分类[204]调节水泥材料碳化的因素

据报道[206,211]，碳化在 50%～70%相对湿度（RH）下更迅速。高湿度增加了填充水的孔的分数，阻碍了 CO_2 的扩散，低湿度意味着没有足够的水来促进二氧化碳的溶剂化和水化。介于两者之间的湿度，对于 CO_2 的反应动力学和扩散都是有利的，因而成为碳化的最佳条件[204]。

近年来，混凝土中矿物掺合料（SCM）急剧增加。因此目前市场上的大多数水泥均为混合胶凝材料。然而，对于此类材料的碳化有较少的关注。与不含任何矿物混合物的水泥相比，混合胶凝材料的酚酞指示剂的碳化测试方法[212-215]表明其具有更高的碳化易感性。火山灰反应降低了胶凝材料中 $Ca(OH)_2$ 的含量。

因此，矿物掺合料或其他替代水泥用于硅酸盐水泥生产混凝土时，是否因其不存在 $Ca(OH)_2$ 而产生腐蚀过程。事实上，混合水泥的使用对混凝土耐久性具有显著的积极影响，可显著降低包括炉渣[216,217]或其他火山灰的混凝土钢筋的腐蚀速率[218,219]。因此酚酞指示剂测试方法作为碳化进程的唯一量度对于评价"现代"水泥的碳化性能有点过于简单，其化学和微观结构不同于通常使用的水泥在 20 年前的碳化性能。第 3、第 4 和第 5 章表明，AAM 胶凝材料中 Ca(OH)$_2$ 不能被识别为反应产物以及化学成分差异性对于理解 AAM 中的碳化是至关重要的。

9.6.3 碳化测试方法

大气中低的 CO_2 相对浓度（0.03%～0.04%）和硬化胶凝材料（混凝土和砂浆）的低气体渗透性导致了现实生活中水泥或混凝土的碳化过程是缓慢的。这就是为什么用于评估水泥质材料中碳酸盐化的试验方法均是基于在受控条件下提高 CO_2 浓度诱导样品加速碳化。

一种方法是在 100% CO_2 的气体环境下暴露试样，同时如标准程序 ASTM E 104-02[220]中所述通过使用饱和盐溶液控制相对湿度[208,210,214,221,222]。这种方法相对流行，但是使用这样高的 CO_2 浓度的科学基础是不确定的。近年来，由于可以完全控制暴露条件（CO_2 浓度、相对湿度和温度），使用气候室诱导加速碳化作用增加[223-225]，这促进了水泥材料碳化评估的三个国际标准的制定：

（1）EN 13295：2004：用于混凝土结构保护和修复的产品和系统测试方法——耐碳化的测定[226]。

该加速测试方法测量建筑产品或体系在加速测试条件下抗碳化的阻力。在这种情况下，将样品暴露于 1% CO_2，(21±2)℃ 的温度和 (60%±10%) 的相对湿度 (RH) 的气体环境中。该标准中的一个基本假设是在这些碳化条件下，形成暴露于大气碳化时在硅酸盐水泥中鉴定的相同反应产物。这在应用加速碳化试验中是必要的，但是仅在硅酸盐水泥的情况下才被验证。

样品根据欧洲标准 EN 196-1 制备，用塑料膜覆盖 24h，然后脱模并再次用塑料膜密封 48h。此后，将样品老化并在用于碳化测试的相同温度和湿度条件下预处理 25d。样品需要预处理以确保均匀的水分含量。碳化深度在预处理期后和在该环境中储存 56d 后测量。还包括与几何效应和大聚合物的结合相关的考虑。

（2）BS ISO/CD 草案 1920-12：混凝土潜在碳化阻力的确定——加速碳酸盐法[227]

该标准目前正在开发中，尚未正式发布。2012 年草案中规定以 4% CO_2、20℃ 和 55% RH 作为基准方法，在热气候地区可选择 27℃ 和 65% RH。样品规定为 100mm 立方体或 100mm×100mm×400mm 的直角棱柱，在与所选测试温度匹配的温度下固化 28d，在 18～29℃ 和 50%～70% RH 下干燥（调节）14d（或者如果优选可以使用和报告不同的条件），然后暴露于升高的 CO_2 条件下 56～70d。通过施用酚酞（1%，在 70/30 乙醇/水混合溶剂中）在分裂（未锯割）样品上显示碳化深度。

（3）EN 14630：2006：用于混凝土结构保护和修复的产品和体系试验方法——通过酚酞法测定硬化混凝土中碳化深度[228]

该方法用于暴露于 CO_2 后的样品分析，并且解释了将 1g 酚酞指示剂溶解在 70mL 乙醇中，用蒸馏水或去离子水稀释至 100mL 的过程。还包括碳化深度测量和报告描述的注意事项。

用于评估混凝土的加速碳化的其他方法如下：

（4）RILEM CPC-18：硬化混凝土碳化深度的测量[229]

这种测试方法包括借助于由 1% 酚酞在 70% 乙醇中的溶液组成的指示剂来确定硬化混凝土表面上的碳化层深度。对于加速碳化，该方法不建议任何特定的暴露条件，但是在测试时需要精确地指示存储的气候条件。对于存储在室内的样品的自然碳化研究，需定义温度和相对湿度（20℃和65%RH）。当样品存放在室外时，需要防雨。空气必须能够在任何时候都不受阻碍地到达试验表面，试样周围的自由空间至少为20mm。

（5）NORDTEST方法：NT Build 357。混凝土、修补材料和保护涂层：耐碳化[230]

该方法规定了加速试验程序，监测暴露于3%CO_2和55%～65%相对湿度的气氛中的样品的碳化速率（使用酚酞指示剂）。严格规定了样品的制备方法：应当用水：胶凝材料为0.60±0.01，坍落度为120±20mm以及最大直径为16mm的骨料制备混凝土。混凝土需要减水剂的情况下可使用三聚氰胺类型。如果工作性不能达到规定的混合参数或三聚氰胺型增塑剂与胶凝材料不相容，则不适用于胶凝材料类型。试样在浇筑后1d脱模，在（20±2）℃的水中养护14d，然后在空气中（50±5）%RH，（20±2）℃环境下养护28d后进行到测试。重要的是，酚酞溶液通过将1g酚酞混合在500mL蒸馏水/离子交换水和500mL乙醇的溶液中制备，所述乙醇是比EN14630：2006中使用的溶液稀得多的溶液。

（6）葡萄牙标准LNEC E391：混凝土-碳化性的测定-稳定性[231]

该方法用于CO_2浓度为（5±0.1）%，RH为55%～65%，温度为（23±3）℃的样品的加速碳化测试。样品需要在（20±2）℃下浸没在水中14d，然后在（50±5）%RH和（20±2）℃的封闭环境中储存直至达到28d龄期。碳化深度的测量根据如前所述的RILEM CPC-18中推荐的方法进行。酚酞指示剂是0.1%醇溶液。

（7）法国测试方法AFPC-AFREM：混凝土耐久性，用于测量相关的可持续性，建议的程序数量推荐的方法，加速碳化试验，测量混凝土碳酸盐的厚度[232]

该加速碳化方法使用在20℃、65%相对湿度和50%CO_2的碳化室。在不同的暴露时间之后使用0.1%的酚酞指示剂的醇溶液测量碳化深度。对于自然碳酸盐化评估，将在常规固化（浸入水中）28d后的样品储存在50%RH和20℃的受控气候条件下直到测试结束。

一般来说，上述测试方法的碳化前端（其被假定为尖锐而不是漫射）的进入作为在特定环境条件下的暴露时间的函数来测量。质量损失、力学性能和渗透性的监测并不是碳化测试方法中的内容。在本报告中，考虑到AAM的特定性质可能影响在这些测试中获得的测量，详细关注了酚酞方法和碳化收缩的问题。

9.6.4 碳化深度——酚酞法

在最初的碱激发建筑材料中的碳化前端通常通过观察作为CO_2暴露结果的pH值指示剂的颜色变化来测量。酚酞（最常用的指示剂）的颜色转变在pH值为10（从紫色/粉红色变为无色，过渡在pH值为8.3时完成）时发生，其大致对应于低于该pH值时，硅酸盐水泥体系中的钢筋钝化膜变得不稳定，并且开始停止保护钢筋免受腐蚀。因此，广泛地假定碳化深度大于膜覆盖深度将导致钢筋表面的腐蚀过程。该方法的主要问题是尽管特定结构处于均匀的CO_2浓度中，但其不同部分的碳化深度不同。由于结构不同部分中的相对湿度、湿/干条件和阳光暴露的差异，导致了混凝土渗透性的变化。混凝土的孔隙溶液的细节也应该考虑，当氯化物和碳酸盐同时存在于钢筋混凝土试件中时具有复杂的效应[233]。然而实际中覆盖物深度在结构中的位置也有所不同。但这种方法的另一个缺点是具有破坏性，不能重复测量，无法识别作为时间函数变化以及不能对应用于关键结构的服役混凝土进行分析。酚酞指示剂方法以一种科学上令人满意的细节水平评价建筑材料的碳化是不可靠[234]。尽管如此，它仍被广泛使用和普遍接受。

此外，如在上述不同测试方法的讨论中所确定的，酚酞指示剂可以在不同浓度和在不同溶液环境中制备，这将完全可能改变在AAM中的碳化深度获得的结果。

9.6.5 碳化收缩率

对于建筑材料中碳化收缩的评估较少有人关注研究，这与水泥浆中由于形成碳化产物而引起的应力有关。在高度碳化条件下，最初在孔隙网络中然后在胶凝材料[235]中的碳化收缩尚无明确的测量方法。但是在评估这种行为的研究[236]中，一般采用类似于 ASTM C596-09[237] 或 ASTM C1090-10[238] 中的方法。测量不同CO_2暴露时间的收缩、碳化深度与样品所示的任何尺寸变化相关联。根据现有文献，AAM 中碳化收缩的程度是完全未知的。然而，Shi[171]确定，在碱激发的 BFS 胶凝材料暴露于 53%RH 的 15%CO_2 的 75d 期间，在开始碳化测试后几天出现裂纹，这是由干燥收缩和碳化的组合效应导致的。

9.6.6 AAM 的碳化

9.6.6.1 暴露条件的影响

关于 AAM 碳化的研究有限。Byfors 等人[239]在加速试验中与普通硅酸盐水泥参考样品进行相似压缩强度比较时，确定了 F-混凝土（增塑碱硅酸盐激发的 BFS）中存在更高的碳化速率。这些结果与 Bakharev 等人[222]的观察结果非常一致，他们还报告了加速碳化条件，硅酸钠和 BFS 制备的 AAM 混凝土中的碳化易感性比基于普通硅酸盐水泥的参考混凝土更高。

相反，Deja[55]认为，碱激发矿渣的砂浆和混凝土的碳化深度与相同环境的硅酸盐水泥无异。随碳化时间的增加，在硬化浆体微孔中因碳化生成大量的碳酸盐沉淀，改善了孔结构的分布，提高了抗压强度。碳酸钠激发加剧了试样的碳化。与碳酸盐激发相比，硅酸钠作为激发剂时此现象更加明显。值得注意的是，本试验所采取的加速条件为湿度 90% 的饱和CO_2碳化室加速碳化。在如此高的相对湿度和饱和CO_2浓度下，AAM 样品碳化反应也与低的相对湿度下不同。Byfors 等人[239]也得到了相同的结果。低的相对湿度环境有利于 AAM 碳化的发生。

AAM 的碳化的判定与其相对湿度条件有关，因为在低湿度条件下可以有利于干燥和后期碳化的收缩，诱发微裂纹，增加碳化的进展。Bernal[194]在碱激发的 BFS/偏高岭土混合混凝土的研究中表明，与 50%RH 或 80%RH 相比，样品在 65%RH 下碳化的进展和随之而来的总孔隙率的增加略高。然而，在更长时间的碳化暴露之后，RH 的影响变得不太相关，并且随着 RH 增加，碳化深度略微增加。

在碱硅酸盐激发的 BFS 和 BFS/偏高岭土混合物的碳酸盐化试验中，样品在测试前并未进行干燥，对未碳化样品的吸水率单独进量测量，较好地表明了在测试期间干燥是否有效，在该试验中，干燥可能会延迟碳化的初始阶段[240]。在高相对湿度下，具有低吸水率的样品（即当孔网络最初是高度饱和的并且重复时）在测试的早期阶段的碳化速率非常低，因为饱和胶凝材料的碳化速率相对较慢，并且随后当干燥前端进入样品并且能够进行碳化时加速碳化过程。

9.6.6.2 样品组成的影响

与常规硅酸盐水泥相比，碱激发 BFS 和 BFS/偏高岭土掺混物混凝土因独特的劣化机理以及对微观结构的影响显示出更高的碳化易感性。特别是对于不存在作为反应产物的氢氧钙石的反应体系[60,107,241,242]，碳化敏感性强烈受到碱激发剂的类型和浓度的影响。

Puertas 等人[241]确定，使用硅酸盐基激发剂制备的碱激发 BFS 样品暴露于完全饱和的 CO_2 环境后呈现高碳化深度，孔隙率显著增加，机械强度降低。另一方面，当用 NaOH 溶液激发 BFS 时，碳化增强了砂浆的密实度，从而提高了抗压强度[243]。这种行为的可能原因是 C-S-H 凝胶的组成和结构的差异。因为其在硅酸钠激发下形成的低的 Ca/Si 比（～0.8）的 C-S-H，而使用 NaOH 生成 Ca/Si 比（～1.2）较高的 C-S-H；然而，其他因素例如孔溶液化学性质以及凝胶孔隙率和稳定性的差异可能影响饱和 CO_2 环境下的 AAM 的行为。

Palacios 和 Puertas[242]分析了暴露在完全饱和 CO_2 气氛中的硅酸钠激发的 BFS 胶凝材料，结果表明 CAST 严重碳化的是其不同多晶型产物中的 Na_2CO_3 和 $CaCO_3$ 的沉淀。Bernal 等人[107]也观察到在硅酸钠激发的 BFS 中形成的 C-A-S-H 结构中 CO_2 暴露的类似效果。方解石被认定为唯一的含钙碳化产物，而天然碱则作为含钠碳化产物。AAM 孔隙溶液化学的热力学模拟[244]预测了在较低 pCO_2 时会形成碳氢钠，在较高 pCO_2 时会形成水化碳酸钠，以及在较高温度及在中间 pCO_2 时天然碱普遍会增加，并且这些趋势也与碱激发 BFS 混凝土的 7 年自然碳化暴露试验[245]相一致。

基于硅酸盐激发的 BFS 混凝土研究也表明，混凝土混合料设计中较高的浆料含量可导致碳化阻力显著增加，使得达到的碳化深度与硅酸盐水泥混凝土相似（图 9-8）。这突出了对混凝土的设计参数进行操纵的可能性，以实现 AAM 中期望的性能。

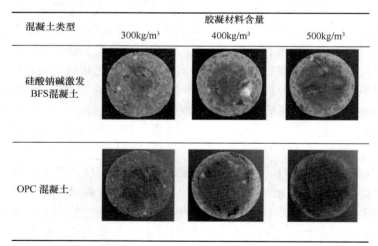

图 9-8　碳酸盐混凝土在暴露于 1% CO_2 环境中 1000h 后的横截面，用酚酞指示剂显示碳化程度（紫色未碳化，无色碳化）。样品直径为 76.2mm（改编自 Bernal 等人[60]）

目前，对于碱激发胶凝材料中的天然碳化的研究是有限的。在苏联和波兰的现役结构中获得的一些数据在第 2、第 11 和第 12 章中给出，证明了在大陆性气候条件下服役的胶凝材料结构一般具有较低的碳化率（<0.5mm/年）[54-56]。如表 9-5 所述，Shi 等[54]报道了位于俄罗斯、乌克兰和波兰的龄期在 12～40 年之间的混凝土结构使用酚酞法测量的自然碳化速率，不超过 1mm/年。

表 9-5 碱激发混凝土的运行中碳化速率的总结

应用	地点	时间	抗压强度(MPa)	平均碳化速率(mm/年)
排水收集池[a]	乌克兰,敖德萨	1966	62(34 年)	<0.1
预制楼板和墙板	波兰,克拉科夫	1974	43(27 年)	0.4
饲料槽[a]	乌克兰,扎波罗热	1982	39(18 年)	0.2~0.4
重载路面	俄罗斯,马格尼托哥尔斯克	1984	86(15 年)	1
高层居民楼	俄罗斯,利佩茨克	1986	35(14 年)	0.4
预应力轨道枕	俄罗斯,楚道伏	1988	88(12 年)	0.7~1

a 材料是在覆盖或地下应用,碳化明显减缓。

Bernal 及其同事[224,244,245]认为,与硅酸盐水泥材料相比,AAM 的加速诱导碳化比在自然碳化条件变化更严重。在 AAM 和 OPC 材料的自然和加速碳化结果之间没有相同的相关性(图 9-9),这表明应用于硅酸盐水泥样品的加速测试在 AAM 体系碳化过程测试是不准确的。这与在较高 CO_2 浓度下诱导的孔溶液化学性质的差异有关,特别是碳酸盐/碳酸氢盐比例的差异[244]。在 AAM 的碳化期间发生的相互作用和/或竞争机制需进一步研究。

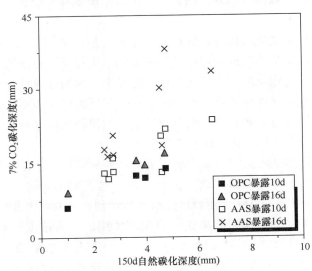

图 9-9 AAM 混凝土的自然和加速(不同暴露时间)碳化深度之间的
关系(来自 Bernal 和 Rodríguez 的数据[224])

如第 2 章所讨论的,Xu 等[56]评估了碱碳酸盐激发的 BFS 硬化混凝土(超过 35 年,在苏联开发和利用)的自然碳化,发现其碳化深度低于 8mm,表明其对天然碳化的良好耐久性。在碱激发 BFS 混凝土的加速碳化试验中,发现碳化深度约为 8mm。相对于 CO_2 速率扩散,碳化过程似乎主要是由化学反应控制[60]。超过碳化过程这一点,扩散控制似乎更加显著。这表明在碳化反应的时间内,AAM 孔结构的致密化使得 CO_2 的扩散受到碳化反应的限制。

在文献中也有关于碳化对碱金属硅酸盐激发BFS混凝土的力学性能的影响的不同意见。Bernal等人[86]发现,在加速碳化后其压缩强度显著降低,但是Hakkinen[104]在22个月的加速碳化后其抗压强度提高了40%。

基于低Ca硅酸铝的AAM的碳化仅在迄今为止公开的单一研究中进行了详细的评估。Criado等人[246]在评估碱激发粉煤灰不同固化条件的影响时,确定了在大气条件下固化的样品中的碳酸氢钠的形成与孔溶液中碱的碳化有关。相比于高温度(85℃)下的硬化,常温硬化显示出低的反应程度和力学性能。这可能部分与碳化反应期间碱的消耗有关,导致溶液的pH值减小,从而降低未反应粉煤灰的溶解度。

9.6.7　碱激发材料试验方法的评价

传统水泥碳化性能试验方法的评价标准比较适用。与自然碳化相比,当材料暴露在标准条件下时两者形成了相似的反应产物。这为假想提供了足够的证据,证明这些协议所提出的试验和条件准确地可再现在自然碳化情况下发生的情况。

没有任何标准或方法来评估AAM的碳化性能。在某些情况下,学者须控制用于碳化评估的样品不良的测试条件。这就是为什么少数报告检查AAM碳化的结果需要解释这些材料在特定碳化条件下的性能指标。与自然碳化条件不同,这些指标不适用于所有AAMs。同时干燥和碳化过程的控制和科学解释,即其中侵蚀性干燥条件可能导致材料的开裂,但在测试方法方面尚未完全成熟。

苏联和波兰几十年来老化结构的良好稳定性是现有最佳证据。AAM可以抵抗时间的流逝,而存在碳化问题与文献中报道的加速碳化试验的结果相反。在加速碳化条件下分析样品时,自然条件下发生的碳化反应平衡的变化很可能发生。因为加速碳化试验之前样品中的渗透性低,机械强度较好,导致性能差。这说明,在该领域的进一步研究,不仅仅是要了解这些体系中劣化过程如何进行,而且还要明白测试条件如何影响所进行测试的结果,以便可以给出当暴露于CO_2时AAM的耐久性的"真实"证明。

材料加速碳酸盐化测试评估的主要影响因素有多种。例如碳酸化收缩和孔隙溶液的化学性质以及酚酞指示剂的制备等因素可改变测试结果。这表明需要进一步研究开发用于测量AAM中碳化前端进程发展的方法,用以解释酚酞指示剂在AAM中获得的结果。

碱激发混凝土中碳化的进程与其加速试验期间使用的CO_2浓度有直接关系。高浓度的CO_2可诱导孔结构的差异[240]。同时在较高CO_2浓度[240]和孔溶液中的碳酸盐/碳酸氢盐比率下,孔隙溶液碳酸盐化(由酚酞显示)和碳化引起的凝胶劣化(在孔结构中显而易见)的速率特别明显[244]。因此,不建议在CO_2浓度高于1%CO_2的情况下对碱激发胶凝材料进行加速碳化试验,需要进一步的工作来为确定关于试验程序及其他方面提出建议。

9.7　风化

9.7.1　测试方法

盐霜是水通过多孔材料的运动导致当水蒸发时在材料表面沉积白色沉积物的现象。当混凝土构件与潮湿的土壤接触时，水通过毛细管作用向上通过混凝土并从其表面蒸发。这留下了富含存在于孔溶液中的可溶性阳离子的混凝土表面。沉积的阳离子（碱金属或碱土金属）可以与存在于孔溶液或地下水中的二氧化碳和/或硫酸盐反应，导致形成白色表面沉积物。其影响混凝土美观的同时对材料的性能影响不大。

在国际上描述建筑产品的风化测试标准是为砖和砌体产品专门设计的，而非针对一般的混凝土。例如 ASTM C67[247]，AS/NZS 4456.6：2003[248]、ČSN73 1358[249] 以及 ČSN（捷克）标准，还包括蒸养混凝土的讨论。在砌筑方面，有关风化试验的规定最近已从英国标准（BS 3921）撤回。实验室也采用非标准化检测方案，以碱浸出为代表来表征风化潜力[111]。在大多数标准化风化试验中，将试样（立方体、梁或圆柱体）部分浸入指定量的水中。水通过毛细管被吸入混凝土中，通过材料上升并从孔溶液中携带盐析出。在浸泡和干燥之后，析出的盐保留在混凝土的表面上，可进行目测评估，也可以把表面的析出盐除去后称重来进行评价。

风化取决于混凝土的微观结构及其组成。因为盐霜产物通常是碱或碱土碳酸盐或硫酸盐，其必须由胶凝材料孔隙溶液提供或由大气碳化所形成[250,251]。为了能够形成这些产物，需要溶解的孔溶液组分具有足够的移动性，以使产物能够到达材料的表面。这意味着风化测试基本上是微结构和孔溶液化学分析的组合。含水硫酸钠的膨胀性质引起的孔隙压力在这里也是重要的。

9.7.2 AAM 的风化

如在本章的前面部分和第 4 章中所述，具有低钙含量和高碱含量的 AAM（这在实验室中由粉煤灰或偏高岭土合成的材料中是常见的）具有多孔和开放的微结构。高 Na_2O/Al_2O_3 比硅铝酸钠胶凝材料[111,252-257]易遭受在材料中未反应的过量氧化钠引起的起泡。未反应的氧化钠在孔网络内移动，与大气中 CO_2 接触时易于形成白色晶体（即风化）。这不同于第 9.6 节所述胶凝材料的大气碳化过程。碳化通常导致胶凝材料裂化，pH 值降低和碳酸盐反应产物在样品主体中的沉积，其可以是肉眼可见的或可以不是肉眼可见的，而风化导致形成可见的表面沉积物。

高碱浓度的孔溶液和 Na 在硅铝酸盐凝胶结构中弱的结合能力造成了低钙体系 AAM 中的开放微观结构，加剧了风化[258-260]。钾替代钠作为激发剂中碱源可有效地减少泡沫[258,259]。因为钾更强地结合到铝硅酸盐凝胶框架[261]，所以碳酸钾晶体通常比它们的钠对应物在视觉上更不明显。补充的铝源（如铝酸钙水泥）添加到低反应程度的胶凝材料中，通过提供可以与碱结合的反应性 Al 在降低碱迁移率方面是有意义的[111]。然而，降低碱迁移率最重要的因素是总渗透率的降低，与本章中讨论的大多数其他参数相同，这意味着良好配制的抗侵蚀的 AAM，例如氯化物渗透和碳化也将抵抗风化。

9.8 小结

碱激发胶凝材料与传输相关的耐久性非常强地依赖于孔结构，其通过胶凝材料化学性（反应性钙倾向于降低渗透性）和成熟度来确定（提供充分固化对于形成不可渗透的和耐用 AAM 胶凝材料）。似乎没有高渗透率的特定区域接近聚集颗粒的边缘以形成明显的"界面过渡区"，如硅酸盐水泥基材料的情况，这可能提供性能优点，但仍然需要详细探讨。

碱激发砂浆和混凝土的氯化物渗透性测试显示出广泛的性能，其结果强烈依赖于所选择的测试方法的细节。广泛使用的 ASTM C1202 测试方法主要是孔溶液化学，因此它使得碱激发胶凝材料有时表现出非常好的抗传质性，有时表现相当差。提供氯离子迁移到胶凝材料中的进程的更直接测量的替代方法（例如池测试或 NordTest NT Build 492 加速测试）将提供更有效的比较，其相对独立于胶凝材料的孔溶液化学。

在 AAM 混凝土中，钢筋腐蚀和碳化的问题在确定服役性能和耐久性中被认为是非常重要的。两者都受到孔溶液化学的强烈影响，在这两个领域都需要仔细解释通过广泛使用的测试方案获得的结果。钢筋腐蚀化学被认为是由于 BFS 基胶凝材料中存在硫化物而受到强烈影响，因此通过应用为硅酸盐水泥混凝土设计的指导而获得的预测可能是不可靠的。在较高 CO_2 浓度下测试 AAM，改变了孔溶液中碳酸钠-碳酸氢钠的相平衡，会影响胶凝材料的化学变化，并且与碳化前端的切入速率相关，但这并不能代表其服役性能。从可用于长期（>20年）老化材料的有限数据集，已经显示 AAM 混凝土在使用中提供相对良好的耐碳化性，这与从加速测试获得的结果相反。

一般来说，与第 8 章中讨论的化学耐久性过程一样，似乎有必要通过与服役性能比较来详细验证 AAM 的加速测试方法。碱激发胶凝材料的化学（特别是孔溶液化学）和微观结构的细节似乎导致控制这些材料与传输相关耐久性的机制的显著差异，并且可能几种广泛使用和标准化的测试及协议提供的结果在预测材料长期性能方面是不完全可靠的。

参考文献

[1] Powers T. C., Brownyard T. L.. Studies of the physical properties of hardened Portland cement paste. Part 7. Permeability and absorptivity. J. Am. Concr. Inst. Proc, 1947, 18(7): 865-880.

[2] Garboczi E. J.. Permeability, diffusivity, and microstructural parameters: a critical review. Cem. Concr. Res, 1990, 20(4): 591-601.

[3] Ollivier J. P., Maso J. C., Bourdette B.. Interfacial transition zone in concrete. Adv. Cem. Based Mater, 1995, 2(1): 30-38.

[4] Lu S., Landis E., Keane D.. X-ray microtomographic studies of pore structure and permeability in Portland cement concrete. Mater. Struct, 2006, 39(6): 611-620.

[5] Kropp J., Hilsdorf H. K. (eds.). Performance criteria for concrete durability: RILEM Report REP12. E&FN Spon. London, UK, 1995.

[6] Torrent R., Fernández Luco L. (eds.). Non-destructive Evaluation of the Penetrability and Thickness of the Concrete Cover: State of the Art Report of RILEM Technical Committee 189-NEC. RILEM Publications, Bagneux, 2007.

[7] Harris A. W., Atkinson A., Claisse P. A.. Transport of gases in concrete barriers. Waste Manag, 1992, 12(2-3): 155-178.

[8] Houst Y. F., Wittmann F. H.. The diffusion of carbon dioxide and oxygen in aerated concrete. In: Wittmann, F. H. (ed.) 2nd International Colloquium: Materials Science and Restoration, 1986: 629-634. WTA, Esslingen, Germany.

[9] Jung S. H., Lee M. K., Oh B. H.. Measurement device and characteristics of diffusion coefficient of carbon dioxide in concrete. ACI Mater. J, 2011, 108(6): 589-595.

[10] Brunauer S., Emmett P. H., Teller E.. Adsorption of gases in multimolecular layers. J. Am. Chem. Soc, 1938, 60: 309-319.

[11] Barrett E. P., Joyner L. G., Halenda P. P.. The determination of pore volume and area distributions in porous substances. I. computations from nitrogen isotherms. J. Am. Chem. Soc, 1951, 73(1): 373-380.

[12] Diamond S.. Mercury porosimetry. An inappropriate method for the measurement of pore size distributions in cement-based materials. Cem. Concr. Res, 2000, 30: 1517-1525.

[13] Kaufmann J., Loser R., Leemann A.. Analysis of cement-bonded materials by multi-cycle mercury intrusion and nitrogen sorption. J. Colloid Interf. Sci, 2009, 336: 730-737.

[14] Kaufmann J.. Characterization of pore space of cement-based materials by combined mercury and Wood's metal intrusion. J. Am. Ceram. Soc, 2009, 92(1): 209-216.

[15] Lloyd R. R., Provis J. L., Smeaton K. J., Van Deventer J. S. J.. Spatial distribution of pores in fly ash-based inorganic polymer gels visualised by Wood's metal intrusion. Microporous Mesoporous Mater, 2009, 126(1-2): 32-39.

[16] Deutsches Institut für Normung. Bestimmung der Porengrößenverteilung und der spezifischen Oberfläche mesoporöser Feststoffe durch Stickstoffsorption; Verfahren nach Barrett, Joyner und Halenda (BJH) (Determination of the pore size distribution and specific surface area of mesoporous solids by means of nitrogen sorption-Method of Barrett, Joyner and Halenda (BJH)) (DIN 66134). Berlin, Germany, 1997.

[17] International Organization for Standardization: Pore size distribution and porosity of solid materials by mercury porosimetry and gas adsorption-Part 2: Analysis of Mesopores and Macropores by Gas Adsorption (ISO 15901-2: 2006). Geneva, Switzerland, 2006.

[18] Neimark A. V., Ravikovitch P. I.. Capillary condensation in MMS and pore structure characterization. Microporous Mesoporous Mater, 2001, 44-45(1): 697-707.

[19] Metroke T., Thommes M., Cychosz K.. Porosity characteristics of geopolymers: influence of synthesis conditions. In: 36th International Conference and Exposition on Advanced Ceramics and Composites, Daytona Beach, FL. American Ceramic Society, 2012.

[20] Zheng L., Wang W., Shi Y.. The effects of alkaline dosage and Si/Al ratio on the immobilization of heavy metals in municipal solid waste incineration fly ash-based geopolymer. Chemosphere, 2010, 79(6): 665-671.

[21] Sindhunata, Provis J. L., Lukey G. C., Xu H., Van Deventer J. S. J.. Structural evolution of fly ashbased geopolymers in alkaline environments. Ind. Eng. Chem. Res, 2008, 47(9): 2991-2999.

[22] Sindhunata P. K., Van Deventer J. S. J., Lukey G. C., Xu H.. Effect of curing temperature and silicate concentration on fly-ash-based geopolymerization. Ind. Eng. Chem. Res, 2006, 45(10): 3559-3568.

[23] Deutsches Institut für Normung: Mikroporenanalyse mittels Gasadsorption (Micropore analysis by gas adsorption) (DIN 66135). Berlin, Germany, 2001.

[24] International Organization for Standardization: Pore size distribution and porosity of solid materials by mercury porosimetry and gas adsorption-Part 3: Analysis of Micropores by Gas Adsorption (ISO 15901-3: 2007). Geneva, Switzerland, 2007.

[25] Sazama P., Bortnovsky O., Dědeček J., Tvarůžková Z., Sobalík Z.. Geopolymer based catalysts-New group of catalytic materials. Catal. Today, 2011, 164 (1): 92-99.

[26] ASTM International: Standard Test Method for Determination of Pore Volume and Pore Volume Distribution of Soil and Rock by Mercury Intrusion Porosimetry (ASTM D4404 -10). West Conshohocken, PA, 2010.

[27] ASTM International: Standard Test Method for Determining Pore Volume Distribution of Catalysts by Mercury Intrusion Porosimetry (ASTM D4284-07). West Conshohocken, PA, 2007.

[28] Deutsches Institut für Normung: Bestimmung der Porenvolumenverteilung und der spezifischen Oberfläche von Feststoffen durch Quecksilberintrusion (Determination of pore volume distribution and specific surface area of solids by mercury intrusion) (DIN 66133). Berlin, Germany, 1993.

[29] ASTM International: Automated Pore Volume and Pore Size Distribution of Porous Substances by Mercury Porosimetry (UOP578-11). West Conshohocken, PA, 2011.

[30] British Standards Institution: Porosity and Pore Size Distribution of Materials. Method of Evaluation by Mercury Porosimetry (BS 7591-1: 1992). London, UK, 1992.

[31] International Organization for Standardization: Pore size distribution and porosity of solid materials by mercury porosimetry and gas adsorption-Part 1: Mercury Porosimetry (ISO 15901-1: 2005). Geneva, Switzerland, 2005.

[32] Häkkinen T.. The influence of slag content on the microstructure, permeability and mechanical properties of concrete: Part 1. Microstructural studies and basic mechanical properties. Cem. Concr. Res, 1993, 23(2): 407-421.

[33] Bell J. L., Gordon M., Kriven W. M.. Nano- and microporosity in geopolymer gels. Microsc. Microanal, 2006, 12(S02): 552-553.

[34] Bell J. L., Kriven W. M.. Nanoporosity in aluminosilicate, geopolymeric cements. In: Microscopy and Microanalysis '04 (Proceedings of 62nd Annual Meeting of the Microscopy Society of America), 2004, 10. Microscopy Society of America. Reston, VA.

[35] Kriven W. M., Bell J. L., Gordon M.. Microstructure and nanoporosity in as-set

[36] Zhang Z., Yao X., Zhu H.. Potential application of geopolymers as protection coatings for marine concrete: II. Microstructure and anticorrosion mechanism. Appl. Clay Sci, 2010, 49(1-2): 7-12.

[37] Wong H. S., Buenfeld N. R., Head M. K.. Estimating transport properties of mortars using image analysis on backscattered electron images. Cem. Concr. Res, 2006, 36(8): 1556-1566.

[38] Brough A. R., Atkinson A.. Automated identification of the aggregate-paste interfacial transition zone in mortars of silica sand with Portland or alkali-activated slag cement paste. Cem. Concr. Res, 2000, 30(6): 849-854.

[39] Ben Haha M., Le Saout G., Winnefeld F., Lothenbach B.. Influence of activator type on hydration kinetics, hydrate assemblage and microstructural development of alkali activated blast-furnace slags. Cem. Concr. Res, 2011, 41(3): 301-310.

[40] Ben Haha M., Lothenbach B., Le Saout G., Winnefeld F.. Influence of slag chemistry on the hydration of alkali-activated blast-furnace slag-Part I: effect of MgO. Cem. Concr. Res, 2011, 41(9): 955-963.

[41] Ben Haha M., Lothenbach B., Le Saout G., Winnefeld F.. Influence of slag chemistry on the hydration of alkali-activated blast-furnace slag-Part II: effect of Al_2O_3. Cem. Concr. Res, 2012, 42(1): 74-83.

[42] Le Saoût G., Ben Haha M., Winnefeld F., Lothenbach B.. Hydration degree of alkaliactivated slags: a 29 Si NMR study. J. Am. Ceram. Soc, 2011, 94(12): 4541-4547.

[43] Willis K. L., Abell A. B., Lange D. A.. Image-based characterization of cement pore structure using Wood's metal intrusion. Cem. Concr. Res, 1998, 28(12): 1695-1705.

[44] Nemati K. M.. Preserving microstructure of concrete under load using the Wood's metal technique. Int. J. Rock Mech. Min Sci, 2000, 37(1-2): 133-142.

[45] Diamond S., Landis E. N.. Microstructural features of a mortar as seen by computed microtomography. Mater. Struct, 2007, 40(9): 989-993.

[46] Gallucci E., Scrivener K., Groso A., Stampanoni M., Margaritondo G.. 3D experimental investigation of the microstructure of cement pastes using synchrotron X-ray microtomography (μCT). Cem. Concr. Res, 2007, 37(3): 360-368.

[47] Nakashima Y., Kamiya S.. Mathematica programs for the analysis of three-dimensional pore connectivity and anisotropic tortuosity of porous rocks using X-ray computed tomography image data. J. Nucl. Sci. Technol, 2007, 44(9): 1233-1247.

[48] Promentilla M. A. B., Sugiyama T., Hitomi T., Takeda N.. Quantifi cation of tortuosity in hardened cement pastes using synchrotron-based X-ray computed microtomography. Cem. Concr. Res, 2009, 39: 548-557.

[49] Rattanasak U., Kendall K.. Pore structure of cement/pozzolan composites by X-ray

microtomography. Cem. Concr. Res, 2005, 35(4): 637-640.

[50] Provis J. L., Myers R. J., White C. E., van Deventer J. S. J.. Linking structure, performance and durability of alkali-activated aluminosilicate binders. In: Palomo, A. (ed.) 13th International Congress on the Chemistry of Cement, Madrid, Spain. CD-ROM, 2011.

[51] Provis J. L., Myers R. J., White C. E., Rose V., Van Deventer J. S. J.. X-ray microtomography shows pore structure and tortuosity in alkali-activated binders. Cem. Concr. Res, 2012, 42(6): 855-864.

[52] Van Deventer J. S. J., Provis J. L., Duxson P.. Technical and commercial progress in the adoption of geopolymer cement. Miner. Eng, 2012, 29: 89-104.

[53] Provis J. L., Rose V., Winarski R. P., Van Deventer J. S. J.. Hard X-ray nanotomography of amorphous aluminosilicate cements. Scripta Mater, 2011, 65(4): 316-319.

[54] Shi C., Krivenko P. V., Roy D. M.. Alkali-Activated Cements and Concretes. Taylor & Francis, Abingdon, 2006.

[55] Deja J.. Carbonation aspects of alkali activated slag mortars and concretes. Silic. Ind, 2002, 67(1): 37-42.

[56] Xu H., Provis J. L., Van Deventer J. S. J., Krivenko P. V.. Characterization of aged slag concretes. ACI Mater. J, 2008, 105(2): 131-139.

[57] Provis J. L., Muntingh Y., Lloyd R. R., Xu H., Keyte L. M., Lorenzen L., Krivenko P. V., Van Deventer J. S. J.. Will geopolymers stand the test of time? Ceram. Eng. Sci. Proc, 2007, 28(9): 235-248.

[58] Adam A. A.. Strength and durability properties of alkali activated slag and fly ash-based geopolymer concrete. Ph. D. thesis, RMIT University, 2009.

[59] Rodríguez E., Bernal S., Mejía de Gutierrez R., Puertas F.. Alternative concrete based on alkali-activated slag. Mater. Constr, 2008, 58(291): 53-67.

[60] Bernal S. A., Mejía de Gutierrez R., Pedraza A. L., Provis J. L., Rodríguez E. D., Delvasto S.. Effect of binder content on the performance of alkali-activated slag concretes. Cem. Concr. Res, 2011, 41(1): 1-8.

[61] Dhir R. K., Hewlett P. C., Chan Y. N.. Near surface characteristics of concrete: intrinsic permeability. Mag. Concr. Res, 1989, 41(147): 87-97.

[62] Parrott L. J.. Influence of cement type and curing on the drying and air permeability of cover concrete. Mag. Concr. Res, 1995, 47(171): 103-111.

[63] Tsivilis S., Chaniotakis E., Batis G., Meletiou C., Kasselouri V., Kakali G., Sakellariou A., Pavlakis G., Psimadas C.. The effect of clinker and limestone quality on the gas permeability, water absorption and pore structure of limestone cement concrete. Cem. Concr. Res, 1999, 21(2): 139-146.

[64] Monlouis-Bonnaire J. P., Verdier J., Perrin B.. Prediction of the relative permeability to gas flow of cement-based materials. Cem. Concr. Res, 2004, 34(5): 737-744.

[65] Henderson G. D., Basheer P. A. M., Long A. E.. Pull-off test and permeation tests. In: Malhotra, V. M., Carino, N. J. (eds.) Handbook on Nondestructive Testing of Concrete, 2004: 6.1-6.2. CRC Press, Boca Raton.

[66] Romer M., RILEM TC 189-NEC. Recommendation of RILEM TC 189-NEC: 'Non-destructive evaluation of the concrete cover'-Comparative test-Part I-Comparative test of 'penetrability' methods. Mater. Struct, 2005, 38(10): 895-906.

[67] Kollek J.. The determination of the permeability of concrete to oxygen by the Cembureau method—a recommendation. Mater. Struct, 1989, 22(3): 225-230.

[68] Alarcon-Ruiz L., Brocato M., Dal Pont S., Feraille A.. Size effect in concrete intrinsic permeability measurements. Transp. Porous Media, 2010, 85(2): 541-564.

[69] Torrent R.. A two-chamber vacuum cell for measuring the coefficient of permeability to air of the concrete cover on site. Mater. Struct, 1992, 25(6): 358-365.

[70] Schweizerisches Ingenieur and Architektenverein (SIA): Betonbau-Ergänzende Festlegungen (SIA 262/1). Zürich, Switzerland, 2003.

[71] Jacobs F., Leemann A., Denarié E., Teruzzi T.. Empfehlungen zur Qualitätskontrolle von Beton mit Luftpermeabilitätsmessungen (VSS Report 641). Zurich, Switzerland, 2009.

[72] Materials Advanced Services Ltd: Annotated bibliography related to testing the air-permeability of the concrete cover according to the "Torrent" method (Swiss Standard SIA 162/1-E), 2009. http://www.tfb.ch/htdocs/Files/Annotated_Bibliography_on_TM_090616.pdf.

[73] Romer M.. Effect of moisture and concrete composition on the Torrent permeability measurement. Mater. Struct, 2005, 38(5): 541-547.

[74] Basheer P. A. M., Long A. E., Montgomery F. R.. The Autoclam-a new test for permeability. Concrete, 1994, 28(4): 27-29.

[75] RILEM Technical Committee 189-NEC: Update of the recommendation of RILEM TC 189-NEC 'Non-destructive evaluation of the concrete cover' "Comparative test-Part I-Comparative Test of Penetrability Methods", Materials & Structures, 2005, 38: 895-906. Mater. Struct, 2008, 41(3): 443-447.

[76] Häkkinen T.. Durability of alkali-activated slag concrete. Nord. Concr. Res, 1987, 6(1): 81-94.

[77] Sagoe-Crentsil K., Brown T., Yan S.. Medium to long term engineering properties and performance of high-strength geopolymers for structural applications. Adv. Sci. Technol, 2010, 69: 135-142.

[78] Hearn N., Hooton R. D., Nokken M. R.. Pore structure, permeability, and penetration resistance characteristics of concrete. In: Lamond J. F., Pielert J. H. (eds.) Significance of Tests and Properties of Concrete and Concrete-Making Materials, 2006: 238-252. ASTM International, West Conshohocken.

[79] Banthia N., Mindess S.. Water permeability of cement paste. Cem. Concr. Res, 1989,

19(5): 727-736.
[80] ASTM International. Standard Test Method for Density, Absorption, and Voids in Hardened Concrete (ASTM C642-13). West Conshohocken, PA, 2013.
[81] Standards Australia. Methods of Testing Concrete-Determination of Water Absorption and Apparent Volume of Permeable Voids in Hardened Concrete (AS 1012.21). Sydney, Australia, 1999.
[82] Andrews-Phaedonos F.. VicRoads technical note 89: Test methods for the assessment of durability of concrete(2007).
[83] De Schutter G., Audenaert K.. Evaluation of water absorption of concrete as a measure for resistance against carbonation and chloride migration. Mater. Struct, 2004, 37(9): 591-596.
[84] Ismail I., Bernal S. A., Provis J. L., Hamdan S., Van Deventer J. S. J.. Drying-induced changes in the structure of alkali-activated pastes. J. Mater. Sci, 2013, 48(9): 3566-3577.
[85] Wimpenny D., Duxson P., Cooper T., Provis J. L., Zeuschner R.. Fibre reinforced geopolymer concrete products for underground infrastructure. In: Concrete 2011, Perth, Australia. CD-ROM proceedings. Concrete Institute of Australia, 2011.
[86] Bernal S. A., Mejía de Gutiérrez R., Provis, J. L.. Engineering and durability properties of concretes based on alkali-activated granulated blast furnace slag/metakaolin blends. Constr. Build. Mater, 2012, 33: 99-108.
[87] British Standards Institution. Testing Concrete. Method for Determination of Water Absorption (BS 1881-122: 2011). London, UK, 2011.
[88] European Committee for Standardization (CEN). Testing Hardened Concrete. Depth of Penetration of Water Under Pressure (EN 12390-8). Brussels, Belgium, 2009.
[89] U. S. Army Engineer Research and Development Center. Test Method for Water Permeability of Concrete Using Triaxial Cell (CRD-C 163-92). Vicksburg, Mississippi, 1992.
[90] Figg J. W.. Methods of measuring the air and water permeability of concrete. Mag. Concr. Res, 1973, 25(85): 213-219.
[91] International Organization for Standardization. Concrete, Hardened-Determination of the Depth of Penetration of Water Under Pressure (ISO 7301). Geneva, Switzerland, 1963.
[92] Deutsches Institut für Normung. Prüfverfahren für Beton; Festbeton, gesondert hergestellte Probekörper (Testing concrete; hardened concrete, specially prepared specimens) (DIN 1048-5). Berlin, Germany, 1991.
[93] Olivia M., Nikraz H., Sarker P.. Improvements in the strength and water penetrability of low calcium fly ash based geopolymer concrete. In: Uomoto T., Nga T. V. (eds.) The 3rd ACF International Conference- ACF/VCA 2008, Ho Chi Minh City, Vietnam, 2008: 384-391. Vietnam Institute for Building Materials.

[94] Shi C.. Strength, pore structure and permeability of alkali-activated slag mortars. Cem. Concr. Res, 1996, 26(12): 1789-1799.

[95] Wongpa J., Kiattikomol K., Jaturapitakkul C., Chindaprasirt P.. Compressive strength, modulus of elasticity, and water permeability of inorganic polymer concrete. Mater. Des, 2010, 31(10): 4748-4754.

[96] Talling B., Krivenko P. V.. Blast furnace slag-the ultimate binder. In: Chandra S. (ed.) Waste Materials Used in Concrete Manufacturing, 1997: 235-289. Noyes Publications, Park Ridge.

[97] Sugama T., Brothers L. E., Van de Putte T. R.. Acid-resistant cements for geothermal wells: sodium silicate activated slag/fly ash blends. Adv. Cem. Res, 2005, 17 (2): 65-75.

[98] Zhang Z., Yao X., Zhu H.. Potential application of geopolymers as protection coatings for marine concrete: I. Basic properties. Appl. Clay Sci, 2010, 49(1-2): 1-6.

[99] Fagerlund G.. On the capillarity of concrete. Nord. Concr. Res, 1982, 1: 6.1-6.20.

[100] ASTM International. Standard Test Method for Measurement of Rate of Absorption of Water by Hydraulic-Cement Concretes (ASTM C1585-11). West Conshohocken, PA, 2011.

[101] European Committee for Standardization (CEN). Methods of Test for Mortar for Masonry. Determination of Water Absorption Coefficient due to Capillary Action of Hardened Mortar (EN 1015-18). Brussels, Belgium, 2002

[102] Schweizerisches Ingenieur and Architektenverein (SIA). Determination of Water Infiltration Rate (porosity) (SIA 262/1 Appendix A). Zürich, Switzerland, 2003.

[103] RILEM TC 116-PCD. Test for gas permeablity of concrete. C. Determination of the capillary absorption of water of hardened concrete. Mater. Struct, 1999, 32 (3): 178-179.

[104] Häkkinen T.. The permeability of high strength blast furnace slag concrete. Nord. Concr. Res, 1992, 11 (1): 55-66.

[105] Bernal S., De Gutierrez R., Delvasto S., Rodriguez E.. Performance of an alkali-activated slag concrete reinforced with steel fibers. Constr. Build. Mater, 2010, 24 (2): 208-214.

[106] Adam A. A., Molyneaux T. C. K., Patnaikuni I., Law D. W.. Strength, sorptivity and carbonation of geopolymer concrete. In: Ghafoori N. (ed.) Challenges, Opportunities and Solutions in Structural Engineering and Construction, 2009: 563-568. CRC Press, Boca Raton.

[107] Bernal S. A., Mejía de Gutierrez R., Rose V., Provis J. L.. Effect of silicate modulus and metakaolin incorporation on the carbonation of alkali silicate-activated slags. Cem. Concr. Res. 2010, 40 (6): 898-907.

[108] Collins F., Sanjayan J.. Unsaturated capillary flow within alkali activated slag concrete. J. Mater. Civil Eng, 2008, 20 (9): 565-570.

[109] Collins F., Sanjayan J.. Capillary shape: influence on water transport within unsaturated alkali activated slag concrete. J. Mater. Civil Eng, 2010, 22 (3): 260-266.

[110] Collins F., Sanjayan J.. Prediction of capillary transport of alkali activated slag cementitious binders under unsaturated conditions by elliptical pore shape modeling. J. Porous Mater, 2010, 17 (4): 435-442.

[111] Najafi Kani E., Allahverdi A., Provis J. L.. Efflorescence control in geopolymer binders based on natural pozzolan. Cem. Concr. Compos, 2012, 34 (1): 25-33.

[112] Okada K., Ooyama A., Isobe T., Kameshima Y., Nakajima A., MacKenzie K. J. D.. Water retention properties of porous geopolymers for use in cooling applications. J. Eur. Ceram. Soc, 2009, 29 (10): 1917-1923.

[113] British Standards Institution. Testing Concrete. Recommendations for the Determination of the Initial Surface Absorption of Concrete (BS 1881-208: 1996). London, UK, 1996.

[114] European Committee for Standardization (CEN). Methods of Test for Masonry Units. Determination of Water Absorption of Aggregate Concrete, Autoclaved Aerated Concrete, Manufactured Stone and Natural Stone Masonry Units due to Capillary Action and the Initial Rate of Water Absorption of Clay Masonry Units (EN 772-11). Brussels, Belgium, 2011.

[115] Vicat L.-J., Smith J. T.. A practical and scientific treatise on calcareous mortars and cements, artificial and natural; containing, directions for ascertaining the qualities of the different ingredients, for preparing them for use, and for combining them together in the most advantageous manner; with a theoretical investigation of their properties and modes of action. The whole founded upon an extensive series of original experiments, with examples of their practical application on the large scale. John Weale, Architectural Library, London, UK, 1837.

[116] Nmai C. K.. Freezing and thawing. In: Lamond J. F., Pielert J. H. (eds.). Significance of Tests and Properties of Concrete and Concrete-Making Materials, 2006: 154-163. ASTM International, West Conshohocken.

[117] Garrabrants A. C., Sanchez F., Kosson D. S.. Leaching model for a cement mortar exposed to intermittent wetting and drying. AIChE J, 2003, 49 (5): 1317-1333.

[118] Puertas F., Amat T., Fernández-Jiménez A., Vázquez T.. Mechanical and durable behavior of alkaline cement mortars reinforced with polypropylene fibres. Cem. Concr. Res, 2003, 33 (12): 2031-2036.

[119] Slavík R., Bednařík V., Vondruška M., Nemec A.. Preparation of geopolymer from fluidized bed combustion bottom ash. J. Mater. Proc. Technol, 2008, 200 (1-3): 265-270.

[120] Häkkinen T.. The influence of slag content on the microstructure, permeability and mechanical properties of concrete: Part 2. Technical properties and theoretical examinations. Cem. Concr. Res, 1993, 23 (3): 518-530.

[121] Breton D., Carles-Gibergues A., Ballivy G., Grandet J.. Contribution to the formation mechanism of the transition zone between rock-cement paste. Cem. Concr. Res, 1993, 23 (2): 335-346.

[122] Struble L., Skalny J., Mindess S.. A review of the cement-aggregate bond. Cem. Concr. Res, 1980, 10 (2): 277-286.

[123] Monteiro P. J. M., Mehta P. K.. Improvement of the aggregate-cement paste transition zone by grain refinement of hydration products. In: 8th International Congress on the Chemistry of Cement, 1986, 3: 433-437. Rio de Janeiro, Brazil.

[124] Scrivener K. L., Bentur A., Pratt P. L.. Quantitative characterization of the transition zone in high-strength concretes. Adv. Cem. Res, 1988, 1: 230-237.

[125] Mehta P. K., Monteiro P. J. M.. Concrete: Microstructure, Properties and Materials, 3rd edn. McGraw-Hill, New York, 2006.

[126] Mitsui K., Li Z., Lange D. A., Shah S. P.. Relationship between microstructure and mechanical properties of the paste-aggregate interface. ACI Mater. J, 1994, 91: 30-39.

[127] Trende U., Büyüköztürk O.. Size effect and influence of aggregate roughness in interface fracture of concrete composites. ACI Mater. J. 1998, 95: 331-338.

[128] Brough A. R., Atkinson A.. Sodium silicate-based, alkali-activated slag mortars: Part I. Strength, hydration and microstructure. Cem. Concr. Res, 2002, 32 (6): 865-879.

[129] Shi C., Xie P.. Interface between cement paste and quartz sand in alkali-activated slag mortars. Cem. Concr. Res, 1998, 28 (6): 887-896.

[130] Pacheco-Torgal F., Castro-Gomes J. P., Jalali S.. Investigations of tungsten mine waste geopolymeric binder: strength and microstructure. Constr, Build. Mater 2008, 22 (11): 2212-2219.

[131] Škvára F., Doležal J., Svoboda P., Kopecký L., Pawlasová S., Lucuk M., Dvořáček K., Beksa M., Myšková L., Šulc R.. Concrete based on fly ash geopolymers. In: Proceedings of 16th IBAUSIL, 2006, 1: 1079-1097. Weimar, Germany.

[132] San Nicolas R., Provis J. L.. Interfacial transition zone in alkali-activated slag concrete. In: 12th International Conference on Recent Advances in Concrete Technology and Sustainability Issues, ACI SP 289. Supplementary papers CD-ROM. American Concrete Institute, Prague, Czech Republic, 2012.

[133] Lee W. K. W., Van Deventer J. S. J.. The interface between natural siliceous aggregates and geopolymers. Cem. Concr. Res, 2004, 34 (2): 195-206.

[134] Lee W. K. W., Van Deventer J. S. J.. Chemical interactions between siliceous aggregates and low-Ca alkali-activated cements. Cem. Concr. Res, 2007, 37 (6): 844-855.

[135] Zhang J. X., Sun H. H., Wan J. H., Yi Z. L.. Study on microstructure and mechanical property of interfacial transition zone between limestone aggregate and Sial-

ite paste. Constr. Build. Mater, 2009, 23 (11): 3393-3397.

[136] Zhang Y., Sun W., Li Z.. Hydration process of interfacial transition in potassium polysialate (K-PSDS) geopolymer concrete. Mag. Concr. Res, 2005, 57 (1): 33-38.

[137] Krivenko P. V., Gelevera A. G., Petropavlovsky O. N., Kavalerova E. S.. Role of metakaolin additive on structure formation in the contact zone "cement-alkali-susceptible aggregate". In: Bilek V. (ed.) 2nd International Conference on Non-Traditional Cement & Concrete, Brno, Czech Republic. Brno University of Technology & ZPSV A. S, 2005.

[138] Alonso C., Andrade C., Castellote M., Castro P.. Chloride threshold values to depassivate reinforcing bars embedded in a standardized OPC mortar. Cem. Concr. Res, 2000, 30 (7): 1047-1055.

[139] Angst U., Elsener B., Larsen C. K., Vennesland O.. Critical chloride content in reinforced concrete - a review. Cem. Concr. Res., 2009, 39 (12): 1122-1138.

[140] Stanish K. D., Hooton R. D., Thomas M. D. A.. Testing the chloride penetration resistance of concrete: a literature review, FHWA Contract Report DTFH61-97-R-00022. Toronto, Canada, 1997.

[141] Tang L.. CHLORTEST - EU funded research project under 5FP growth programme, Final Report: Resistance of Concrete to Chloride Ingress - From Laboratory Tests to In-Field Performance, SP Swedish National Testing and Research Institute, 2005.

[142] Andrade C., Kropp J. (eds.). Testing and Modelling Chloride Ingress into Concrete: Proceedings of the 3rd International RILEM Workshop. RILEM Proceedings PRO38, Madrid, Spain, 2005.

[143] Castellote M., Andrade C.. Round-Robin test on methods for determining chloride transport parameters in concrete. Mater. Struct, 2006, 39 (10): 955-990.

[144] European Committee for Standardization (CEN). Products and Systems for the Protection and Repair of Concrete Structures. Test Methods. Measurement of Chloride Ion Ingress (EN 13396: 2004). Brussels, Belgium, 2004.

[145] ASTM International. Standard Test Method for Determining the Penetration of Chloride Ion into Concrete by Ponding (ASTM C1543-10a). West Conshohocken, PA, 2010.

[146] McGrath P. F., Hooton R. D.. Re-evaluation of the AASHTO T259 90-day salt ponding test. Cem. Concr. Res, 1999, 29 (8): 1239-1248.

[147] Nordtest. Concrete, Hardened: Accelerated Chloride Penetration (NT BUILD 443). Espoo, Finland, 1995.

[148] ASTM International. Standard Test Method for Determining the Apparent Chloride Diffusion Coefficient of Cementitious Mixtures by Bulk Diffusion (ASTM C1556-11). West Conshohocken, PA, 2011.

[149] Yang C. C., Cho S. W., Huang R.. The relationship between charge passed and the chlorideion concentration in concrete using steady-state chloride migration test. Cem. Concr. Res, 2002, 32 (2): 217-222.

[150] Yang C. C., Chiang S. C.. The chloride ponding test and its correlation to the accelerated chloride migration test for concrete. J. Chin. Inst. Eng, 2006, 29 (6): 1007-1015.

[151] Yang C.. The relationship between charge passed and the chloride concentrations in anode and cathode cells using the accelerated chloride migration test. Mater. Struct, 2003, 36 (10): 678-684.

[152] Nordtest. Concrete, Mortar and Cement-Based Repair Materials: Chloride Diffusion Coefficient from Migration Cell Experiments (NT BUILD 355), 2nd Edition. Espoo, Finland, 1997.

[153] Nordtest. Concrete, Mortar and Cement-Based Repair Materials: Chloride Migration Coefficient from Non-Steady State Migration Experiments (NT BUILD 492). Espoo, Finland, 1999.

[154] Tang L., Nilsson L.-O.. Rapid determination of the chloride diffusivity in concrete by applying an electrical field. ACI Mater. J, 1992, 89 (1): 49-53.

[155] Tang L., Sørensen H.. Precision of the Nordic test methods for measuring the chloride diffusion/migration coefficients of concrete. Mater. Struct, 2001, 34 (8): 479-485.

[156] ASTM International. Standard Test Method for Electrical Indication of Concrete's Ability to Resist Chloride Ion Penetration (ASTM C1202-10). West Conshohocken, PA, 2010.

[157] Whiting D.. Rapid determination of the chloride permeability of concrete, Report No. FHWA RD-81-119, Federal Highway Administration. Washington DC, 1981.

[158] Shi C.. Another Look at the Rapid Chloride Permeability Test (ASTM C1202 or ASSHTO T277), FHWA Resource Center. Federal Highway Administration, Baltimore, MD, 2003.

[159] Andrade C.. Calculation of chloride diffusion coefficients in concrete from ionic migration measurements. Cem. Concr. Res, 1993, 23 (3): 724-742.

[160] Pfeifer D. W., McDonald D. B., Krauss P. D.. The rapid chloride permeability test and its correlation to the 90-day chloride ponding test. PCI J, 1994, 39 (1): 38-47.

[161] Wee T. H., Suryavanshi A. K., Tin S. S.. Evaluation of rapid chloride permeability test (RCPT) results for concrete containing mineral admixtures. ACI Mater. J, 2000, 97 (2): 221-232.

[162] Shi C., Stegemann J. A., Caldwell R. J.. Effect of supplementary cementing materials on the specific conductivity of pore solution and its implications on the rapid chloride permeability test (AASHTO T277 and ASTM C1202) results. ACI Mater. J. 1998, 95 (4): 389-394.

[163] Shi C. J.. Effect of mixing proportions of concrete on its electrical conductivity and the rapid chloride permeability test (ASTM C1202 or ASSHTO T277) results. Cem. Concr. Res, 2004, 34 (3): 537-545.

[164] Douglas E., Bilodeau A., Malhotra V. M.. Properties and durability of alkali-activated slag concrete. ACI Mater. J, 1992, 89 (5): 509-516.

[165] Roy D. M.. Hydration, microstructure, and chloride diffusion of slag-cement pastes and mortars. In: Malhotra, V. M. (ed.) 3rd International Conference on Fly Ash, Silica Fume, Slag and Natural Pozzolans in Concrete, ACI SP114, 1989, 2: 1265-1281. American Concrete Institute, Trondheim.

[166] Roy D. M., Jiang W., Silsbee M. R.. Chloride diffusion in ordinary, blended, and alkali- activated cement pastes and its relation to other properties. Cem. Concr. Res, 2000, 30: 1879-1884.

[167] Mejía R., Delvasto S., Gutiérrez C., Talero R.. Chloride diffusion measured by a modified permeability test in normal and blended cements. Adv. Cem. Res, 2003, 15 (3): 113-118.

[168] Husbands T. B., Malone P. G., Wakeley L. D.. Performance of concretes proportioned with Pyrament blended cement, U. S. Army Corps of Engineers Construction Productivity Advancement Research Program, Report CPAR-SL-94-2. Vicksburg, MS, 1994.

[169] Zia P., Ahmad S. H., Leming M. L., Schemmel J. J., Elliott R. P.. Mechanical Behavior of High Performance Concretes, Volume 3: Very High Early Strength Concrete. SHRP-C-363, Strategic Highway Research Program, National Research Council. Washington DC, 1993.

[170] Al-Otaibi S.. Durability of concrete incorporating GGBS activated by water-glass. Constr. Build. Mater, 2008, 22 (10): 2059-2067.

[171] Shi C.. Corrosion resistance of alkali-activated slag cement. Adv. Cem. Res, 2003, 15 (2): 77-81.

[172] Bertolini L., Elsener B., Pedeferri P., Polder R.. Corrosion of Steel in Concrete - Prevention, Diagnosis, Repair. Wiley-VCH. Weinheim, Germany, 2004.

[173] Lloyd R. R., Provis J. L., Van Deventer J. S. J.. Pore solution composition and alkali diffusion in inorganic polymer cement. Cem. Concr. Res, 2010, 40 (9): 1386-1392.

[174] ASTM International. Standard Test Method for Corrosion Potentials of Uncoated Reinforcing Steel in Concrete (ASTM C876 - 09). West Conshohocken, PA, 2009.

[175] Gu P., Beaudoin J. J.. Construction Technology Update No. 18, Obtaining Effective Half- Cell Potential Measurements in Reinforced Concrete Structures, Institute of Research in Construction. National Research Council of Canada, Ottawa, Canada, 1998.

[176] Alonso C., Sánchez M., Andrade C., Fullea J.. Protection capacity of inhibitors a-

gainst the corrosion of rebars embedded in concrete. In: Brillas E., Cabot P.-L. (eds.) Trends in Electrochemistry and Corrosion at the Beginning of the 21st Century, 2004: 585-598. Universitat de Barcelona, Barcelona.

[177] ASTM International: Standard Test Method for Determining Effects of Chemical Admixtures on Corrosion of Embedded Steel Reinforcement in Concrete Exposed to Chloride Environments (ASTM G109-07). West Conshohocken, PA, 2007.

[178] European Committee for Standardization (CEN). Admixtures for Concrete, Mortar and Grout. Test Methods. Determination of the Effect on Corrosion Susceptibility of Reinforcing Steel by Potentiostatic Electro-Chemical Test (EN 480-14: 2006). Brussels, Belgium, 2006.

[179] Trejo D., Halmen C., Reinschmidt K.. Corrosion performance tests for reinforcing steel in concrete: Technical Report FHWA/TX-09/0-4825-1. Texas Transportation Institute, 2009.

[180] Poursaee A., Hansson C. M.: Potential pitfalls in assessing chloride-induced corrosion of steel in concrete. Cem. Concr. Res, 2009, 39 (5): 391-400.

[181] Fratesi R.. Galvanized reinforcing steel bars in concrete. In: Working Group A2, Project I2, Final Report, COST 521 Workshop, 2002: 33-44. Luxembourg.

[182] Wheat H. G.. Corrosion behavior of steel in concrete made with Pyrament® blended cement. Cem. Concr. Res, 1992, 22: 103-111.

[183] Miranda J. M., Fernández-Jiménez A., González J. A., Palomo A.. Corrosion resistance in activated fly ash mortars. Cem. Concr. Res, 2005, 35 (6): 1210-1217.

[184] Bastidas D., Fernández-Jiménez A., Palomo A., González J. A.. A study on the passive state stability of steel embedded in activated fl y ash mortars. Corros. Sci, 2008, 50 (4): 1058-1065.

[185] Criado M., Fernández-Jiménez A., Palomo A.. Corrosion behaviour of steel embedded in activated fl y ash mortars. In: Shi C., Shen X. (eds.) First International Conference on Advances in Chemically-Activated Materials, Jinan, China, 2010: 36-44. RILEM. Bagneux, France.

[186] Glasser F. P.. Mineralogical aspects of cement in radioactive waste disposal. Miner. Mag, 2001, 65 (5): 621-633.

[187] Fernández-Jiménez A., Miranda J. M., González J. A., Palomo A.. Steel passive state stability in activated fly ash mortars. Mater. Constr, 2010, 60 (300): 51-65.

[188] Castro-Borges P., Troconis de Rincón O., Moreno E. I., Torres-Acosta A. A., Martínez-Madrid M., Knudsen A.. Performance of a 60-year-old concrete pier with stainless steel reinforcement. Mater. Perform, 2002, 41: 50-55.

[189] Criado M., Bastidas D. M., Fajardo S., Fernández-Jiménez A., Bastidas J. M.. Corrosion behaviour of a new low-nickel stainless steel embedded in activated fly ash mortars. Cem. Concr. Compos, 2011, 33 (6): 644-652.

[190] Kukko H., Mannonen R.. Chemical and mechanical properties of alkali-activated

blast furnace slag (F-concrete). Nord. Concr. Res, 1982, 1: 16.1-16.16.

[191] Deja J., Małolepszy J., Jaskiewicz G.. Influence of chloride corrosion on durability of reinforcement in the concrete. In: Malhotra V. M. (ed.) 2nd International Conference on the Durability of Concrete, 1991: 511-521. Montreal, Canada. American Concrete Institute.

[192] Małolepszy J., Deja J., Brylicki, W. : Industrial application of slag alkaline concretes. In: Krivenko, P. V. (ed.) Proceedings of the First International Conference on Alkaline Cements and Concretes, 1994, 2: 989-1001. Kiev, Ukraine. VIPOL Stock Company.

[193] Holloway M., Sykes J. M.. Studies of the corrosion of mild steel in alkali-activated slag cement mortars with sodium chloride admixtures by a galvanostatic pulse method. Corros. Sci, 2005, (12): 3097-3110.

[194] Bernal S. A.. Carbonatación de Concretos Producidos en Sistemas Binarios de una Escoria Siderúrgica y un Metacaolín Activados Alcalinamente. Ph. D. thesis, Universidad del Valle. Cali, 2009.

[195] Aperador W., Mejía de Gutierrez R., Bastidas D. M.. Steel corrosion behaviour in carbonated alkali-activated slag concrete. Corros. Sci, 2009, 51 (9): 2027-2033.

[196] Montoya R., Aperador W., Bastidas D. M.. Influence of conductivity on cathodic protection of reinforced alkali-activated slag mortar using the finite element method. Corros. Sci, 2009, 51 (12): 2857-2862.

[197] Krivenko P. V.. Alkaline cements. In: Krivenko P. V. (ed.) Proceedings of the First International Conference on Alkaline Cements and Concretes, 1994, 1: 11-129. Kiev, Ukraine. VIPOL Stock Company.

[198] Krivenko P. V., Pushkaryeva E. K.. Durability of the Slag Alkaline Cement Concretes. Budivelnik, Kiev, 1993.

[199] International Atomic Energy Agency. Safety Reports Series No. 49: Assessing the Need for Radiation Protection Measures in Work Involving Minerals and Raw Materials, . Vienna, 2006.

[200] Singh D. D. N., Ghosh R., Singh B. K.. Fluoride induced corrosion of steel rebars in contact with alkaline solutions, cement slurry and concrete mortars. Corros. Sci, 2002, 44 (8): 1713-1735.

[201] Hobbs D. W.. Concrete deterioration: causes, diagnosis, and minimising risk. Int. Mater. Rev, 2001, 46 (3): 117-144.

[202] Poonguzhali A., Shaikh H., Dayal R. K., Khatak H. S.. A review on degradation mechanism and life estimation of civil structures. Corros. Rev, 2008, 26 (4): 215-294.

[203] Glasser F. P., Marchand J., Samson E.. Durability of concrete — degradation phenomena involving detrimental chemical reactions. Cem. Concr. Res, 2008, 38 (2): 226-246.

[204] Fernández-Bertos M., Simons S. J. R., Hills C. D., Carey P. J.. A review of accelerated carbonation technology in the treatment of cement-based materials and sequestration of CO_2. J. Hazard. Mater, 2004, B112: 193-205.

[205] Bary B., Sellier A.. Coupled moisture—carbon dioxide-calcium transfer model for carbonation of concrete. Cem. Concr. Res, 2001, 34: 1859-1872.

[206] Papadakis V. G., Vayenas C. G., Fardis M. N.. Experimental investigation and mathematical modeling of the concrete carbonation problem. Chem. Eng. Sci, 1991, 46: 1333-1338.

[207] Johannesson B., Utgenannt P.. Microstructural changes caused by carbonation of cement mortar. Cem. Concr. Res., 2001, 31: 925-931.

[208] Gonen T., Yazicioglu S.. The influence of mineral admixtures on the short and long-term performance of concrete. Build. Environ., 2007, 42 (8): 3080-3085.

[209] Rasheeduzzafar. Influence of cement composition on concrete durability. ACI Mater. J, 1992, 89 (6): 574-586.

[210] Anstice D. J., Page C. L., Page M. M.. The pore solution phase of carbonated cement pastes. Cem. Concr. Res, 2005, 35 (2): 377-383.

[211] Houst Y. F.. The role of moisture in the carbonation of cementitious materials. Int. Z. Bauinstandsetzen, 1996, 2 (1): 46-66.

[212] Litvan G. G., Meyer A.. Carbonation of granulated blast furnace slag cement concrete during twenty years of field experience. In: ACI SP91, Proceedings of the Second International Conference on Fly ash, Silica Fume, Slag, and Other Natural Pozzolans in Concrete, 1986: 1445-1462. CANMET/ACI, Detroit, MI.

[213] Tumidajski P. J., Chan G. W.. Effect of sulfate and carbon dioxide on chloride diffusivity. Cem. Concr. Res, 1996, 26 (4): 551-556.

[214] Papadakis V. G.. Effect of supplementary cementing materials on concrete resistance against carbonation and chloride ingress. Cem. Concr. Res, 2000, 30 (2): 291-299.

[215] Chindaprasirt P., Rukzon S., Sirivivatnanon V.. Effect of carbon dioxide on chloride penetration and chloride ion diffusion coefficient of blended Portland cement mortar. Constr. Build. Mater, 2008, 22 (8): 1701-1707.

[216] Song H.-W., Saraswathy V.. Studies on the corrosion resistance of reinforced steel in concrete with ground granulated blast-furnace slag—an overview. J. Hazard. Mater, 2006, 138 (2): 226-233.

[217] Topçu, İB., Boğa A. R.. Effect of ground granulate blast-furnace slag on corrosion performance of steel embedded in concrete. Mater. Des, 2010, 31 (7): 3358-3365.

[218] Fajardo G., Valdez P., Pacheco J.. Corrosion of steel rebar embedded in natural pozzolan based mortars exposed to chlorides. Constr. Build. Mater, 2009, 23 (2): 768-774.

[219] Parande A. K., Babu B. R., Karthik M. A., Kumaar, K. K. D., Palaniswamy N.. Study on strength and corrosion performance for steel embedded in metakaolin blended concrete/mortar. Constr. Build. Mater. 2008, 22 (3): 127-134.

[220] ASTM International. Standard Practice for Maintaining Constant Relative Humidity by Means of Aqueous Solutions (ASTM E104-02). West Conshohocken, PA, 2007.

[221] Henry B. M., Kilmartin B. A., Groves G. W.. The microstructure and strength of carbonated aluminous cements. J. Mater. Sci, 1997, 32: 6249-6253.

[222] Bakharev T., Sanjayan J. G., Cheng Y. B.. Resistance of alkali-activated slag concrete to carbonation. Cem. Concr. Res, 2001, 31 (9): 1277-1283.

[223] Duran Atiş C.. Accelerated carbonation and testing of concrete made with fly ash. Constr. Build. Mater. 2003, 17 (3): 147-152.

[224] Bernal S. A., Rodríguez E.. Durability and Mechanical Properties of Alkali-Activated Slag Concretes. B. Eng. thesis, Universidad del Valle, 2004.

[225] Jerga J.. Physico-mechanical properties of carbonated concrete. Constr. Build. Mater, 2004, 18 (9): 645-652.

[226] European Committee for Standardization (CEN). Products and Systems for the Protection and Repair of Concrete Structures -Test Methods-Determination of Resistance to Carbonation (EN 13295: 2004). Brussels, Belgium, 2004.

[227] International Organization for Standardization. Determination of the Potential Carbonation Resistance of Concrete — Accelerated Carbonation Method (ISO/CD 1920-12). Geneva, Switzerland, 2012.

[228] European Committee for Standardization (CEN). Products and Systems for the Protection and Repair of Concrete Structures - Test Methods - Determination of the Carbonation Depth in a Hardened Concrete Through the Phenolphthalein Method (EN 14630: 2006). Brussels, Belgium, 2006.

[229] RILEM TC 56-MHM. CPC-18 Measurement of hardened concrete carbonation depth. Mater. Struct, 1988, 21 (6): 453-455.

[230] NORDTEST. Concrete, Repairing Materials and Protective Coating: Carbonation resistance (NT Build 357). Espoo, Finland, 1989.

[231] Laboratório Nacional de Engenharia Civil: Betões. Determinação da resistência à carbonatação. Estacionário (LNEC E391). Lisbon, Portugal, 1993.

[232] AFPC-AFREM. Durabilité des bétons, méthodes recommandées pour la mesure des grandeurs associées à la durabilité: Mode opératoire recommandé, essai de carbonatation accéléré, mesure de l'épaisseur de béton carbonaté, 1997: 153-158. Toulouse, France.

[233] Melchers R. E., Li C. Q., Davison M. A.. Observations and analysis of a 63-year-old reinforced concrete promenade railing exposed to the North Sea. Mag. Concr. Res. 2009, 61 (4): 233-243.

[234] Vassie P. R.. Measurement techniques for the diagnosis, detection and rate estimation of corrosion in concrete structures. In: Dhir R. K., Newlands M. D. (eds.) Controlling concrete degradation. Proceedings of the International Seminar, 1999: 215-229. Thomas Telford, Dundee.

[235] Alexander K. M., Wardlaw J.. A possible mechanism for carbonation shrinkage and crazing, based on the study of thin layers of hydrated cement. Austr. J. Appl. Sci, 1959, 10 (4): 470-483.

[236] Houst Y. F.. Carbonation shrinkage of hydrated cement paste. In: Malhotra V. M. (ed.) Proceedings of the 4th CANMET/ACI International Conference of Durability of Concrete, Supplementary Papers, 1997: 481-491. Sydney, Australia. American Concrete Institute.

[237] ASTM International. Standard Test Method for Drying Shrinkage of Mortar Containing Hydraulic Cement (ASTM C596 - 09). West Conshohocken, PA, 2009.

[238] ASTM International. Standard Test Method for Measuring Changes in Height of Cylindrical Specimens of Hydraulic-Cement Grout (ASTM C1090 - 10). West Conshohocken, PA, 2010.

[239] Byfors K., Klingstedt G., Lehtonen H. P., Romben L.. Durability of concrete made with alkali-activated slag. In: Malhotra V. M. (ed.) 3rd International Conference on Fly Ash, Silica Fume, Slag and Natural Pozzolans in Concrete, ACI SP114, 1989: 1429-1444. Trondheim, Norway. American Concrete Institute.

[240] Bernal S. A., Mejía de Gutierrez R., Provis J. L.. Carbonation of alkali-activated GBFS-MK concretes. In: Justnes H., et al. (eds.) International Congress on Durability of Concrete, Trondheim, Norway. CD-ROM. Norsk Betongforening, 2012.

[241] Puertas F., Palacios M., Vázquez T.. Carbonation process of alkali-activated slag mortars. J. Mater. Sci, 2006, 41: 3071-3082.

[242] Palacios M., Puertas F.. Effect of carbonation on alkali-activated slag paste. J. Am. Ceram. Soc, 2006, 89 (10): 3211-3221.

[243] Puertas F., Palacios M.. Changes in C-S-H of alkali-activated slag and cement pastes after accelerated carbonation. In: Beaudoin J. J., Makar J. M., Raki L. (eds.) 12th International Congress on the Chemistry of Cement, Montreal, Canada. CD-ROM Proceedings, 2007.

[244] Bernal S. A., Provis J. L., Brice D. G., Kilcullen A., Duxson P., Van Deventer J. S. J.. Accelerated carbonation testing of alkali-activated binders significantly underestimates service life: the role of pore solution chemistry. Cem. Concr. Res. 2012, 42 (10): 1317-1326.

[245] Bernal S. A., San Nicolas R., Provis J. L., Mejía de Gutiérrez R., Van Deventer J. S. J.. Natural carbonation of aged alkali-activated slag concretes. Mater. Struct. (2013, in press). doi 10.1617/s11527-013-0089-2.

[246] Criado M., Palomo A., Fernández-Jiménez A.. Alkali activation of fly ashes. Part

1: effect of curing conditions on the carbonation of the reaction products. Fuel, 2005, 84 (16): 2048-2054.

[247] ASTM International. Standard Test Methods for Sampling and Testing Brick and Structural Clay Tile (ASTM C67 - 11). West Conshohocken, PA, 2011.

[248] Standards Australia. Masonry units, segmental pavers and flags - Methods of test. Method 6: Determining potential to effl oresce (AS/NZS 4456.6: 2003). Sydney, Australia, 2003.

[249] Czech Office for Standards Metrology and Testing. Stanovení náchylnosti pórobetonu k tvorbě primárních výkvětů(Determination of susceptibility to the formation of primary efflorescence) (Č SN73 1358). Prague, Czech Reublic, 2010.

[250] Dow C., Glasser F. P.. Calcium carbonate efflorescence on Portland cement and building materials. Cem. Concr. Res, 2003, 33 (1): 147-154.

[251] Brocken H., Nijland T. G.. White efflorescence on brick masonry and concrete masonry blocks, with special emphasis on sulfate efflorescence on concrete blocks. Constr. Build. Mater, 2004, 18 (5): 315-323.

[252] Najafi Kani E., Allahverdi A.. Effect of chemical composition on basic engineering properties of inorganic polymeric binder based on natural pozzolan. Ceram.-Silik, 2009, 53 (3): 195-204.

[253] Allahverdi A., Mehrpour K., Najafi Kani E.. Investigating the possibility of utilizing pumice- type natural pozzolan in production of geopolymer cement. Ceram.-Silik, 2008, 52 (1): 16-23.

[254] Škvára, F., Kopecký L., Myšková L., Šmilauer V., Alberovská L., Vinšová L.. Aluminosilicate polymers -influence of elevated temperatures, effl orescence. Ceram.-Silik, 2009, 53 (4): 276-282.

[255] Temuujin J., Van Riessen A.. Effect of fly ash preliminary calcination on the properties of geopolymer. J. Hazard. Mater, 2009, 164 (2-3): 634-639.

[256] Pacheco-Torgal F., Jalali S.. Influence of sodium carbonate addition on the thermal reactivity of tungsten mine waste mud based binders. Constr. Build. Mater, 2010, 24 (1): 56-60.

[257] Smith M. A., Osborne G. J.. Slag/fly ash cements. World Cem. Technol, 1977, 1 (6): 223-233.

[258] Szklorzová H., Bílek V.. Influence of alkali ions in the activator on the performance of alkali activated mortars. In: Bílek V., Keršner Z. (eds.) Proceedings of the 3rd International Symposium on Non-Traditional Cement and Concrete, 2008. 777-784. ZPSV A. S, Brno, Czech Republic.

[259] Škvara F., Pavlasová S., Kopecký L., Myšková L. and Alberovská L.. High-temperature properties of fly ash-based geopolymers. In: Bílek V. and Keršner, Z. (eds.) Proceedings of the 3rd International Symposium on Non-Traditional Cement and Concrete, 2008: 741-750. ZPSV A. S., Brno, Czech Republic.

[260] Bortnovsky O., Dĕdeček, J., Tvarůžková Z., Sobalík Z., Šubrt J.. Metal ions as probes for characterization of geopolymer materials. J. Am. Ceram. Soc,2008,91(9): 3052-3057.

[261] Duxson P., Provis J. L., Lukey G. C., Van Deventer J. S. J., Separovic F., Gan Z. H.. 39 KNMR of free potassium in geopolymers. Ind. Eng. Chem. Res,2006,45(26): 9208-9210.

第 10 章 耐久性和测试——物理性能

John L. Provis，Vlastimil Bílek，Anja Buchwald，
Katja Dombrowski-Daube and Benjamin Varela

10.1 引言

众所周知，混凝土的抗压强度较高，抗折强度和抗拉强度较低。然而，结合预拉技术、钢筋增韧、配合适当的结构工程设计方法来补偿这些弱点，可以使混凝土的胶凝材料和骨料经受最小的拉伸载荷。这就意味着混凝土的抗压强度、抗折强度和其他力学性能之间的关系是土木和结构工程设计的基本依据。实际上，随着硅酸盐水泥基混凝土普遍应用于市政基础设施中，这些相关性能指标已经通过经验公式和定律写在了标准中。如 28d 抗压强度，有时作为唯一参数用到预测方程里，有时也和其他一些性能参数一起使用。又如，美国混凝土协会[1]指出，弹性模量是抗压强度和混凝土密度的函数，但是也提供了仅基于抗压强度的方程，并且在实际工程中广泛应用。其实还有更复杂详细的理论模型，或者含有大量参数的经验关系式，经常发表在学术文献中，但是没有被广泛应用。Neville[2]提出了很好的硅酸盐水泥混凝土的理论模型，读者可以参考获取更多信息。

这些标准和通用经验关系都是基于几十年来大量硅酸盐水泥混凝土测试的数据得出的。但是，使用硅酸盐水泥混凝土经验关系来预测碱激发混凝土的性能容易出现误差，因为碱激发胶凝材料的物理化学性质及其与骨料的相互作用与硅酸盐水泥明显不同，本书前一章已做讨论。与碱激发混凝土性能相关的数据较少，所以碱激发混凝土各种性能参数之间的可用经验关系或标准也很少。所以，很有必要为 AAMs（和/或验证当前这些材料的标准化关系，如果适当）建立系列经验关系，用于指导 AAM 混凝土的大规模工程应用，这就意味着需要建立一套全面适用的测试方法，用来预测碱激发材料的力学性能。

10.2 力学性能测试

10.2.1 测试方法

有大量的测试方法和尺寸要求用于测试硅酸盐水泥基材料的机械强度，这些都在 ASTM 标准[3]中有详细的描述。碱激发胶凝材料、砂浆及混凝土的力学性能和弹性通常也跟水泥基材料的性能相似，技术人员能够很好地混料、脱模制备高质量材料，并且没有大的缺陷、离析或分层现象，这就没有任何理由不让 AAMs 首先使用现有的检测方法。标准测试方法 ASTM C39[4]，在许多国家标准和国际标准中有等同标准，被广泛应用于混凝土圆柱试样的检测分析。虽然样品尺寸在标准中没有被严格规定，但是大多数试样都是依据 ASTM C31[5]制备成直径 100mm 或 150mm（在美国，直径相当于 4 英寸或 6 英寸），长宽

比为 2，粗骨料粒径不大于圆柱直径的三分之一。测试长宽比低于 1.75 的圆柱体在高强度测量中会出现误差[4]，标准给出校正因子以解决已有的偏差。标准 ASTM C31 是在 1921 年制定发布的，强调抗压强度是混凝土的基本属性，其测试方法能够标准量化混凝土的性能，该标准自使用以来已做过多次修订。圆柱体的顶端涂抹硫黄砂浆、石膏灰泥或者可重复使用的氯丁橡胶垫（氯丁橡胶垫一般仅限于低强度试件使用），测试的加载速率控制在 0.20～0.30MPa/s 之间[4]。

对同种材料，小样品测试会加载更高的应变速率，通常导致测量强度更高，小样品也会表现出更高的断裂韧性[6]，这就意味着在一个测试程序中应尽可能保持这些参数的一致性。近期对 AAMs 的测试发现，力学性能与应变速率有关[7,8]，类似的变化规律也出现在硅酸盐水泥混凝土中。这就表明，在对 AAMs 的力学性能测试时，可以照搬大部分传统混凝土的知识来理解 AAMs 的材料性能。研究显示，碱激发矿渣混凝土比普通水泥混凝土的尺寸效应更强[9]，这主要是 AAM 混料困难，导致样品均质性不佳。

欧洲标准 EN 12390-3[10] 还给出了硬化混凝土的测试方法，除了圆柱体[11]外，还可以使用立方体（100mm 或 150mm）的试样，有更多封顶化合物可以使用，加载速率范围放宽到 0.20～1.0MPa/s。与相应尺寸的立方体相比，本标准文件中圆柱测试数据的再现性和重复性更好，但是两种样品模式都是可用的。由于长宽比的差异，立方体试块将给出比 ASTM 标准圆柱体试块[6]高 10%～15%的测量强度。当比较不同测试方法获得的数据时，应考虑到试块强度间的差异。

在国际上砂浆抗压强度的测试同样也有大量如 ASTM C109 的标准、规范[12]，样块使用的是 50mm 的立方体，标准砂作料。这些标准主要用于砂浆研究的实验室测试，对于大体积混凝土的混料与浇筑并不适用。ASTM C109 根据水灰比（硅酸盐水泥）或者流动度（其他胶凝材料）提出了样块的配比，如果 AAMs 浆体的流变性能与硅酸盐水泥明显不同，就会出现问题。此外，标准中指定的养护条件（20～72h 后脱模，然后将样品浸入石灰水中，直到测试）可能不适合碱激发胶凝材料体系。众所周知，如果想要把混凝土强度数据和砂浆的强度数据建立联系时，就需要仔细考虑测试条件和制样方式（含气量、骨料含水率、养护条件和样品检测龄期等参数），通过精细的实验设计，才有可能得到合理的经验关系式。抗压强度要比抗拉/抗折强度的相关性更好。

如 EN196-1[16] 或者 ASTM C349[17]标准给出，使用 40mm×40mm×160mm 长方形模具用于砂浆测试，长方体样块在弯曲应力下破坏（或者通过适当的方法让一半的样块不受有害应力），然后用半截试块做抗压强度测定。这种方法的优势是单个样品可以同时给出抗折强度和抗压强度数据，但也有可能测完抗折强度后样品的边角破损，或者样品的微观结构有所损坏，导致抗压强度测试出现问题。该方法的前提假设是，该方法导致的弯曲破坏对试块两边的损伤很小，但 ASTM 标准则具体说明该方法可供参考，并不能直接代替立方体测试（ASTM C109），表明数据的可靠性较低。

类似情况如使用工业陶瓷的测试方法对不添加骨料的、小棱柱状（通常是 10mm 或者是更小的尺寸）的硬化胶凝样块，或者小圆柱体，进行实验室测试[18]。前面所提到的测试方式很少应用于实际混凝土强度的测试，因为砂浆的水胶比跟混凝土[15]有很大区别，但可以用作胶凝材料设计优化研究。同样，ASTM C773 标准用于测试白陶瓷的抗压强度，规定每次测试使用至少 10 个小柱状样块[18]，以说明这种材料的脆性破坏程度。相比之下，在

ASTM C109[12]中只需要2~3个砂浆样块，或在 ASTM C39 中使用2个混凝土圆柱样块[4]。在实验室测试砂浆样块往往不能满足标准要求的数量，存在一定不确定性和偏差，所以只能用于优化配方，不能用来预测混凝土性能。

混凝土的弹性特性可以在载荷下直接测量，也可以使用基于振动传播的速率和传播能力的其他技术测量，如超声技术和谐振频率分析技术[19]。弹性测试可用于新浇混凝土或硬化混凝土，数据对结构设计计算和服役混凝土评估很有用。现在普遍使用超声波和谐振频率技术测试服役混凝土样块的抗压强度，但还是有些争议[19]，AAMs 这一领域的数据也有发表，将在 10.2.4 节中介绍。其实，测试方法对静态和动态的弹性模量有很大影响[20]，弹性模量测试的经验关系式也很少。此外，还有许多不同的测试方法和标准主导超声波测试[20]（在文献 [21] 中报道了20种）的应用，这些测试方法[21]总结的结论是"存在固有的不确定性……如此之高以至于不适用于许多实际的工程测试"。ASTM 测试方法 C597[22]明确警告不能使用上述方法测试强度和弹性模量，但也指出，可以给出一些特定混凝土的参考数据来评价该混凝土。因此，要对比不同测试方法获得的数据，通过对不同样品和测试方法的研究，科学地选择使用相关数据。

样品制备过程对碱激发材料力学性能影响很大。在测试混凝土抗压强度时，湿样品比干样品强度更低，对这种现象的解释纷说不一。AAMs 在测试过程中要特别考虑材料对干燥的敏感性，这会导致材料后期的开裂和/或碳化，这在本书第9章及下面一节做了讨论。

10.2.2 碱激发材料的机械强度

几乎所有学术刊物涉及 AAMs 工程性能时，就会涵盖净浆、砂浆或者混凝土的机械性能，砂浆和混凝土的机械强度从低到高都会涉及。碱硅酸盐激发 BFS 砂浆样块的抗压强度超过了95MPa；25℃下养护 28d 的碱硅酸盐激发粉煤灰砂浆样块的抗压强度达 70MPa[23-26]；30mm 见方的碱硅激发磷矿渣砂浆试块，20℃养护 28d，抗压强度达到 120MPa[27]；40℃养护 20h，室温（20±5）℃养护 28d，碱硅酸盐激发偏高岭土的砂浆圆柱试样（直径 25mm×高 50mm）的抗压强度超过 90MPa[28]。这些数据否定了一直以来重复的说法，即碱激发材料需要长时间的加热蒸养才能达到满意的机械强度。相反，在室温或接近室温的条件下，恰当控制和设计 AAM 配合比，在某种程度上就能够硬化并表现出极优的力学性能。

在实验室制备的 BFS 和/或粉煤灰基混凝土样品抗压强度都能超过 50MPa[9,29-34]，强度等级达到 110MPa 的混凝土在乌克兰也能标准化制备[34]。工业化产品要普遍实现这种强度也是比较难的，但是通过选择适当的原材料，碱激发制备商用"高性能"混凝土是有可能的，如第2章和第11章的阐述和论证。

众所周知，硅酸盐水泥基材料的强度与孔隙率有关，这也是所有脆性材料的普遍规律；从不锈钢到煅烧的氧化铝，都遵循指数关系[2,6]。对于混凝土，这种关系在一定程度上受到材料的孔径影响，碱激发混凝土其实也有类似规律。一些碱激发胶凝材料，特别是偏高岭土基材料，由于新拌浆体需水量较大，相对更加多孔。Smilauer 等人建立了孔隙率和其他胶凝材料参数与强度的微观机理模型[35]，证明模型可有效应用于未来碱激发材料的分析研究。在低水胶比下，使用高压压实碱激发材料，减小孔隙率所得到的样品具有更高的抗压强度；500MPa 压实的碱激发 BFS，其抗压强度超过 250MPa[36]。

用 AAM 混凝土做预制构件，经蒸养后抗压强度发展得很好[34,37-40]。碱硅酸盐激发粉煤的细长柱状（截面 175mm², 长 1500mm）试样，经纵向和横向增韧，热养护后的力学性能能很好地满足标准规范 AS 3600[41] 和 ACI 318[43] 结构混凝土的要求。最近，Yost 等人[44,45] 设计制备了过度增韧、欠增韧和临界剪切态的碱激发粉煤灰混凝土，其性能可分别满足服役和实效时 ACI 标准的相关规定。

10.2.3 抗折强度和抗压强度关系

越来越多的证据表明，在相同抗压强度下，根据标准经验公式预测，AAMs 的抗折强度和抗拉强度远高于硅酸盐水泥混凝土[30,32,43,46-48]。这些标准经验公式在不同国家略有不同，但通常抗压强度和抗折强度之间呈指数关系（指数 0.5~0.7）；美国混凝土协会给出的公式是 $\sigma_f = 0.6\sigma_c^{0.5}$，其中 σ_f 是抗折强度（断裂模数，单位 MPa），σ_c 是混凝土的抗压强度（也以 MPa 计）[42]。图 10-1 是由不同文献的数据总结出来的抗压强度与抗折强度的关系图[9,30,32,48-54]，几乎所有 AAM 混凝土的抗折强度都高于指数公式计算的硅酸盐水泥混凝土的抗折强度。值得注意的是，一些作者认为 ACI 公式计算的数据是硅酸盐水泥基材料[55]断裂模数的下限，而这里碱激发混凝土的数据是遵循这一要求给出的。

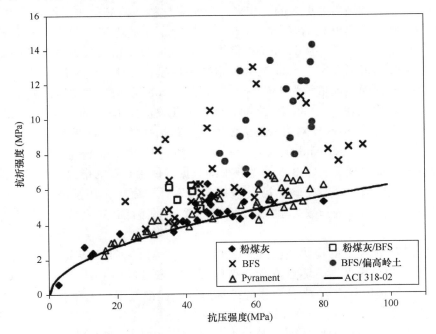

图 10-1 不同原料制备碱激发混凝土的抗折强度和抗压强度关系（龄期从 4h 到 1 年）和 ACI 318-02 中 OPC 混凝土计算关系的对照图（用于 AAM 混凝土的数据取自文献 [9, 30, 32, 48-54]）

Sumajouw、Rangan[39,47]和 Dattatreya 等人[56]测试了各种载荷下钢筋增韧碱激发粉煤灰混凝土的梁和柱，发现与澳大利亚标准 AS 3600 规定的硅酸盐水泥混凝土的失效模式和挠度类似[41]，印度标准 IS 456: 2000[57] 和 ACI 318−02[42] 也能用于检测 AAM 混凝土的力学性能。通常用于 OPC 混凝土的关系式，测试碱激发天然火山灰混凝土[58]和碱激发粉煤灰混凝土[37]的劈裂抗拉强度，也会得到与OPC混凝土类似的数据变化规律。

10.2.4 弹性模量和泊松比

在结构设计中必不可少的两个参数是泊松比和弹性模量。当使用AAM混凝土结构作力学建模时，需要对以上参数有更深的理解。一种方法是通过应力-应变曲线得到弹性模量，通过应变测量仪得到泊松比，画正交曲线得到关系式；另一种方法是使用超声技术获得力学数据。对于碱激发偏高岭土胶凝材料，Lawson[59]使用超声技术，弹性模量随着Si/Al比的增加而降低，当Si/Al比从1.5增加到5.0，弹性模量从9.1GPa降至5.5GPa。但是，Duxson等人[60]研究发现碱激发偏高岭土样品的Si/Al比1.15增加到1.90时，弹性模量（直接从应力-应变曲线获得数据）从2.3GPa增加到5.2GPa。在这些数据中，只有添加Si使Si/Al比超过2.0时，弹性模量才降低。对于碱激发粉煤灰混凝土，Wongpa等人[61]发现弹性模量随着养护时间的增加而降低，Talling和Krivenko也发现蒸养高强度AAS早期，弹性模量是降低的[34]。然而，Douglas等人[62]发现，龄期从28d到91d，碱激发BFS混凝土的弹性模量略有增加，并且样品的弹性模量在30~35GPa之间，这很好地吻合了ACI 318模型，作为碱激发BFS"F混凝土"[33,46]的参考数据。

很明显，在这一领域需要更多的工作来确定胶凝材料结构演化与弹性性能之间的关系，近期有人使用细观力学建模和纳米压痕方式[35,63]取得了一些进展，但还需要逐步构建微观模型与宏观混凝土性能关系。从Němeček等人[64]研究碱激发粉煤灰和碱激发偏高岭土胶凝材料的纳米压痕数据得出，N-A-S-H型凝胶本征弹性模量为17~18GPa（相比之下，沸石结构的晶态方钠石的弹性模量约为43GPa[65]），而Puertas等人[63]发现碱硅酸盐激发BFS的C-A-S-H凝胶弹性模数在28~50GPa之间，碱氢氧化物激发BFS的值在12~42GPa之间。Oh等人[66]提出Al取代C-S-H对凝胶的力学性能，包括弹性模量，影响不大。

Diaz-Loya等人[30]、Fernández-Jiménez等人[50]和Sofi等人[48]研究表明，碱激发粉煤灰混凝土的弹性模量都要低于ACI 318模型获得的数据（使用0.5的指数关系式，给定硅酸盐水泥相应的抗压强度值）。Diaz-Loya等人[30]提出AAM混凝土抗压强度和弹性模量是线性关系，但是也提出了0.5的指数关系式添加一个混凝土密度的校正系数，也能很好地与实际数据吻合。

碱激发粉煤灰混凝土的应力-应变曲线也与公式计算的OPC混凝土相应曲线做了对比[67]，测量和预测曲线直到破坏点都吻合得很好[43]。Sarker[68]提出一个高性能OPC的修正公式，可以很好地解释应力-应变曲线的破坏后区，但把数据套入原有标准公式中也能给出合理解释。

在文献中关于AAMs泊松比的实验数据相对较少。对于碱硅酸盐激发偏高岭土浆体，Lawson[59]也使用超声波技术测定泊松比，随着Si/Al比的增加而减小，从Si/Al=1.5处的0.221减小到Si/Al=3处的0.111。Diaz-Loya等人[30]提出碱硅酸盐激发粉煤灰混凝土的测量值在0.07~0.23之间，尽管数据看起来相当分散，但在较高的抗压强度下趋于增加（图10-2）。

一般来说，AAM混凝土基本的力学性能与硅酸盐水泥混凝土标准里的关系曲线能很好吻合。然而，有必要在工程背景下对可能偏离"预期"（即硅酸盐水泥）行为进行理论和机理研究，如徐变和收缩（见第10.3节和第10.4节）。很有必要探究OPC标准中公式背后的

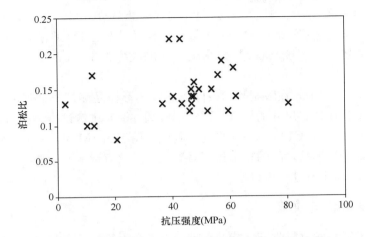

图 10-2 碱激发粉煤灰混凝土的泊松比随抗压强度的
变化关系（来自参考文献[30]的数据）

基本原理，看能否修改后用于 AAM 混凝土基础化学力学性能的研究。这就需要我们继续努力做大量工作，来完成上述目标。

10.3 收缩和开裂

10.3.1 收缩测试

有许多测试标准用于检测新拌合硬化水泥净浆、砂浆和混凝土尺寸的稳定性，在 ASTM 体系中就有 10 种测试方法[69]。测试一般都具体到某一收缩性能（干缩或自收缩经常单独考虑），或者有时测量多种收缩的叠加引起在不同收缩方式和收缩速率下尺寸的变化。从相对基础的化学和热力学角度看，它预示着碱激发 BFS 胶凝材料中 C-A-S-H 相比 OPC 的水化反应形成的低 Al C-S-H 相[70,71]更容易收缩。然而，碱激发砂浆和碱激发混凝土选择适度的 w/b 比和良好级配的骨料后，其收缩要低于 OPC 基混凝土，这表明在混凝土服役期间有其他的因素影响材料的性能。因此，了解一些用于评价收缩性能的具体测试模型至关重要，以便准确反映材料的服役性能。

用于检测硬化材料尺寸稳定性基本测试方法是 ASTM C157[72]检测使用长方体试样，长 285mm、截面积 25mm×25mm（用于砂浆），截面积 75mm×75mm（用于 25mm 以下骨料的混凝土），截面积 100mm×100mm（用于含有大于 25mm 骨料的混凝土）。横截面表面嵌入了测量螺栓。将样品湿养 24h，测量其"初始"长度，然后在石灰水中浸泡至 28d 再次测量。此时，样品从石灰水中取出，干态下测试，或者把样品放在 23℃、相对湿度 50% 的环境中，湿态下测试尺寸稳定性，通常长度测量要持续 64 周。石灰水浸渍是为了提供 Ca^{2+} 饱和环境以防止通过软水侵蚀 OPC 胶凝材料脱钙，而碱激发胶凝材料有未结合入凝胶中的碱，希望通过早期将孔溶液中的碱脱除，如果使用石灰石水会有很大问题。

ASTM C596 中砂浆干缩试验[73]与 C157 类似，但样品只需要放置于相对湿度 50% 环境下 3d，意味着 C596 样品比 C157 龄期更短。这个标准期望通过砂浆的收缩性能与相应混凝土的性能关联，但实际效果也被质疑[69]。澳大利亚标准 AS1012.13[74]，湿养护龄期比 C596 长一些（AS1012.13 是养护 7d），对于大掺量矿渣胶凝材料也不太理想，因为这些材料在一定龄期会发生微膨胀（与传统相反）。如测试 56d 的收缩性能[75]，膨胀会有效补偿实际的收缩。

在 ASTM C1581[76]中，在限制收缩条件下通过环形样品测试早期尺寸的稳定性。一种是在环中裂纹，或是在特定尺寸的模具（圆锥形或圆柱形）中测试凝结前后的尺寸变化（如 ASTM C 827[77]或 ASTM C1090[78]）。将这些测试应用于 AAM 时，AAM 浆体往往比 OPC 浆体更黏一些，试样和模具之间的粘附作用对测试效果会有影响。这些测试也通常用于研究抗裂性，这将在 10.3.4 节进一步讨论。

10.3.2 碱激发材料的收缩

在发表的关于碱激发材料收缩性能的文献中大家众说纷纭，有些文献说收缩很小，有些说收缩很大。特别是，碱激发 BFS 胶凝材料的收缩在一些环境下养护会有问题，特别是在没有充分养护后就直接放置于干燥环境中。碱激发偏高岭土胶凝材料收缩现象明显；碱激发粉煤灰材料在不同配合比设计条件下有时收缩非常少，有时收缩明显。一般来说，加过量水到 AAM 中，或早期置于干燥条件下养护，都会出现干缩（和/或开裂）问题，因为水在 AAM 凝胶结构中的化学结合程度比硅酸盐水泥基材料小得多。

Kukko 和 Mannonen 的早期研究[46]表明，（在相对湿度 40% 以下）碱激发 BFS "F 混凝土" 的干缩略低于（～10%）在相同配合比、工作性和两年龄期强度等级下的硅酸盐水泥混凝土，干燥收缩率随着 w/b 比的增加而增加。相反，Häkkinen[79,80] 发现在相对湿度～70%（过饱和亚硝酸钠溶液），F 混凝土早期收缩（<2 个月）低于 OPC 混凝土，但是 AAM 混凝土的收缩持续时间更长，4 个月龄期的收缩量超过了 OPC 样品总的收缩量。这可能由于 AAM 中 BFS 在早期膨胀引起的，如 10.3.1 节中所提到的，70%BFS/30%快硬性硅酸盐水泥混凝土也有类似现象（相对于快硬性硅酸盐水泥，早期收缩慢，但后期收缩延长）[79,80]。暴露于加速碳化环境（5%CO_2）时，可观察到收缩程度略有降低[79]。

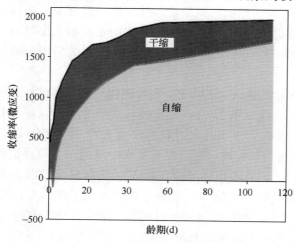

图 10-3　在 24℃、相对湿度 50% 下干燥，碱激发 BFS 砂浆中的干缩和自缩性能曲线（来自 Melo Neto 等[81]的数据）

Kukko 和 Mannonen[46]研究发现，在 40% 和 65% 相对湿度下的收缩行为不同，较高相对湿度与 40% 相对湿度的纯干燥状态相比，同时进行的自缩/干缩速率是不同的。Melo Neto 等人[81]使用平行干燥/自收缩测试来确定每种收缩模式的影响（图 10-3），并且发现碱激发 BFS 在持续收缩中的自收缩比干燥更有影响，与 OPC 相比，收缩与自身干燥相关程度更大，这与 Chen 和 Brouwers 建立的模型一致[70]。然而，碱激发 BFS 砂浆样品在水中放置 28d 后观察到没有明显收缩。相反，与微收缩的硅酸盐水泥基材料相比，有略微膨胀[80]。Douglas 等人[62]也

第10章 耐久性和测试——物理性能

观察到膨胀现象,并发现碱激发 BFS 砂浆存放在水中持续 9 个月,样块有微膨胀。Sakulich 和 Bentz[82]使用内养护方式,在混凝土中加入饱和的膨胀黏土颗粒来减轻碱激发 BFS 混凝土的收缩,影响了自缩和干缩程度。

Wang 等人[83]和 DuranAtiş 等人[84]注意到通过实验发现 NaOH 或 Na_2CO_3 激发 BFS 与 OPC 有相似的收缩性质,而在干燥条件下水玻璃激发 BFS 胶凝材料比 OPC 水泥收缩严重。Collins 和 Sanjayan[85]将这种收缩归因于 AAM 胶凝材料内部存在大量中孔,而 Krizan 和 Zivanovic[86]提出了一种凝胶脱水收缩的机制,这是从 Glukhovsky 早期的研究工作衍生出来的。Douglas 等人[62]和 Cincotto 等人[87]发现随着 BFS 基胶凝材料中激发剂含量的增加干缩程度也增加,以及 Krizan 和 Zivanovic[86]观察到随着激发剂模数增加收缩程度增加。前边每项研究都表明 AAM 的收缩要高于 OPC 基浆体。

Collins 和 Sanjayan 研究制备了抑制收缩和开裂的碱激发 BFS 混凝土[88,89],发现当 AAM 和 OPC 基样品同时养护 24h,然后做抑制收缩处理,AAM 混凝土更易收缩,这就更容易开裂。然而,AAM 混凝土养护 3d 的收缩性能与同龄期养护的 OPC 混凝土性能相当 (OPC 混凝土的养护 1~3d 收缩性能差别不大)。在 AAM 混凝土中引入多孔和低活性的矿渣粗骨料代替天然岩石骨料,也发现由于内养护效应[90]减少了收缩。

在碱激发粉煤灰和粉煤灰-BFS 砂浆中,Yong 研究了配合比设计和制备条件对砂浆收缩性能的影响[91],研究了这些砂浆的自缩和干缩性能。40%BFS 胶凝材料早期没有微膨胀现象,而会有一定程度的收缩。用水量会特别影响早期的干燥收缩,但在养护后期,由于凝胶生成会产生更多孔隙,所以收缩性能与激发剂(以及凝胶形成的程度,铝硅酸盐凝胶的组成)的特性有关[91]。60%粉煤灰/40%BFS 胶凝材料使用低模数硅酸钠溶液($SiO_2/Na_2O=0.5$)激发,7 个月的收缩应变约为 5×10^{-4},偏硅酸钠溶液($SiO_2/Na_2O=1.0$)激发的收缩应变为 6×10^{-3}[91]。文献报道碱激发粉煤灰混凝土浇筑的小路没有明显的收缩开裂,即使浇筑长度达到 12m 都没有收缩裂纹[92]。

Yong[91]同样研究了密封养护时间对碱激发砂浆干缩的影响,如图 10-4 所示。样品在测试开始之前养护较长时间,使凝胶生长充分,更能抵制尺寸变化。Rangan[37]研究发现,粉煤灰基 AAM 在热养护条件下比室温养护的干缩程度低很多,这也可能与胶凝材料的凝胶成熟度有关。但对于强度和微观结构发展完全又放置于室温的体系,使用过高的养护温度或过度延长高温养护时间可能或多或少会导致收缩[93]。

煅烧黏土基 AAM 需水量较高,如在书中第 4 章所讨论的,很容易收缩。N-A-S-H 凝胶的显微结构和化学性质本身具有以下特性[25,94],因为缺乏结合水和凝胶密度低,材料内部的微观结构可能会逐渐塌陷,转成沸石晶体。用 20%方解石或白云石取代的黏土基铝硅酸盐前驱体的收缩程度更大[95],使用石灰石作骨料也有同样的

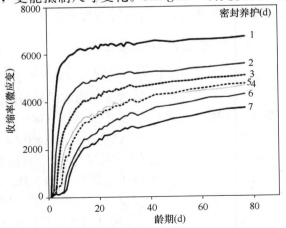

图 10-4 使用硅酸钠溶液(模数为 1.5)激发的 60%粉煤灰/40%BFS 砂浆不同龄期的收缩率(样品放置在密封袋中养护)(数据来自文献[91])

现象[96]。Elimbi 等人[97]也发现不同温度下煅烧黏土制备的样品，抗压强度和尺寸稳定性直接相关，表明反应活性越高的前驱体，反应过程中微结构演变越明显，因此会进一步形成稳定的凝胶。

鉴于 AAM 在干燥过程中容易导致收缩和开裂，人们尝试了各种方法来减少这些问题。为了开发高品质的材料，除了使用有效的养护制度，也可能使用减缩的化学外加剂（如第 6 章所述）。第 12 章详细讨论的使用纤维增韧来控制 AAM 中的裂纹开裂，研究已经证明纤维增韧可以有效控制尺寸变化和载荷引起的开裂。

10.3.3 开裂分析

建筑材料普遍都不希望出现开裂问题，因为它会导致力学性能的下降，加速潜在有害离子侵入结构内部。混凝土开裂一般由机械化学（由于晶态变化、化学反应或放热引起的胶凝材料的尺寸或骨料的变化）或机械物理（载荷或温升循环）过程引起。混凝土中出现微裂纹的根本原因是硬化后部分抑制了凝胶相的化学收缩，这也是从热力学角度提出碱激发胶凝材料和相关（火山灰）体系固有的化学特性[70,98]。几乎所有用于建材的胶凝材料都显示出收缩或膨胀，在硬化前后发生相变而产生强度，并逐渐发展[99]。因此，在水泥和混凝土技术领域控制尺寸稳定是一项重要难题。

应用于评价混凝土开裂最常用的测试是 ASTM 1581 的受限环试验[76]，在较长龄期内监测由于抑制收缩引起的开裂。也有其他类似方法，如双同心环测试法，能够检测膨胀过程以及收缩[100]，但这些方法尚未标准化。这些方法的缺点是具有较低收缩率或较高拉伸强度的双环样品需要花几个月去破裂，而对椭圆环（在中国的标准[101]），埋有应力放大器的约束梁[89]，或部分受限的板[102]更实用。开裂可以通过视觉，或通过声、电或超声方法观察，这些方法更适合于检测微裂纹。

在较小的（薄截面的混凝土、砂浆或净浆）试样中，同样可以使用荧光染色的光学显微镜或通过扫描电镜来分析微裂纹[103]，而 AAM 尚未有相应的文献公开。在裂纹中浸渍熔融金属如伍德合金，然后在室温下凝固，保留的裂纹模式可用作后续观察研究[104,105]，但在 AAMs 中，似乎也没有以这种方式做裂纹研究。

10.3.4 碱激发材料的微裂纹

通过 $NaOH/Na_2CO_3$[46,106]或 $NaOH$[9,80]激发的 F 混凝土样品中可看到明显的微裂纹。Hakkinen[79,107]研究发现随着碳化速率的增加，强度减小，并且高掺量的激发剂会出现更多微裂纹。Collins 和 Sanjayan[89,108]在这一领域进行了详细研究，使用约束环和约束梁测试方法，加上毛细管吸入法、显微镜和压汞法测试，结果发现养护期间的干燥效应是碱激发 BFS 开裂的主要原因，但是使用多孔粗 BFS 骨料能抑制开裂，性能优于 OPC 混凝土。Shen 等人[109]采用国标《水泥胶砂干缩试验方法》JC/T 603—2004 约束椭圆环试验[101]研究发现，加入粉煤灰或活性 MgO 可以降低碱激发 BFS 开裂的趋势。

Bernal 等人[31]也通过毛细管吸入法测量显示硅酸盐激发 BFS 混凝土出现微裂纹程度明显取决于混凝土配合比设计中的胶凝材料含量；在养护混凝土样品过程中，过量的胶凝材料会使放热量增加，早期热诱导引起混凝土的微裂纹。然而，在这种情况下，没有观察到微裂纹对混凝土的碳化速率有任何显著影响，表明过量的胶凝材料能够提高基体致密度来补偿裂缝传输引起的碳化。

碱激发混凝土的微裂纹程度与胶凝材料含量和养护制度直接相关，这就需要理解胶凝材料和骨料之间的相互作用，养护放热，养护中热和湿传输的影响，进而减少微裂纹对混凝土性能和耐久性的影响。

10.4 徐变

徐变是混凝土在长期载荷下尺寸的缓慢变化过程。在实际中，尽管拉伸徐变试验方法已经被开发和应用[112,113]，但在 ASTM C512[110] 或 ISO 1920-9[111] 中通常都是用抗压载荷。徐变是混凝土材料内应力释放的一种特性，也是混凝土设计制备过程时结构破坏的潜在原因。所以，适当的徐变可以补偿任何混凝土材料的收缩，但不会引起建筑中其他结构或非结构部件的开裂。

徐变试验需要几个月的时间或者更长，但通常不可能等待这么长时间才决定给定的混凝土是否适用于所需的应用，所以一般都选择一个合适的标准模型来预测徐变。徐变方面的文献很多，也有很多模型可以使用。通常把实验室测试的 28d 抗压强度作为主要（通常是唯一的）输入参数，代入经验公式推出徐变数据。抗压强度、混合料配合比、骨料类型和载荷龄期等信息，可以为硅酸盐水泥徐变提供很好的模型数据，使其用于世界各地的结构工程。现有大量模型都存在争议，大家就哪个模型适用于特定类型和强度等级的混凝土，以及这些争议的细节和结果（如果有）远远超出了本报告讨论的范围。这里最重要的是对于 AAM 的分析，而所有的模型都是基于硅酸盐水泥及其相关材料的物理化学性质（特别是 28d 抗压强度和弹性模数之间的关系）得出的。

正如在 10.2 节中所讨论的，在 AAMs 中 28d 抗压强度和弹性模量不同于硅酸盐水泥混凝土，意味着标准徐变模型不可能直接用于 AAMs。此外，AAMs 和硅酸盐水泥基混凝土的微观结构、凝胶结构以及强度发展机制之间的差异可能导致材料对施加载荷的时间响应存在差异。另外 AAM 养护期间强度发展曲线与硅酸盐水泥有很大差异，也可能导致需要重新考虑使用 28d 抗压强度作为主要测量变量来评价 AAM 徐变。Kukko 和 Mannonen 发现，在长龄期，甚至是 90d 内，硅酸盐水泥混凝土和 AAM 混凝土的徐变差异明显。对于碱激发 BFS "F 混凝土"[46]，AAM 混凝土的徐变程度超过 OPC 混凝土，AAM 混凝土随着龄期增长徐变增大，而 OPC 混凝土的徐变减少。100d 的短龄期徐变试验数据显示[52]，并没有以上规律，因此可能不需要考察长期服役下的徐变行为。早龄期加载的 AAM 混凝土具有更高的徐变系数，在硅酸盐水泥混凝土中也是类似情况。只有很少的数据是关于配合比设计对碱激发 BFS 混凝土徐变的影响，但也有文献报道，硅酸盐激发混凝土比氢氧化物激发混凝土徐变趋势低，碳酸盐激发胶凝材料徐变最高[34]。

当前，学者对粉煤灰基 AAM 混凝土徐变的研究较少。Gourley 和 Johnson[114] 报告了当粉煤灰基材料用于预制构件时具有较低的徐变。Wallah 和 Rangan[115] 研究了两种不同的热养护粉煤灰基 AAM 混凝土的徐变行为，根据澳大利亚标准 AS1012.16[116]，7d 后用 40% 的抗压强度加载圆柱形试样，监测 12 个月。此时所有样品的总徐变大约为 1.4×10^{-3} 左右，几乎不受配合比和养护制度的影响。徐变比（徐变与抗压强度的比率）低于相同抗压强度的混凝土[115]，但因为总徐变在所有测试样品中相似，较低强度样品显示出较高的徐变比。澳

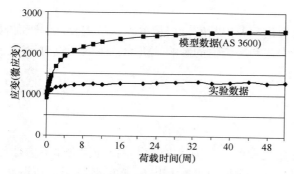

图 10-5　碱激发粉煤灰混凝土徐变的实验数据与 AS 3600 模型数据的比较（改编自文献[115]）

大利亚标准 AS 3600[41]有一个适用于硅酸盐水泥混凝土徐变经验模型，Gilbert 使用该模型数据与 12 个月龄期碱激发粉煤灰混凝土的徐变行为作对比[117,118]，图 10-5 显示 AAM 混凝土的徐变量远低于标准预测模型的数据。

Lee[92]使用标准 AS 1012.16[116]进行测试室温养护的粉煤灰/BFS 碱激发混凝土的徐变速率低于 OPC 混凝土，在 AAM 混凝土强度范围内的徐变比也有类似结果，较密封养护而言，室外养护 AAM 混凝土样品的徐变增加了 3 倍。

也有文献根据 ASTM C512[49]测试了 Pyrament AAM 混凝土的干缩和徐变，该标准中描述的徐变速率数据的线性化计算公式不适用于 AAM 材料，但没有给出偏差方向或程度。

基于以上讨论，显然需要合适的模型来解释 AAM 的徐变（如果硅酸盐水泥混凝土的现有经验模型恰好符合 AAM 具体数据中的趋势，那就不用考虑力学或微观力学方面的原因，可直接使用），这就需要使用在良好控制条件下获得的高性能 AAM 混凝土的大量实验数据来推导这些模型。迄今为止，通过各种测试方法，各种配合比设计条件获得的数据还远远不足以建立 AAM 徐变的模型。因此，这一领域非常重要，也是 AAMs 体系未来研究的重点。确定影响 AAM 混凝土徐变性能的载荷压力也很重要，标准试验和其他实验室研究使用了多种载荷压力，通常为无侧限抗压强度的 25%～40%。由于 AAM 混凝土和硅酸盐水泥混凝土的微观结构和纳米结构不同，载荷压力对徐变数据的影响也不同。

10.5　冻融和抗冻性

10.5.1　测试方法

材料的冻融循环是混凝土的耐冻性评价重要指标，相关参考标准包括 ASTMC 666[119]、CSA 231[120,121]、ASTM C1645[122]，或诸如用于瓷砖测试的 ASTM C1026[123]或 ISO/NP 15045—12[124]。ASTM C67[125]通常用于混凝土试样冻融测试，特别是预制路面，最初用于黏土砖墙。ASTM C6 用作冻融试验测试还不够严格，因为在北美环境下，冷却速度较缓，冷冻温度适中。用于混凝土每种冻融条件和要求都在 EN 206—1 给了定义[127]。

在这些标准中，混凝土样品（通常是棱柱状）做冻融循环时，不同标准中的冻融温度、循环次数和循环时间各有不同。ASTM C666 的冻融实验可以在空气或水中进行，但也注意到"从硬化混凝土切割的样品和实验室制备的样品没有建立冻融循环之间的关系[119]。" RILEM TC 176—IDC[128]特别区分了在去离子水接触的循环所引起的冻融损伤和含盐（测试方法中使用 NaCl）引起的结垢损伤。最早 RILEM 在 TC 117—FDC[129]提出了 "CDF" 测试方法；后来优化方法，在 TC 176—IDC 中给出 "CIF" 测试方法和板材测试方法[128]。不同的国家标准都测试阻垢；CEN/TS 12390—9[130]或现在取消的 ASTM C672[131]把样品置于除冰盐水中会结垢。NaCl 溶液在这些试验中是最常用的盐溶液。

在每个测试中，在特定次数循环后，要把标况储存样品或未循环前样品与循环后样品的外观、单位面积的质量损失和/或力学性能作对比。另一种表征初始和循环后样品的冻融循环的便捷方法是使用超声波检测动态弹性模量。一些规范和标准也会测量梁的弯曲强度和抗压强度，样品的尺寸稳定性或质量变化（特别是在设计用于分析结垢或与受损样品中的水分吸收有关的测试中）。大多数国际标准都没有规定接受或生效的标准限值；虽然一些国家的标准给了详细的结垢限值，但通常都是用于材料性能的比较。

在 AAM 的冻融测试中特别重要的一点是在检测之前样品的养护。Boos 等人[132]对几种复合水泥混凝土做了研究，证明在北欧环境下能提供很好的抗冻性，但是当根据 CEN12390-9 中的三种方法检测时，发现性能较差。作者建议复合水泥应该在 56d（而不是 28d）后测试冻融性能，相比前边测试结果，复合水泥和纯硅酸盐水泥混凝土在 56d 养护后测试的结果相当。鉴于一些 AAMs，特别是 BFS 混凝土中获取的冻融数据较少，建议直接与这些材料分析结果关联，获取有用的数据。

10.5.2　碱激发材料的性能

冻融和结垢测试广泛用于硅酸盐水泥混凝土，也应能够类推应用于碱激发材料。Janssen 和 Snyder[133]提供了硅酸盐水泥混凝土的抗冻融性的详细综述，Valenza 和 Scherer[134]已经阐明了盐析的机理。基于现有理论，硅酸盐水泥混凝土的抗冻性（包括孔结构、孔隙水饱和度、力学性能和含气孔隙）的控制机制和参数在很大程度上是物理机械作用，而不是化学作用，因此应该也是 AAM 混凝土的控制参数。然而，孔溶液的冻结温度也非常重要[135]，这可能对 AAM 和 OPC 混凝土有所不同，因为离子强度和临界孔径下限制孔隙流体的差异。这可能是两类材料冻融性能差异的主要原因。Krivenko[136]指出，在－50℃下发生毛细管水分冻结，BFS 基 AAM 会发生冷冻破坏，并且该冻结点被孔溶液的高离子强度强烈抑制。

在冻融或霜盐攻击下，AAMs 的性能表现不一。某些情况下，碱激发材料比普通硅酸盐水泥混凝土表现出更好的抗冻性[33,137]。Rostovskaya 等人[138]提出很多实际应用碱激发混凝土的良好抗冻性的实例。有趣的是，上述混凝土在服役期间的强度并不高，而它们的耐冻性很好。Kukko、Mannonen[46]和 Bin、Pu[139]也注意到 BFS 基 AAM 有很好的抗冻性，这可能是因为 AAM 基体的总孔隙率较低、孔径较小。在 Kukko 和 Mannonen 的测试中，与空白样品相比，AAM 混凝土经 100 次冻融循环（＋20℃/－20℃）后，抗折强度增加；而 Hakkine[140]经过 700 次冻融循环后抗压强度略有下降，其中残余抗压强度为 87%～97%，抗折强度为 66%～87%。

Douglas 等人[62]对大量 AAM 做冻融测试，经 500 次冻融循环后，抗折强度损失（约 60%残余强度）的程度相似，只有极低激发剂含量的混凝土易于破损。然而，动态弹性模量在绝大部分样品中（不大于±7%）中几乎观察不到变化，除了极低激发剂含量的样品以外。抗折强度降低 40%且弹性模量没有变化的事实表明，需要进一步研究 AAM 混凝土中与冻融循环相关和不相关的参数。Gifford 和 Gillott[141]发现碱硅酸盐激发的 BFS 混凝土的冻融耐久性主要取决于含气量和气泡分布，并且某些样品的性能差是因为难以实现期望的气孔分布。然而，他们得出结论，AAM 或 OPC 胶凝材料制成的混凝土可以得到类似的孔隙率，AAM 混凝土的抗冻与 OPC 混凝土相比"至少是相当的"。

Talling 和 Krivenko[34]还指出，使用硅酸钠作为 BFS 的激发剂比 NaOH 激发具有更好的抗冻融性，硅酸盐激发的样品耐受 1000 次冻融循环（+20℃/-15℃），而 NaOH 激发的样品在 200 次和 700 次循环之间失效，并且显示出比相同强度的 OPC 混凝土更低的性能。Byfors 等人[106]发现碱激发 BFS 混凝土的耐冻性似乎与含气量无关（虽然这是难以控制的，即使大量的引气剂加入到混凝土中），而主要取决于水胶比。他们还发现，只有当材料强度降低至 5MPa 时，早期（至-20℃）冻结才会显著有害，否则 AAM 混凝土在早期冷冻和再融化中比 OPC 混凝土损害小[106]。

粉煤灰基 AAM 混凝土的抗冻融性也被证明是可接受的，150 次冻融循环后抗压强度保留 70%[142]。Husbands 等人[49]还发现 Pyrament 复合 AAM 混凝土具有高的抗冻融性能（根据 ASTM C666 和 C672 程序）。

根据生命周期分析原则，将碱激发混凝土与硅酸盐水泥混凝土在技术、经济和生态水平上进行了比较[143-147]。设计和测试了几种粉煤灰基胶凝材料和 BFS 基胶凝材料，对两类材料的冻融性能作比较。使用硅酸钠作激发剂，根据样品要求做配合比设计，使样品具有相似的力学性能。基准硅酸盐水泥混凝土（320kg/m³ 水泥，w/b=0.5，无外加剂），与碱激发混凝土同时测试，比较冻融和霜/除冰盐攻击（冻融等级 XF2 和 XF4，根据 EN206-1/DIN 1045-2[127,148]）的性能数据。AAM 样品在 40℃、100% 相对湿度下养护 24h，然后在 20℃ 下置于水中直接测试。这偏离了 DIN EN206 中规定的测试方法（其指定在水下的储存），但限制了激发剂从样品中浸出。

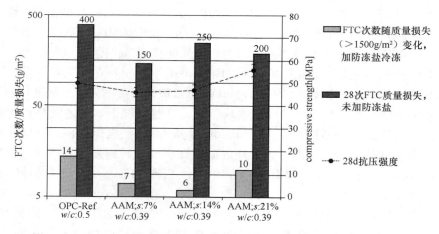

图 10-6　CDF 冻融循环测试方法检测粉煤灰基胶凝材料混凝土的实验结果
s 为硅酸盐激发剂含量；w/c 为水灰比；c 为粉煤灰+BFS

通过 CF 和 CDF 测试来确定冻融循环是否添加除冰盐[149]。110mm×150mm×75mm 混凝土样品养护 28d 后，放入测试溶液浸泡 7d 做毛细管吸入，然后做 28 次冻融循环。从毛细管吸入开始到 28 次冻融循环结束，将一些样品置于去离子水中（用于冻融测试），其余样品置于 3% 氯化钠溶液中（用于霜/除冰盐攻击），粉煤灰基胶凝材料的冻融试验结果如图 10-6 所示，BFS 基胶凝材料的冻融试验结果如图 10-7 所示。

如果不使用除冰盐，碱激发粉煤灰基胶凝材料（图 10-6）具有比基准硅酸盐水泥混凝土更好的抗冻性（更少的质量损失）。然而，如果存在除冰盐溶液，碱激发混凝土（主要是粉煤灰）比参考混凝土的早期抗冻性较差。为了满足 XF4 分类的标准，必须达到 28 次冻融循环（FTC），且质量损失不超过 1500g/m² 极限。基准混凝土在 14 次冻融循环（FTC）之后超过 1500g/m² 的质量损失，而由碱激发材料（粉煤灰占多数）制成的混凝土较早地超过该限度（6～10FTC）。

在含除冰盐和不含除冰盐的情况下，碱激发 BFS 基胶凝材料（图 10-7）两种试验都显示出较好的抗冻性。所有 AAM 混凝土满足 XF4 级别的标准。

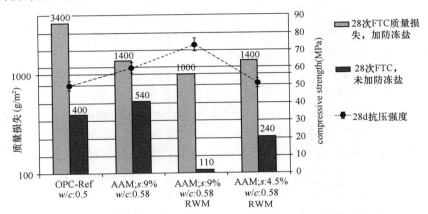

图 10-7　CDF 冻融循环测试方法检测 BFS 基胶凝材料混凝土的实验结果
s 为硅酸盐激发剂含量；w/c 为水灰比；c 为粉煤灰＋矿渣；RWM 为添加 16% 二次碱性材料

其他一些研究人员发现，AAM 的抗冻性比 OPC[150] 的差，这归因于碱激发混凝土结构中存在更多被冻结的游离水。硬化硅酸盐水泥的水大部分都是晶相中的化学结合水（氢氧化钙、钙矾石、单硫型和/或相关相）。碱激发材料中不存在以上化合物，因此当混凝土中的水达到饱和，碱激发材料的孔道中就会含有更多水，吸附于凝胶上，这部分水就会被冻结。

碱激发材料的抗冻性和抗结垢性也可能受到碳化（如第 9 章所述）以及高收缩影响（如第 10.3 节所述）而产生微裂纹[151]。碳化是影响抗结垢性最重要的参数之一，它会影响混凝土表层的力学性能，使材料更脆，特别是降低抗拉强度会影响抗结垢性能[134]。碱激发材料的收缩，如 10.3 节所述，可能高于硅酸盐水泥混凝土。收缩相关过程引起的微裂纹会强烈地影响抗结垢性。

碱激发材料代表具有各种性质（孔隙率和组成）的一大类材料。不可能就抗冻性得出一般性结论，但即使是硅酸盐水泥混凝土也不可能。然而，除了需要改进用于硅酸盐水泥混凝土样品的一些养护方案外，广泛使用的测试方法似乎通用于碱激发材料。

10.6　小结

当前，混凝土的力学性能的分析检测方法大部分直接用于 AAM 的分析中。最重要的差异是与在测试之前使用的养护和配合比设计方案不同，这就需要详细研究比较，提出合理的建议。一些碱激发混凝土事实上比硅酸盐水泥混凝土更容易收缩和开裂，但在大多数情况下，通过适当的养护和/或配合比设计，收缩和开裂是可控的。过量水在 AAM 材料中很成问题。对于良好养护、低孔隙率的 AAM 材料，其冻融试验结果是好的。但如果使用硅酸盐水泥混凝土试样的标准养护方式,AAM 就会出现适应性问题。

本章中指出的 AAMs 和硅酸盐水泥混凝土之间最显著的区别是徐变。人们对 AAM 混凝土的徐变行为了解不足,但不能使用硅酸盐水泥混凝土在载荷下的长期服役的标准经验模型来预测 AAM 混凝土性能。这一领域需要做进一步研究工作,开发 AAM 混凝土的经验模型,为未来 AAMs 在结构领域的应用提供理论支撑。

参考文献

[1] American Concrete Institute. Building Code Requirements for Structural Concrete and (ACI318-08). Farmington Hills, 2008.

[2] Neville A. M.. Properties of Concrete, 4th edn. Wiley, Harlow, 1996.

[3] Ozyildirim C., Carino N. J.. Concrete strength testing. In: Lamond J. F., Pielert J. H. (eds.) Significance of Tests and Properties of Concrete and Concrete-Making Materials, 2006: 125-140. ASTM International, West Conshohocken.

[4] ASTM International. Standard Test Method for Compressive Strength of Cylindrical Concrete Specimens (ASTM C39/C39M-10). West Conshohocken, 2010.

[5] ASTM International. Standard Practice for Making and Curing Concrete Test Specimens in the Field (ASTM C31/C31M-09). West Conshohocken 2009.

[6] Mehta P. K., Monteiro P. J. M.. Concrete: Microstructure, Properties and Materials, 3rd edn. McGraw-Hill, New York, 2006.

[7] Pernica D., Reis P. N. B., Ferreira J. A. M., Louda P.. Effect of test conditions on the bending strength of a geopolymer-reinforced composite. J. Mater. Sci, 2010, 45 (3): 744-749.

[8] Khandelwal M., Ranjith P., Pan Z., Sanjayan J.. Effect of strain rate on strength properties of low-calcium fly-ash-based geopolymer mortar under dry condition. Arabian J. Geosci, 2013, 6 (7): 2383-2389.

[9] Häkkinen T.. The influence of slag content on the microstructure, permeability and mechanical properties of concrete: Part 2. Technical properties and theoretical examinations. Cem. Concr. Res, 1993, 23 (3): 518-530.

[10] European Committee for Standardization (CEN). Testing Hardened Concrete. Compressive Strength of Test Specimens (EN 12390-3). Brussels, 2002.

[11] European Committee for Standardization (CEN). Testing Hardened Concrete. Making and Curing Specimens for Strength Tests (EN 12390-2). Brussels, 2000.

[12] ASTM International. Standard Test Method for Compressive Strength of Hydraulic cement Mortars (Using 2-in. or [50-mm] Cube Specimens) (ASTM C109/C109M-11). West Conshohocken, 2011.

[13] Gaynor R. D.. Cement strength and concrete strength - an apparition or a dichotomy? Cem. Concr. Agg, 1993, 15 (2): 135-144.

[14] Struble L. J.. Hydraulic cements - physical properties. In: Lamond J. F., Pielert J. H. (eds.) Significance of Tests and Properties of Concrete and Concrete-Making

Materials, 2006: 435-449. ASTM International, West Conshohocken.

[15] Popovics S.. Strength and Related Properties of Concrete: A Quantitative Approach. Wiley, New York, 1998.

[16] European Committee for Standardization (CEN). Methods of Testing Cement - Part 1: Determination of Strength (EN 196-1). Brussels, 2005.

[17] ASTM International. Standard Test Method for Compressive Strength of Hydraulic-Cement Mortars (Using Portions of Prisms Broken in Flexure) (ASTM C349 - 08). West Conshohocken, 2008.

[18] ASTM International. Standard Test Method for Compressive (Crushing) Strength of Fired Whiteware Materials (ASTM C773 - 88, reapproved 2011). West Conshohocken, 2011.

[19] Malhotra V. M., Carino N. J. (eds.). Handbook on Nondestructive Testing of Concrete, 2nd edn. CRC Press, Boca Raton, 2004.

[20] Popovics J. S., Zemajtis J., Shkolni I.. ACI-CRC Final Report: A Study of Static and Dynamic Modulus of Elasticity of Concrete, 2008.

[21] Komlos K., Popovics S., Nürnbergerová T., Babál B., Popovics J. S.. Ultrasonic pulse velocity test of concrete properties as specified in various standards. Cem. Concr. Compos. 1996, 18 (5): 357-364.

[22] ASTM International. Standard Test Method for Pulse Velocity Through Concrete (ASTM C597 - 09). West Conshohocken, 2009.

[23] Wang S.-D., Scrivener K. L., Pratt P. L.. Factors affecting the strength of alkali-activated slag. Cem. Concr. Res, 1994, 24 (6): 1033-1043.

[24] Fernández-Jiménez A., Palomo J. G., Puertas F.. Alkali-activated slag mortars. Mechanical strength behaviour. Cem. Concr. Res, 1999, 29: 1313-1321.

[25] Lloyd R. R.. Accelerated ageing of geopolymers. In: Provis J. L., Van Deventer J. S. J. (eds.). Geopolymers: Structures, Processing, Properties and Industrial Applications, 2009: 139-166. Woodhead, Cambridge.

[26] Lloyd R. R.. The durability of inorganic polymer cements. Ph. D. thesis, University of Melbourne, Australia, 2008.

[27] Shi C., Li Y.. Investigation on some factors affecting the characteristics of alkali-phosphorus slag cement. Cem. Concr. Res, 1989, 19 (4): 527-533.

[28] Duxson P., Mallicoat S. W., Lukey G. C., Kriven W. M., Van Deventer J. S. J.. The effect of alkali and Si/Al ratio on the development of mechanical properties of metakaolin-based geopolymers. Colloids Surf, 2007 A 292 (1): 8-20.

[29] Wang S. D.. Review of recent research on alkali-activated concrete in China. Mag. Concr. Res, 1991, 43 (154): 29-35.

[30] Diaz-Loya E. I., Allouche E. N., Vaidya S.. Mechanical properties of fly-ash-based geopolymer concrete. ACI Mater. J. 2011, 108 (3): 300-306.

[31] Bernal S. A., Mejía de Gutierrez R., Pedraza A. L., Provis J. L., Rodríguez E. D.,

Delvasto S.. Effect of binder content on the performance of alkali-activated slag concretes. Cem. Concr. Res. 2011, 41(1): 1-8.

[32] Bernal S. A., Mejía de Gutiérrez R., Provis J. L.. Engineering and durability properties of concretes based on alkali-activated granulated blast furnace slag/metakaolin blends. Constr. Build. Mater, 2012, 33: 99-108.

[33] Shi C., Krivenko P. V., Roy D. M.. Alkali-Activated Cements and Concretes. Taylor & Francis, Abingdon, 2006.

[34] Talling B., Krivenko P. V.. Blast furnace slag-the ultimate binder. In: Chandra S. (ed.) Waste Materials Used in Concrete Manufacturing, 1997: 235-289. Noyes Publications, Park Ridge, 1997.

[35] Šmilauer V., Hlaváček P., Škvára F., Šulc R., Kopecký L., Němeček J.. Micromechanical multiscale model for alkali activation of fly ash and metakaolin. J. Mater. Sci, 2011, 46 (20): 6545-6555.

[36] Xu Z., Deng Y., Wu X., Tang M., Beaudoin J. J.. Influence of various hydraulic binders on performance of very low porosity cementitious systems. Cem. Concr. Res, 1993, 23 (2): 462-470.

[37] Rangan B. V.. Engineering properties of geopolymer concrete. In: Provis J. L., van Deventer J. S. J. (eds.). Geopolymers: Structure, Processing, Properties and Industrial Applications, 2009: 213-228. Woodhead, Cambridge.

[38] Hardjito D., Wallah S. E., Sumajouw D. M. J., Rangan B. V.: On the development of fly ash-based geopolymer concrete. ACI Mater. J, 2004, 101(6): 467-472.

[39] Sumajouw D. M. J., Hardjito D., Wallah S. E., Rangan B. V.. Fly ash-based geopolymer concrete: study of slender reinforced columns. J. Mater. Sci, 2007, 42 (9): 3124-3130.

[40] Bakharev T., Sanjayan J. G., Cheng Y. B.. Effect of elevated temperature curing on properties of alkali-activated slag concrete. Cem. Concr. Res, 1999, 29 (10): 1619-1625.

[41] Standards Australia. Concrete Structures (AS 3600-2009). Sydney, 2009.

[42] American Concrete Institute. Building Code Requirements for Structural Concrete (ACI 318-02) and Commentary (ACI 318R-02). Farmington Hills, 2002.

[43] Hardjito D., Rangan B. V.. Development and Properties of Low-Calcium Fly Ash-Based Geopolymer Concrete. Curtin University of Technology. Perth, Australia, 2005.

[44] Yost J. R., Radlińska A., Ernst S., Salera M.. Structural behavior of alkali activated fly ash concrete. Part 1: mixture design, material properties and sample fabrication. Mater. Struct, 2013, 46 (3): 435-447.

[45] Yost J. R., Radlińska A., Ernst S., Salera M., Martignetti N. J.. Structural behavior of alkali activated fly ash concrete. Part 2: structural testing and experimental findings. Mater. Struct, 2013, 46 (3): 449-462.

[46] Kukko H., Mannonen R.. Chemical and mechanical properties of alkali-activated blast furnace slag (F-concrete). Nord. Concr. Res, 1982, 1: 16.1-16.16.

[47] Sumajouw D. M. J., Rangan B. V.. Low-Calcium Fly Ash-Based Geopolymer Concrete: Reinforced Beams and Columns. Curtin University of Technology. Perth, Australia, 2006.

[48] Sofi M., Van Deventer J. S. J., Mendis P. A., Lukey G. C.. Engineering properties of inorganic polymer concretes (IPCs). Cem. Concr. Res, 2007, 37 (2): 251-257.

[49] Husbands T. B., Malone P. G., Wakeley L. D.. Performance of Concretes Proportioned with Pyrament Blended Cement, U. S. Army Corps of Engineers Construction Productivity Advancement Research Program, Report CPAR-SL-94-2, 1994.

[50] Fernández-Jiménez A. M., Palomo A., López-Hombrados C.. Engineering properties of alkali-activated fly ash concrete. ACI Mater. J, 2006, 103 (2): 106-112.

[51] Bernal S., De Gutierrez R., Delvasto S., Rodriguez E.. Performance of an alkali-activated slag concrete reinforced with steel fibers. Constr. Build. Mater, 2010, 24 (2): 208-214.

[52] Collins F. G., Sanjayan J. G.. Workability and mechanical properties of alkali activated slag concrete. Cem. Concr. Res, 1999, 29 (3): 455-458.

[53] Douglas E., Bilodeau A., Brandstetr J., Malhotra V. M.. Alkali activated ground granulated blast-furnace slag concrete: preliminary investigation. Cem. Concr. Res, 1991, 21 (1): 101-108.

[54] Wu Y., Cai L., Fu Y.. Durability of green high performance alkali-activated slag pavement concrete. Appl. Mech. Mater, 2011, 99-100: 158-161.

[55] Paultre P., Mitchell D.. Code provisions for high-strength concrete - an international perspective. Concr. Int, 2003, 25 (5): 76-90.

[56] Dattatreya J. K., Rajamane N. P., Sabitha D., Ambily P. S., Nataraja M. C.. Flexural behaviour of reinforced geopolymer concrete beam. Int. J. Civil. Struct. Eng, 2011, 2 (1): 138-159.

[57] Bureau of Indian Standards. Plain and Reinforced concrete - Code of Practice (IS 456: 2000). New Delhi, 2000.

[58] Bondar D., Lynsdale C. J., Milestone N. B., Hassani N., Ramezanianpour A. A.. Engineering properties of alkali-activated natural pozzolan concrete. ACI Mater. J, 2011, 108 (1): 64-72.

[59] Lawson J. L.. On the determination of the elastic properties of geopolymeric materials using non-destructive ultrasonic techniques. MSc thesis, Rochester Institute of Technology, 2009.

[60] Duxson P., Provis J. L., Lukey G. C., Mallicoat S. W., Kriven W. M., Van Deventer J. S. J.. Understanding the relationship between geopolymer composition, microstructure and mechanical properties. Colloids Surf, 2005 A269 (1-3): 47-58.

[61] Wongpa J., Kiattikomol K., Jaturapitakkul C., Chindaprasirt P.. Compressive

strength, modulus of elasticity, and water permeability of inorganic polymer concrete. Mater. Des, 2010, 31 (10): 4748-4754.

[62] Douglas E., Bilodeau A., Malhotra V. M.. Properties and durability of alkali-activated slag concrete. ACI Mater. J, 1992, 89 (5): 509-516.

[63] Puertas F., Palacios M., Manzano H., Dolado J. S., Rico A., Rodríguez J.. A model for the C-A-S-H gel formed in alkali-activated slag cements. J. Eur. Ceram. Soc, 2011, 31 (12): 2043-2056.

[64] Němeček J., Šmilauer, V., Kopecký L.. Nanoindentation characteristics of alkali-activated aluminosilicate materials. Cem. Concr. Compos, 2011, 33 (2): 163-170.

[65] Oh J. E., Moon J., Mancio M., Clark S. M., Monteiro P. J. M.. Bulk modulus of basic sodalite, $Na_8[AlSiO_4]_6(OH)_2 \cdot 2H_2O$, a possible zeolitic precursor in coal-fly-ash-based geopolymers. Cem. Concr. Res, 2011, 41 (1): 107-112.

[66] Oh J. E., Clark S. M., Monteiro P. J. M.. Does the Al substitution in C-S-H(I) change its mechanical property? Cem. Concr. Res, 2011, 41 (1): 102-106.

[67] Collins M. P., Mitchell D., Macgregor J. G.. Structural design considerations for high strength concrete. Concr. Int, 1993, 15 (5): 27-34.

[68] Sarker P. K.. Analysis of geopolymer concrete columns. Mater. Struct, 2009, 42 (6): 715-724.

[69] Goodwin F.. Volume change. In: Lamond J. F., Pielert J. H. (eds.). Signifi cance of Tests and Properties of Concrete and Concrete-Making Materials, 2006: 215-225. ASTM International, West Conshohocken.

[70] Chen W., Brouwers H.. The hydration of slag, part 1: reaction models for alkali-activated slag. J. Mater. Sci, 2007, 42 (2): 428-443.

[71] Thomas J. J., Allen A. J., Jennings H. M.. Density and water content of nanoscale solid C-S-H formed in alkali-activated slag (AAS) paste and implications for chemical shrinkage. Cem. Concr. Res, 2012, 42 (2): 377-383.

[72] ASTM International. Standard Test Method for Length Change of Hardened Hydraulic-Cement Mortar and Concrete (ASTM C157/C157M-08). West Conshohocken, 2008.

[73] ASTM International. Standard Test Method for Drying Shrinkage of Mortar Containing Hydraulic Cement (ASTM C596 - 09). West Conshohocken, 2009.

[74] Standards Australia. Methods of Testing Concrete - Determination of the Drying Shrinkage of Concrete for Samples Prepared in the Field or in the Laboratory (AS 1012.13). Sydney, 1992.

[75] Sanjayan J.. Non-Portland based cements and concretes. Concr. Austr, 2012, 38 (1): 34-39.

[76] ASTM International. Standard Test Method for Determining Age at Cracking and Induced Tensile Stress Characteristics of Mortar and Concrete under Restrained Shrinkage (ASTM C1581/C1581M-09a). West Conshohocken, 2009.

[77] ASTM International. Standard Test Method for Change in Height at Early Ages of

Cementitious Mixtures (ASTM C827/C827M-10). West Conshohocken.

[78] ASTM International. Standard Test Method for Measuring Changes in Height of Cylindrical Specimens of Hydraulic-Cement Grout (ASTM C1090-10). West Conshohocken, 2010.

[79] Häkkinen, T.. The microstructure of high strength blast furnace slag concrete. Nord. Concr. Res, 1992, 11 (1): 67-82.

[80] Häkkinen, T.. The influence of slag content on the microstructure, permeability and mechanical properties of concrete: Part 1. Microstructural studies and basic mechanical properties. Cem. Concr. Res, 1993, 23 (2): 407-421.

[81] Melo Neto A. A., Cincotto M. A., Repette W.. Drying and autogenous shrinkage of pastes and mortars with activated slag cement. Cem. Concr. Res, 2008, 38: 565-574.

[82] Sakulich A. R., Bentz D. P.. Mitigation of autogenous shrinkage in alkali activated slag mortars by internal curing. Mater. Struct, 2013, 46 (8): 1355-1367.

[83] Wang S.-D., Pu X.-C., Scrivener K. L., Pratt P. L.. Alkali-activated slag cement and concrete: a review of properties and problems. Adv. Cem. Res, 1995, 7 (27): 93-102.

[84] Duran Atiş C., Bilim C., Çelik Ö., Karahan O.. Influence of activator on the strength and drying shrinkage of alkali-activated slag mortar. Constr. Build. Mater, 2009, 23 (1): 548-555.

[85] Collins F., Sanjayan J. G.. Effect of pore size distribution on drying shrinking of alkaliactivated slag concrete. Cem. Concr. Res, 2000, 30 (9): 1401-1406.

[86] Krizan*D., Zivanovic B.. Effects of dosage and modulus of water glass on early hydration of alkali-slag cements. Cem. Concr. Res, 2002, 32 (8): 1181-1188.

[87] Cincotto M. A., Melo A. A., Repette W. L.. Effect of different activators type and dosages and relation to autogenous shrinkage of activated blast furnace slag cement. In: Grieve G., Owens G. (eds.) Proceedings of the 11th International Congress on the Chemistry of Cement, Durban, South Africa, 2003: 1878-1887. Tech Books International, New Delhi, India, 2003.

[88] Collins F., Sanjayan J. G.. Numerical modeling of alkali-activated slag concrete beams subjected to restrained shrinkage. ACI Mater. J, 2000, 97 (5): 594-602.

[89] Collins F., Sanjayan J. G.. Cracking tendency of alkali-activated slag concrete subjected to restrained shrinkage. Cem. Concr. Res, 2000, 30 (5): 791-798.

[90] Collins F., Sanjayan J. G.. Strength and shrinkage properties of alkali-activated slag concrete containing porous coarse aggregate. Cem. Concr. Res, 1999, 29 (4): 607-610.

[91] Yong C. Z.. Shrinkage behaviour of geopolymers. M. Eng. Sci. thesis, University of Melbourne, 2010.

[92] Lee N. P.. Creep and Shrinkage of Inorganic Polymer Concrete, BRANZ Study Report

175, BRANZ, 2007.

[93] Guo X. L., Shi H. S., Dick W. A.. Compressive strength and microstructural characteristics of class Cfly ash geopolymer. Cem. Concr. Compos, 2010, 32 (2): 142-147.

[94] Rüscher C. H., Mielcarek E., Lutz W., Ritzmann A., Kriven W. M.. Weakening of alkaliactivated metakaolin during aging investigated by the molybdate method and infrared absorption spectroscopy. J. Am. Ceram. Soc, 2010, 93 (9): 2585-2590.

[95] Yip C. K., Lukey G. C., Provis J. L., Van Deventer J. S. J.. Carbonate mineral addition to metakaolin-based geopolymers. Cem. Concr. Compos, 2008, 30 (10): 979-985.

[96] Pacheco-Torgal F., Castro-Gomes J., Jalali S.. Investigations about the effect of aggregates on strength and microstructure of geopolymeric mine waste mud binders. Cem. Concr. Res, 2007, 37 (6): 933-941.

[97] Elimbi A., Tchakoute H. K., Njopwouo D.. Effects of calcination temperature of kaolinite clays on the properties of geopolymer cements. Constr. Build. Mater, 2011, 25 (6): 2805-2812.

[98] Justnes H., Ardoullie B., Hendrix E., Sellevold E. J., Van Gemert D.. The chemical shrinkage of pozzolanic reaction products. In: Malhotra V. M. (ed.) 6th CANMET Conference on Fly Ash, Silica Fume, Slag, and Natural Pozzolans in Concrete, Bangkok, Thailand. 1998, 1: 191-205. ACI SP178. Detroit, MI.

[99] Brouwers H.. The work of Powers and Brownyard revisited: Part 1. Cem. Concr. Res, 2004, 34 (9): 1697-1716.

[100] Schlitter J. L., Senter A. H., Bentz D. P., Nantung T., Weiss W. J.. A dual concentric ring test for evaluating residual stress development due to restrained volume change. J. ASTM Int, 2010, 7 (9): JAI103118.

[101] Standardization Administration of the People's Republic of China. Standard Test Method for Drying Shinkage of Mortar (JC/T 603—2004). Beijing, 2004.

[102] Raoufi K., Pour-Ghaz M., Poursaee A., Weiss W. J.. Restrained shrinkage cracking in concrete elements: role of substrate bond on crack development. J. Mater. Civil. Eng, 2011, 23 (6): 895-902.

[103] Bisschop J., Van Mier J. G. M.. How to study drying shrinkage microcracking in cementbased materials using optical and scanning electron microscopy? Cem. Concr. Res, 2002, 32 (2): 279-287.

[104] Nemati K. M.. Preserving microstructure of concrete under load using the Wood's metal technique. Int. J. Rock Mech. Mining Sci, 2000, 37 (1-2): 133-142.

[105] Nemati K. M., Monteiro P. J. M., Cook N. G. W.. A new method for studying stress-induced microcracks in concrete. J. Mater. Civil. Eng, 1998, 10 (3): 128-134.

[106] Byfors K., Klingstedt G., Lehtonen H. P., Romben L.. Durability of concrete

made with alkali-activated slag. In: Malhotra V. M. (ed.) 3rd International Conference on Fly Ash, Silica Fume, Slag and Natural Pozzolans in Concrete, ACI SP114, Trondheim, Norway, 1989: 1429-1444. American Concrete Institute. Detroit, MI.

[107] Häkkinen T.. The permeability of high strength blast furnace slag concrete. Nord. Concr. Res, 1992, 11 (1): 55-66.

[108] Collins F., Sanjayan J. G.. Microcracking and strength development of alkali activated slag concrete. Cem. Concr. Compos, 2001, 23 (4-5): 345-352.

[109] Shen W., Wang Y., Zhang T., Zhou M., Li J., Cui X.. Magnesia modification of alkaliactivated slag fly ash cement. J. Wuhan Univ. Technol. Mater. Sci. Ed, 2011, 26 (1): 121-125.

[110] ASTM International. Standard Test Method for Creep of Concrete in Compression (ASTM C512/C512M-10). West Conshohocken, 2010.

[111] International Organization for Standardization. Testing of Concrete - Part 9: Determination of Creep of Concrete Cylinders in Compression (ISO 1920-9: 2009), Geneva, 2009.

[112] Bissonnette B., Pigeon M., Vaysburd A. M.. Tensile creep of concrete: study of its sensitivity to various parameters. ACI Mater. J, 2007, 104 (4): 360-368.

[113] Garas V. Y., Kahn L. F., Kurtis K. E.. Tensile creep test of fiber-reinforced ultra-high performance concrete. J. Testing Eval, 2010, 38 (6): JTE102666.

[114] Gourley J. T., Johnson G. B.. Developments in geopolymer precast concrete. In: Davidovits J. (ed.). Proceedings of the World Congress Geopolymer 2005 - Geopolymer, Green Chemistry and Sustainable Development Solutions, Saint-Quentin, France, 2005: 139-143. Institut Géopolymère, 2005.

[115] Wallah S. E., Rangan B. V.. Low-Calcium Fly Ash-Based Geopolymer Concrete: Long-Term Properties, Curtin University of Technology, Research Report GC2, 2006.

[116] Standards Australia. Methods of Testing Concrete-Determination of Creep of Concrete Cylinders in Compression (AS 1012.16). Sydney, 1996.

[117] Gilbert R. I.. AS3600 creep and shrinkage models for normal and high strength concrete. In: Gardner N. J., Weiss W. J. (eds.) Shrinkage and Creep of Concrete, ACI SP 227, Farmington Hills, MI, 2005, 21-40. American Concrete Institute.

[118] Gilbert R. I.. Creep and shrinkage models for high strength concrete - proposals for inclusion in AS3600. Aust. J. Struct. Eng, 2002, 4 (2): 95-106.

[119] ASTM International. Standard Test Method for Resistance of Concrete to Rapid Freezing and Thawing (ASTM C666/C666M-03(2008)). West Conshohocken, 2008.

[120] Canadian Standards Association. Precast Concrete Pavers (CSA 231.2-06). Mississauga, 2006.

[121] Canadian Standards Association. Precast Concrete Paving Slabs (CSA 231.1-06). Mississauga, 2006.

[122] ASTM International. Standard Test Method for Freeze-Thaw and De-icing Salt Durability of Solid Concrete Interlocking Paving Units (ASTM C1645 - 11). West Conshohocken, 2011.

[123] ASTM International: Standard Test Method for Measuring the Resistance of Ceramic Tile to Freeze-Thaw Cycling (ASTM C1026 - 10). West Conshohocken, 2010.

[124] International Organization for Standardization: Ceramic tiles - Part 12: Determination of frost resistance (ISO/NP 10545-12). Geneva, 2012.

[125] ASTM International: Standard Test Methods for Sampling and Testing Brick and Structural Clay Tile (ASTM C67 - 11). West Conshohocken, 2011.

[126] Ghafoori, N., Smith D. R.. Comparison of ASTM and Canadian freeze-thaw durability tests. In: Fifth International Conference on Concrete Block Paving (Pave Israel 96), Tel-Aviv, Israel, 1996: 93-101. Dan Knassim Limited.

[127] European Committee for Standardization (CEN): Concrete - Part 1: Specification, Performance, Production and Conformity (EN 206-1). Brussels, 2010.

[128] Setzer M., Heine P., Kasparek S., Palecki S., Auberg R., Feldrappe V., Siebel E.. Test methods of frost resistance of concrete: CIF-test: capillary suction, internal damage and freeze thaw test -reference method and alternative methods A and B. Mater. Struct, 2004, 37 (10): 743-753.

[129] Setzer M., Fagerlund G., Janssen D.. CDF test - test method for the freeze-thaw resistance of concrete-tests with sodium chloride solution (CDF). Mater. Struct. 1996, 29 (9): 523-528.

[130] European Committee for Standardization (CEN). Testing Hardened Concrete. Freeze-Thaw Resistance. Scaling (DD CEN/TS 12390-9: 2006). Brussels, 2006.

[131] ASTM International. Standard Test Method for Scaling Resistance of Concrete Surfaces Exposed to Deicing Chemicals (ASTM C672/C672M - 03) (withdrawn 2012). West Conshohocken, 2003.

[132] Boos P., Eriksson B. E., Giergiczny Z., Haerdtl R.. Laboratory testing of frost resistance - do these tests indicate the real performance of blended cements? In: Beaudoin J. J. (ed.) 12th International Congress on the Chemistry of Cement, Montreal, Canada. CD-ROM. National Research Council of Canada. Ottawa, Canada, 2007.

[133] Janssen D. J., Snyder M. B.. Resistance of Concrete to Freezing and Thawing, SHRP-C-391, Strategic Highway Research Program. National Research Council. Washington DC, 1994.

[134] Valenza J. J., Scherer G. W.. Mechanism for salt scaling. J. Am. Ceram. Soc, 2006, 89 (4): 1161-1179.

[135] Powers T. C., Brownyard T. L.. Studies of the physical properties of hardened Portland cement paste. J. Am. Concr. Inst, 1947, 18 (8): 933-992.

[136] Krivenko P. V.. Alkaline cements: Structure, properties, aspects of durability. In: Krivenko P. V. (ed.) Proceedings of the Second International Conference on Alka-

line Cements and Concretes, Kiev, Ukraine, 1999: 3-43. ORANTA.

[137]　Davidovits J.. Geopolymer Chemistry and Applications. Institut Géopolymère, Saint-Quentin, 2008.

[138]　Rostovskaya G., Ilyin V., Blazhis A.. The service properties of the slag alkaline concretes. In: Ertl Z. (ed.) Proceedings of the International Conference on Alkali Activated Materials - Research, Production and Utilization, Prague, Czech Republic, 2007: 593-610. Česká rozvojová agentura.

[139]　Bin X., Pu X.. Study on durability of solid alkaline AAS cement. In: Krivenko P. V. (ed.) Proceedings of the Second International Conference on Alkaline Cements and Concretes, Kiev, Ukraine, 1999: 64-71. ORANTA.

[140]　Häkkinen T.. Durability of alkali-activated slag concrete. Nord. Concr. Res, 1987, 6 (1): 81-94.

[141]　Gifford P. M., Gillott J. E.. Freeze-thaw durability of activated blast furnace slag cement concrete. ACI Mater. J, 1996, 93 (3): 242-245.

[142]　Škvára F., Jílek T., Kopecký, L.. Geopolymer materials based on fly ash. Ceram.-Silik, 2005, 49 (3): 195-204.

[143]　Weil M., Dombrowski, K., Buchwald A.. Sustainable design of geopolymers - evaluation of raw materials by the consideration of economical and environmental aspects in the early phases of material development. In: Weil M., Buchwald A., Dombrowski K., Jeske U., Buchgeister J. (eds.) Materials Design and Systems Analysis, 2007: 57-76. Shaker Verlag, Aachen.

[144]　Weil M., Jeske U., Buchwald A., Dombrowski K.. Sustainable design of geopolymers - evaluation of raw materials by the integration of economic and environmental aspects in the early phases of material development. In: 14th CIRP International Conference on Life Cycle Engineering, Tokyo, 2007: 278-284. Springer.

[145]　Dombrowski K., Buchwald A., Weil M.. Geopolymere Bindemittel. Teil 2: Entwicklung und Optimierung von Geopolymerbetonmischungen für feste und dauerhafte Außenwandbauteile (Geopolymer Binders. Part 2: Development and optimization of geopolymer concrete mixes for strong and durable external wall units). ZKG Int, 2008, 61 (03): 70-82.

[146]　Weil M., Dombrowski-Daube K., Buchwald, A.. Geopolymerbinder - Teil 3: Ökologische und ökonomische Analysen von Geopolymerbeton- Mischungen für Außenbauteile (Geopolymer binders -Part 3: Ecological and economic analyses of geopolymer concrete mixes for external structural elements). ZKG Int, 2011, 7/8: 76-87.

[147]　Buchwald A., Weil M., Dombrowski K.. Life cycle analysis incorporated development of geopolymer binders. Restor. Build. Monum, 2008, 14 (4): 271-282.

[148]　Deutsches Institut für Normung. Tragwerke aus Beton, Stahlbeton und Spannbeton - Teil 2: Beton - Festlegung, Eigenschaften, Herstellung und Konformität - Anwend-

ungsregeln zu DIN EN 206-1 (DIN 1045-2: 2008). Berlin, 2008.

[149] Setzer M., Hartmann V.. Verbesserung der Frost-Tausalz-Widerstandsprüfung. CDF-Test- Prüfvorschrift, 1991, 9: 73-82.

[150] Bilek V., Szklorzova H.. Freezing and thawing resistance of alkali-activated concretes for the production of building elements. In: Malhotra V. M. (ed.) Proceedings of 10th CANMET/ ACI Conference on Recent Advances in Concrete Technology, supplementary papers, Seville, Spain, 2009: 661-670. American Concrete Institute, Detroit, MI.

[151] Copuroglu O.. Freeze thaw de-icing salt resistance of blast furnace slag cement mortars-a factorial design study. In: Walraven J. et al. (eds.) 5th International PhD Symposium in Civil Engineering, London, UK, 2004: 175-182. Taylor & Francis.

第 11 章　建筑与市政基础设施的示范项目和应用

John L. Provis，David G. Brice，Anja Buchwald，Peter Duxson，
Elena Kavalerova，Pavel V. Krivenko，Caijun Shi，
Jannie S. J. van Deventer and J. A. L. M. (Hans) Wiercx

11.1　引言

在这份报告中，能够提供在过去几十年中已经使用的碱激发混凝土的结构和工程应用实例，这是非常有价值的。AAM 混凝土在苏联和中国的应用情况已由 Shi、Craven 和 Roy[1]编写成书的第 12 章进行了介绍。这章将简单给出一些文献（更广领域）中提到的应用，以及在世界上其他地区 AAMs 混凝土规模化用于建筑物和其他的市政基础设施构件的情况。苏联的发展和应用也已经由 Brodkoand[2]和 Craven[3]做了详细综述。在本章中涉及的每个项目至少是中试规模，而且在一些情况下完全是商业推广规模。碱激发混凝土的规模化生产同样使用普通混凝土的标准搅拌、成型设备和施工人员，说明碱激发混凝土是可以根据现有配合比设计，采用当地的原材料直接制备的。特别是在苏联，可以使用不同地区铁厂的矿渣生产混凝土，激发剂大部分来自于当地的工业废碱液。

评价服役的 AAM 混凝土的科学文献现在还很少，但对过去几十年服役的 AAM 构件和建筑分析发现，每个工程的材料都很致密，钢筋保护得很好，强度依然很好（在大多数情况下，明显高于设计强度）[4-7]。AAM 混凝土在不同环境下数十年甚至更长时间的研究数据还较少，这就不能为 AAMs 的耐久性提供有力的证据。但是，应当注意的是现代硅酸盐水泥混凝土有过类似的情况，自从 19 世纪 30 年代以来，在熟料用量、配合比设计、掺合料的使用对长期暴露的混凝土和耐久性的考量也与之前的混凝土相差甚多。70 年前使用水泥与现代硅酸盐水泥不同，现代水泥艾利特含量高，贝利特含量低[8]。在过去的几十年里，还引入了一整套有机外加剂来控制流变性、水胶比和其他工艺参数。使用当时工程项目测试的结果与实验室测试和物理化学性质结合使用，来预测现代混凝土的性能。如果从旧材料获得数据用于预测耐久性，则可以使用相同的方法检测现代碱激发胶凝材料。碱激发化学的相关数据已在第 3～第 6 章做过介绍，劣化机制在第 8～第 10 章中做了详细讨论，前面章节的内容可以与本章论述结合，对现代 AAMs 现场样品分析得到有用的应用信息和耐久性数据。

11.2　碱激发 BFS 混凝土构造

11.2.1　俄罗斯利佩茨克高层住宅楼

1986 年至 1994 年期间，Tsentrmetallurgremont 公司使用碱激发 BFS 混凝土兴建了一批高层住宅楼（超过 20 层），包括图 11-1 所示的 24 层建筑。三栋建筑的外墙在现场浇筑，楼

板、楼梯和其他结构件是预制的,全部使用水胶比为 0.35 的碱碳酸盐激发的 BFS 混凝土,整个配合比设计数据都可以在文献[1]中找到。使用普通混凝土罐车将现场浇筑的混凝土从搅拌站拉到施工现场,再用电加热方式养护。预制构件在预制厂浇筑,蒸汽养护,养护后达到设计强度 25MPa。石英砂作细骨料,白云石灰石作粗骨料,碱激发胶凝材料与骨料结合良好,没有明显的碱-骨料反应。

11.2.2　乌克兰马里乌波尔的砌筑砖

1960 年,Azov Zelezobeton 协会使用 Stroydetal 磨机生产碱氢氧化物激发的 BFS 商品混凝土[1]。这种混凝土做成预制砖,用于建设房屋、库房和其他建筑,包括 2～15 层的公寓楼[1](图 11-2)。AAM 砖也被用于建设卡利米乌斯河银行的外墙。砖厂大约在 1980 年关闭,但是伊里奇钢铁联合公司在 1999 年重新开始生产预拌碱激发 BFS 混凝土和混凝土制品[9]。预拌碱激发 BFS 混凝土主要满足于其自身需要,主要是重型车辆现场浇筑路面或预制路面,以及承重结构件的商业化生产[9]。他们还开发了自流平混凝土,用于道路建设。2002 年以来,该企业已经生产了两万多立方米的混凝土,碳酸钠激发的 BFS 混凝土相比同等强度等级的 OPC 混凝土,在各种强度等级和经济方面都显示出明显的成本优势(节约成本超过 50%)[9]。

图 11-1　用碱激发矿渣水泥混凝土建造的 24 层建筑物,位于俄罗斯利佩茨克[3]

图 11-2　乌克兰马里乌波尔的住宅楼,由碱氢氧化物激发的 BFC 预制砖(外涂石灰装饰)建成

11.2.3　中国湖北省阴山县的办公楼和商业楼以及厂房预制梁和柱

在中国,Na_2SO_4 激发的硅酸盐水泥-矿渣-钢渣复合水泥从 1988 年已经开始商业化生产和应用[1],本产品是由河南省安阳市安阳钢渣水泥厂首先生产和销售的。数月的水泥质检表明,在缺乏碱激发剂的情况下,这种复合水泥比硅酸盐水泥-矿渣-钢渣复合水泥显示出更加稳定的性能。由于硫酸钠是以固体形式与其他组分研磨混合的,而不是作为单独的液体激发

剂加人，所以从材料可操作性和配合比设计角度考虑，这种水泥可以像传统水泥一样用在混凝土里。硫酸钠激发的硅酸盐水泥-矿渣-钢渣复合水泥显示出优异的工作性，相对较短的凝结时间、较高的早期强度和制品表面较好的光洁度，适用于大规模建筑施工。这些水泥在20世纪80年代的中国大量生产和广泛使用。

1988年在湖北省阴山县建成的6层办公楼和商业楼（8.6m×31.5m），就是使用破碎石灰石作粗骨料、硅酸盐水泥-矿渣-钢渣复合水泥作胶凝材料，水胶比为0.44的混凝土（图11-3）。设计的抗压强度为20MPa，28d的平均实际强度为24.1MPa。使用小型混凝土搅拌车，现场混料，现场浇筑，坍落度在30～50mm之间。侧墙模板1d后拆除，楼面底板7d后拆除。拆模后的混凝土表面非常光滑，没有明显裂纹，到现在该建筑都使用得非常好。

图11-3 使用硫酸钠激发的硅酸盐水泥-矿渣-钢渣复合水泥混凝土建造的6层办公楼和商业楼[1]

在同一个县，1988年建了一个面积为3500m²的厂房，混凝土使用42.5级的硅酸盐水泥-矿渣-钢渣复合水泥（设计强度为30MPa，实际28d强度为35.9MPa），水胶比为0.50，也是安阳钢渣水泥厂生产（图11-4）。柱的横截面积为400mm×400mm，梁的横截面积为350mm×450mm，跨度为12.6m。混凝土梁在现场预制，然后组装，预制1d后拆模。该混凝土的服役性能非常好，到现在也不存在耐久性问题。

图11-4 使用硫酸钠激发的硅酸盐水泥-矿渣-钢渣复合水泥混凝土梁和柱建造的厂房[1]

11.2.4 波兰克拉科夫的仓库

1974年，在克拉科夫使用预制钢筋增强、碱碳酸盐激发的BFS混凝土楼板和墙板建造了一个仓库[5,10]。混凝土组分包括磨细高炉矿渣（300kg/m³）、混合骨料（1841kg/m³）、碳酸钠（18kg/m³）和水（140kg/m³）。构件浇筑后，在热的隧道中70℃空气养护6h，然后安装投入使用。

研究人员对这个建筑已经观察了很多年。超过25年后从该结构的外墙板中取出直径为

100mm 的圆柱体样品,测试抗压强度、碳化深度和微观结构。表 11-1 总结了混凝土芯的抗压强度和碳化深度,28d 到 27 年的抗压强度明显增加,所有样品的平均碳化率不足 0.5mm/年。电镜[5]显示主要凝胶相是致密的 C-S-H 相。没有观察到微裂纹和碱-骨料反应,或长期服役后的钢筋锈蚀。

表 11-1　混凝土芯的抗压强度和碳化深度[5]

样品	抗压强度(MPa)		碳化深度(mm)(27 年)
	28d	27 年	
1	22.1	43.4	10.1
2	21.7	41.9	10.3
3	21.2	46.2	9.4
4	22.3	41.1	11.8
5	23.1	45.1	12.9
6	22.8	41.6	11.7
平均	22.6	43.2	11.0

11.3　混凝土路面

11.3.1　俄罗斯马格尼托哥尔斯克采石场的重载路面

1999 年,在俄罗斯的马格尼托哥尔斯克市,研究人员检测了 1984 年修建的两条碱激发混凝土道路。第一条路长约 6km,通向特娜娅山采石场的重载道路,卡车从采石场出来的负载质量在 60~80t,混凝土路面厚度是 45~50cm,混凝土路面厚度是 25~30cm。第二条路长约 5km,靠近 Shuravi 加油站,混凝土路面厚度是 25~30cm。

该路混凝土组成为:500kg/m³ 的磨细矿粉(400m²/kg),25kg/m³ 的硅酸盐水泥,28kg/m³ 的碳酸钠(溶解在混合物的水中)作为激发剂,水胶比为 0.35。用于道路施工的新拌混凝土的坍落度为 9~12cm,使用钢网增强,混凝土的设计强度为 30MPa。使用通用搅拌车运送混凝土,使用振动板压实。

使用 15 年后,到 1999 年,为了检查,取出一批 70mm 见方的混凝土样块。检测的抗压强度平均值达到 86.1MPa(几乎是设计强度的三倍),吸水率为 8%,表面可见碳化深度为 10~15mm(即平均每年小于 1mm)。尽管每年有几个月的冻融载荷,没有发现钢网锈蚀或冻融结垢。

11.3.2　乌克兰捷尔诺波尔混凝土道路和喷泉水池

1984 年到 1999 年之间,在乌克兰捷尔诺波尔市,Trust Ternopol-promstroy 工程公司现场浇筑了一条 330m 长的碱激发 BFS 混凝土马路和一个喷泉水池。在 1999 年,对马路和喷泉水池进行了检测,与同时期旁边修的硅酸盐水泥混凝土路面比较,使用碱激发 BFS 水泥混凝土修建的马路和水池工作性良好,而使用硅酸盐水泥混凝土修建的马路已经严重破坏，如图 11-5 所示[3]。

图 11-5　位于乌克兰的泰尔诺皮尔市，现浇碱激发炉渣混凝土（左侧）和普通硅酸盐水泥混凝土路（右侧）的比较

11.4　地下及沟槽结构

11.4.1　乌克兰奥德萨市排水管（1966 年）

为防止水土流失，1965 年奥德萨市沿海岸线修建了一条长达 33km 的地下排水管（编号为 NO.5）。该排水管的设计与建设与地铁隧道类似，这就意味着对建材的要求非常高，设计强度为 40MPa。Kievmetrostroy 构件厂制作了大概 40 根碱碳酸盐激发 BFS 混凝土管道用于 5 号排水管线，该混凝土的组成材料包括 500kg/m³ 的 BFS，钠钾碳酸盐复合激发剂（30kg/m³ 固态激发剂），水胶比 0.37，坍落度 8~9cm，河砂作骨料。

2000 年（在投入使用的 34 年后）对管道进行检测，结果表明，碱碳酸盐 BFS 水泥混凝土管道耐久性很好，抗压强度增长到 62MPa，pH 值仍保持在 11.5 以上，吸水率低于 5%，尽管钢筋深度只有 3mm，但没有明显锈蚀现象。

11.4.2　乌克兰扎波罗热州奥良克村饲料槽（1982 年）

Zaporozhoblagrodorstory 公司位于乌克兰的扎波罗热州，该公司自 1972 年以来，在当地生产碱激发 BFS 混凝土和混凝土制品，已经有二十多年了[1]。1982 年，公司为瓦西里耶夫区奥良克村奶品农场生产了一批饲料槽。饲料槽之间的过道、底板、侧板都是碱碳酸盐激发的钢筋增强的 BFS 混凝土预制板，板尺寸为 2m×3m×0.2m，蒸汽养护。板的总面积约 20m×60m，这就需要约 2000t 碱激发混凝土。用的激发剂是钠钾碳酸盐固体混合物。该激发剂是工业固废。

众所周知，饲料发酵会形成有机强酸[11,12]，尽管掺矿渣的水泥混凝土具有一定的耐酸腐蚀性，但发酵废液对普通混凝土的腐蚀还是严重的[12]。由于饲料需要重型设备搬运，所以饲料槽容易受到重力载荷。在投入使用的 18 年后发现，用碱激发 BFS 混凝土板制作的饲料槽没有明显损伤的痕迹，而同时间安装的硅酸盐水泥混凝土面板已明显损坏。

11.5 比利时布鲁塞尔 Le purdociment 公司

在比利时 Purdon[13]（专利在比利时、卢森堡、法国、英国、波兰及其他地区授权）的研究基础上，20 世纪 50 年代布鲁塞尔的 Le purdociment 公司（图 11-6）将碱激发水泥混凝土制品批量化生产。该公司于 1952 年到 1958 年在布鲁塞尔由几个合伙人共同经营。该公司并没有获得足够的收益，不仅是因为他们的生产能力比较低，而且他们的市场利润也比较低（现有资料没有确切信息）。Purdociment 公司于 1958 年关门。在图 11-7 中的建筑里，下边 6 层的部分混凝土是 Purdociment 公司建设的，这部分材料可在 SOFINA 公司（A.O. Purdon 被雇佣的公司）档案中查到。2010 年检测该混凝土显示，混凝土稍有腐蚀或碳化，但服役性能良好。

图 11-6　Le purdociment 公司的信笺

图 11-7　比利时布鲁塞尔"58 号停车楼"
6 层以下的部分混凝土是由
Purdociment 提供的

11.6　Pyrament 水泥及相关产品（法国的 Cordi-Géopolymère/美国的 Lone Star）

Davidovits 于 20 世纪 70 年代早期创立了 Cordi-Géopolymère 公司，该公司自创立以来便致力于发展碱激发胶凝材料及相关材料，后来成立了 Geopolymer 研究所来开发碱激发铝硅酸盐的相关技术。第一批商业化产品主要是无机防火材料，替代有机聚合物树脂，因此命名为"地质聚合物"[15,16]。但到 20 世纪 80 年代，是由美国的合资企业 Lone Star 工业公司推动该建材产品[17,18]。

最早专利[19]里边，这种材料是 BFS 作为钙源加入到碱激发黏土中；加入水泥熟料可以提高凝结性和强度，就把这种商业化产品命名为 Pyrament 复合水泥[18]。Pyrament 复合水泥是一种高强高性能的水泥，抗压强度高达 80MPa，即使环境温度很低，在浇模 4~6h 后强度可达到 20MPa，而且在 -5℃ 以下强度也可以稳定增长。Pyrament 系列产品的熟料含量会有所变化，在 PBC-XT 产品（其商业市场很成功）中的熟料含量大约是 60%。PBC-XT

的掺合料是粉煤灰与碳酸钾，外加剂使用柠檬酸和高效减水剂来控制凝结时间和强度[21]。因此该水泥被分类到复合 AAM 体系中，而不是纯碱激发胶凝材料。对该水泥服役后跟踪检测发现，这种高碱胶凝体系具有很好的实用性和耐久性。

Pyrament 产品系列收益巨大，在美国陆军的工程项目中[20,22]中试并全面推广，在美国和加拿大[7,23-26]的多个州和联邦交通部得到应用，特别是作为桥梁面层或路面损伤修补材料。该类材料被认定为"过关"材料，且在多年服役后性能更好。即使到现在，在 Pyrament 水泥停产多年后的今天，Pyrament 水泥仍是美国很多公路局修建公路允许使用的材料。Pyrament 水泥在飞机跑道上[27]使用 25 年后仍性能良好。

为使 Pyrament 水泥在军用[20,22]和民用（高速公路）[28]工程中都能应用，相关政府部门公布了 Pyrament 水泥混凝土详细的技术和工程报告。尽管多年服役后，稍有碱-骨料反应，但根据大多数水泥性能标准，Pyrament 混凝土与 Portland 水泥混凝土性能相当，甚至更好。美国陆军把 Pyrament 样品在缅因州的 Treat 岛上测试（每年进行 100 次冻融循环，每天潮汐都浸没在海水里），而样品没有明显劣化迹象。经过 8 年室外实验后[20,29]，使用超声波脉冲检测，其物理性能衰减不到 10%。实验室检测预埋钢筋也保护得很好[30,31]。然而，也有研究[32]显示 Pyrament 水泥对用水量波动比较敏感。但与 AAM 混凝土在实验室以外批量应用一样，一些异议只是有限范围内的示例。不只是 Pyrament，水胶比变化敏感的问题在AAM 中都比较常见。传统混凝土从业者（他们是最有可能提出这些问题的）撰写 Pyrament 技术报告的数量远多于非 AAM 专家撰写任何 AAM 报告的数量。

11.7 芬兰开发的 F 混凝土以及其他材料

20 世纪 80 年代，芬兰通过研究 F 水泥胶凝材料，开发了一系列预制混凝土制品，包括路面砖板、管道和轨枕[34]。空气养护和水养护的砂浆样品[34]7d 强度就达到 50MPa，1~2 年后强度超过 100MPa。暴露在空气中碳化 15 个月，在盐溶液中浸泡 4 年（多组对比），硬化 F 混凝土制品仍显示出良好的耐久性。

1980 年到 1994 年期间，芬兰最大的建材公司 Partek 启动了关于碱激发混凝土材料的研究。他们在实验室中检测了多种设计方案产品的强度和泵送性质。评估之后，他们开始在多个工厂中试制备预压中空楼板、低高度高强度横梁、废水系统制品、受压管道、铁路轨枕和步道石。产品经过了各种测试，包括强度发展、弹性模量、收缩和徐变、应变和形变、钢筋粘附、冻融、盐冻胀、抗硫酸盐性、孔隙率和吸气性、渗透性、热重分析和防火性能。

对各种碱的组合以及促凝剂和外加剂进行了测试，还有其他的替换原料，如粉煤灰、其他类型的磨细矿渣、天然矿物、浮选渣、煅烧矿物。在常规条件、高温条件、带压和不带压条件下都做了研究。所有混凝土的官方测试都在赫尔辛基理工大学国家研究中心（VTT）进行。

通常来说，这种材料跟普通硅酸盐水泥相比性质更优。对钢筋的粘附性更好，在测试管道、中空楼板和铁路轨枕时，经常会导致钢筋断裂。该材料在德国布伦瑞克大学的防火测试中显示，混凝土不会因剥落而炸裂，残余强度要比普通硅酸盐水泥高。自 1994 年以来，这种碱激发体系已经被成功地应用于复合防火材料以及芬兰的大型巡航舰隔离材料中的防火气密隔离层[35]。

另外，他们开发了一系列的碱激发胶凝材料，如磨细粒状高炉矿渣（GGBFS）、粉煤灰、硅灰、硅酸钠、石粉等，用于固化危险固废。该材料的强度 14d 达到 5MPa，182d 达到

20MPa，364d 达到 35MPa。该材料的透水性特别低，比天然岩石还要低 10000 倍，基本可以认为是不渗透的。

11.7.1 屋面瓦

1988 年末，第一批屋面瓦开始规模化生产。作为胶凝材料，GGBFS 含有 4% 的石灰，用液体水玻璃 $Na_2O \cdot 2.6SiO_2$ 和液体 NaOH 制成的偏硅酸钠激发。一共做了 14 种配方，根据激发剂用量变化（Na_2O 掺量是 BFS 用量的 2.5%~4%），每 $1m^3$ 混凝土加入 240~440kg 的 BFS。为保证产品性能稳定，砂子的级配稍有变化。屋面瓦外涂有氧化铁黑颜料和丙烯酸黑色涂料。有一批瓦只涂上涂料，没涂颜料。屋面瓦用在私人住宅上（图 11-8）。15 年来，每年取样，用多余的瓦替换。

图 11-8 该房位于芬兰赫尔辛基附近，该房屋顶于 1988 年用碱激发的 BFS 屋面瓦覆盖（该图片由 B. Talling 友情提供）

对屋面瓦做常规研究时发现：生坯强度和脱模强度比混合普通硅酸盐水泥高，所有 AAM 产品的渗透性都低，拉伸强度都高。其中一种 AAM 样品的最高拉伸强度比水泥样品平均拉伸强度高两倍。

在观察期内（24 年），屋面瓦的拉伸强度平均增加了 25%。2012 年屋顶又重新上漆，没有发现破裂的屋面瓦，并且涂漆的表面是完整的。没有上颜料的那一批瓦也看不出来了。屋顶上没有发现苔藓，在一些地方仅出现微量的黄色藻类。重新喷漆以后，屋顶的预期寿命将会延长。

11.7.2 芬兰研发成果

通过振动压实的方法，碱激发 BFS 材料非常适合生产硬质混凝土。相对于 OPC 混凝土，硬质混凝土的凝结强度和塑坯强度更优，并且加入少量的石灰可增加早期强度。硅酸钠基激发剂（模数小于 2.8）能给出最高的早期和末期强度。相对于 OPC 混凝土，这种混凝土对于石粉、黏土残渣、细粉和杂质的敏感度很低。在适当的混凝土配合比设计条件下，会与 OPC 混凝土有相同的耐久性，甚至在其他方面比 OPC 混凝土有更好的性能。24 年之后，屋面瓦的抗拉强度和抗渗性仍在增加。未来混凝土配合比设计和应用方面还可以将 BFS 与其他炉渣、工业副产品（普通煤、褐煤、油页岩、生物燃料、尾矿燃烧产生的灰）、造纸和化工产生的富碱废料以及天然的碱性材料结合使用。通过这种方式，可以为许多新的应用提供解决方案。

11.8 ASCEM®水泥（荷兰）

ASCEM® 水泥已在荷兰实现了商业化[36]，该水泥结合了两个主要思想，即铝硅酸盐材料的碱激发和固废原料的再利用，例如粉煤灰。通过冶炼工艺把二次原料转化为质量稳定的高性能材料来使用，这种方法的优点是：一方面是材料（不同组成和性质）选择灵活；另一方面是提供质量稳定的产品。

ASCEM® 水泥是一种包含多种组分的碱激发水泥：
(1) 由原材料（优选二次资源）制得的含 $CaO-Al_2O_3-SiO_2$ 玻璃相的活性玻璃。
(2) 填料成分。
(3) 碱激发剂。

虽然它是从高温工艺得到的，工作温度约 1450℃，但是生产该材料比生产普通硅酸盐水泥会产生更少的 CO_2，在技术性能方面也有竞争优势。该方法是基于不同的废物如粉煤灰和城市建筑废弃物的重新利用[37]。这个技术在 20 世纪 80 年代末已经试验成功了，但那时没有商业化。ASCEM B.V. 是最早开发这种新水泥产品的公司，其他驱动因素如 CO_2 排放量低，成为新水泥研发的契机。

ASCEM 水泥技术的核心是二次资源混合物的熔化，这种混合物（例如粉煤灰＋修正材料）被熔化以满足 $CaO-Al_2O_3-SiO_2$ 三元体系中的特定组成，得到激发玻璃体。在熔融体快速冷却后，将冷却的玻璃体研磨并与填料混合。到这一步，粉煤灰原灰还可以和其他细粉填料一块加入上述混合物中利用。通常，玻璃与填料比约为 1，ASCEM 水泥生产流程如图 11-9 所示。

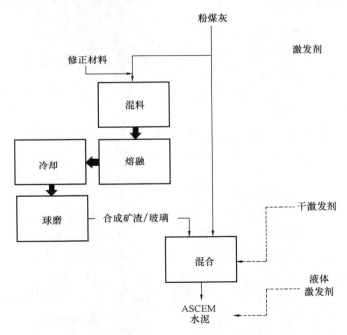

图 11-9　ASCEM 水泥生产的技术流程

根据应用领域不同，可以使用固体激发剂，也可以使用液体激发剂。固体激发剂的优点是可以做成一个化合物产品出厂，只需要在胶凝材料/混凝土中加水即可使用。液体激发剂的优势是混凝土现场施工可以直接加入，掺量可以随时调整，这就需要施工人员具有丰富的施工经验。

11.8.1　改进Ⅰ：水泥生产

传统工艺都是规模化生产的，这就需要对 ASCEM 水泥生产进行工业化改造。因此，2009 年，ASCEM 水泥启动了一个增产项目，目的就是要扩大水泥和商品混凝土的掺量。

事实上，这项计划不是要给 ASCEM 水泥专门新建工厂，而是要使用工厂现有的设备，使用相同工艺或付费服务（例如混料和球磨）的方式生产 ASCEM 水泥。图 11-10 展示了原料熔融出炉时，玻璃水流入冷却槽。中间玻璃产物是使用 1450℃ 的炉温连续熔化产生的。冷却过程在大容量水箱中进行。将冷却的玻璃颗粒从水箱中分批提出。玻璃中间产物在立磨中球磨，细度达到 500~600m²/kg，用于水泥和混凝土生产。普通的干燥粉煤灰作为填料。激发剂仅在混凝土生产中添加，不依赖于液体或固体激发剂，以便在整个测试期间剂量可控。

11.8.2 改进Ⅱ：水泥生产

选择两种不同的产品测试水泥特性。

（1）用于工业领域的 Stelcon® 板材（由 B. V. De Meteoor 在荷兰雷登生产的）：2m×2m，厚度为 14cm，这些制品通常需要使用 CEM Ⅰ 52.5 R 型的硅酸盐水泥生产。

图 11-10 熔化玻璃水进入冷却槽的照片

（2）下水管道（由 B. V. De Hamer 在荷兰阿尔芬生产的）：直径 800mm 的无钢筋混凝土管道。

1. Stelcon® 板的生产

生产的 2m×2m 的混凝土路面板以 Stelcon® 商标命名。这种混凝土是低水泥用量的硬质混凝土（230~250L/m³），根据 EN 206 C1 标准（压实值为 1.4~1.5），先填充后压实（图 11-11）。压实后，将生坯混凝土板翻转，封闭保存约 16h。其中，由于水泥硬化会放热，温度将升至 30℃。封装养护的混凝土板（图 11-12a）放置于室外环境中。

(a)

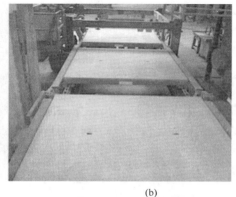
(b)

图 11-11 Stelcon® 板生产
(a) 浆料填充入振动模具；(b) 脱模

2. 下水管道的生产

图 11-13 中的下水管道是使用专门生产管道的设备压实湿混凝土制备的。在管道生产中，均质性和生坯强度是生产的关键。生坯管道脱模后，拉到养护室储存，第二天再把管道放到室外（图 11-12b）。

图 11-12 室外放置产品
（a）板；（b）管道

图 11-13 生产的管道

11.8.3 混凝土配合比设计和组成

本小节的出发点是设计具有相同胶凝材料与骨料体积比的高性能混凝土。混凝土应具有良好的工作性。因此，不同的胶凝材料体系会有不同的需水量，导致有不同的水灰比。这里给出两种不同组成的 ASCEM 水泥。

(1) 液体激发剂：NaOH 溶液（50%）和硅酸钾溶液（13.8% K_2O，26% SiO_2），二者的质量比为 4：3。在这种情况下，活性玻璃和粉煤灰（ASCEM 水泥）进行预混，且要与液体激发剂分开。

(2) 固体激发剂。混凝土组分在表 11-2 中给出。

表 11-2 使用 ASCEM 水泥配制的混凝土组成

成分	单位	板			管道		
		参考值	混合物ⅠA	混合物ⅠB	参考值	混合物ⅡA	混合物ⅡB
参照水泥	kg/m³	320[a]	—	—	390[b]	—	—
ASCEM 水泥（含固体激发剂）	kg/m³	—	—	326	—	—	389
ASCEM 水泥	kg/m³	—	319	—	—	365	—
ASCEM 激发剂（NaOH 和硅酸钾混合物）	kg/m³	—	43.4	—	—	51.26	—
骨料 0/32（天然圆颗粒）	kg/m³	1967	1967	1967			
骨料 0/16（部分破损的）	kg/m³				1823	1800	1807
水（包含液体激发剂中的水）	kg/m³	134.4	115.4	—	144.2	143.8	163.4
水胶比（不含激发剂）	—		0.37	0.46		0.39	0.42
激发剂含量（干重量/水泥质量）	%		6	11.6		6	11.6
Na_2O 在水泥中含量（Na_2O 当量）	%		3.4	2.2		3.4	2.2

a. CEM I：添加超塑化剂 2.08kg/m³；
b. CEM III/B：未添加外加剂。

11.8.4 混凝土和混凝土样块的性质

图 11-14 给出了管道混凝土、实验室水中混凝土试块和参照混凝土的强度发展对比图，图 11-15 给出了几种管道混凝土破坏实验的结果。结果显示参照混凝土性能更高。通过提高活性玻璃的细度才能有效改善强度。

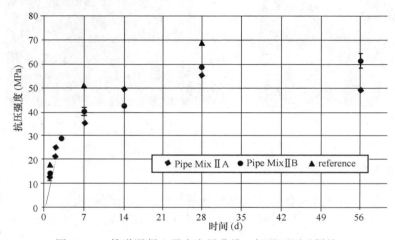

图 11-14 管道混凝土强度发展曲线（标况下测试样块）

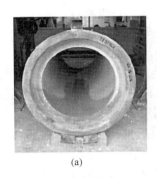
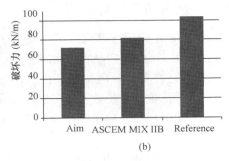

图 11-15　管道的断裂试验结果（见左侧测量设置，与标准混凝土相比）

11.9　E-Crete™ 水泥

11.9.1　E-Crete™ 水泥的技术特点

碱激发材料缺乏工业应用的主要原因已很明确[38,39]：（1）在建材行业以往已经形成了传统做法，已有既定的利益团体；（2）实验制备的碱激发混凝土和工业化生产的混凝土，在粉体和湿混凝土的处理、湿和硬化混凝土的工程性能等方面存在巨大差异；（3）许多研究人员缺乏工业和商业化经验；（4）缺乏供应链动态调控的认识；（5）材料选择的经验不足，只能是小范围的选择原料，对于在不同气候和国家以及在不同的操作条件下进行各种材料选择的经验是缺乏的。

自 2006 年以来，澳大利亚墨尔本的 Zeobond 集团（www.zeobond.com）已经与商业合作伙伴建立了一个团队开展碱激发混凝土的商业化应用。Zeobond 集团开发了 E-Crete™ 水泥胶凝材料，这种胶凝材料由粉煤灰、矿渣和碱激发剂混合生产。把该胶凝材料与砂子、骨料混合，比例与传统水泥相似，做成混凝土。E-Crete™ 水泥可以使用标准碱激发剂和任何原材料生产，例如粉煤灰和矿渣。然而，专利技术通常是把原材料与一系列混合物组合起来，包括一些特殊混合物。E-Crete 水泥使用了特殊的预处理方法，原材料也进行了处理，目的是为了改变它们的性能，并调整它们的反应性以适应某些应用。Zeobond 集团因 E-Crete 水泥的发明收到了 INNOVIC 组织颁发的"未来大事件奖"。

E-Crete™ 技术使用了来自世界各地原材料的测试数据库，这使得新水泥配比得到快速开发。Zeobond 已经培训了几个工业团队，在没有 Zeobond 工作人员直接监督情况下大规模生产 E-Crete™ 水泥，很大程度上扩展了该技术的应用。这表明该技术在工业混凝土应用的苛刻条件下是稳定的。E-Crete™ 与传统水泥具有相同的性能表现，在市场中具有成本竞争优势，并且利用低成本的废弃物材料，这在废弃物资源化利用的市场中具有竞争优势。

E-Crete™ 的商业全生命周期分析由维多利亚州政府委托澳大利亚 NetBalance 基金会执行。2007 年全生命周期分析把地质聚合物水泥与澳大利亚标准的传统水泥做了详细对比，包括水泥与水泥的比较和混凝土与混凝土的比较。水泥间的比较显示 CO_2 排放减少了 80%，而混凝土间的比较显示成本节约高达 60%，但是骨料的生产和运输成本对于两种材料是相同的。

11.9.2 E-Crete™技术的工业化应用

E-Crete™技术在工业化应用方面取得了巨大进展,包含:
(1) 结构和非结构 E-Crete™产品在澳大利亚的维多利亚和昆士兰得到广泛应用。
(2) E-Crete™技术已在美国、阿拉伯联合酋长国和中国进行了工业化示范。
(3) 预拌 E-Crete™混凝土已在墨尔本房地产公司开发的新住宅中应用。
(4) Zeobond 与咨询工程师、建筑师、市议会、政府部门合作,建立新水泥技术使用的协议。
(5) 用于生产混凝土管道的现有设备已经被改装来生产 E-Crete 管道。
(6) 防火纤维增韧的 E-Crete™隧道管片已成功试生产[40]。
(7) 澳大利亚维多利亚州的道路管理局在 2010 年更新的设计规范说明部分的第 703 节当中把地质聚合物混凝土在非结构的应用等同于 OPC 混凝土[41]。
(8) E-Crete™已经被用于维州公路局的基建项目,并且被当地议会和住房开发商在部分工程和路面板中使用。这些大规模应用对于逐渐说服标准局接受地质聚合物混凝土至关重要。
(9) Zeobond 正在与 VicRoads 合作,通过实际工程示范,在《结构混凝土规范》第 620 节预制混凝土单元的章节中接受地质聚合物混凝土在结构工程中的应用[42]。

11.9.3 预制桥面板

根据维州道路管理局规定的 55MPa《结构混凝土规范》(第 620 节[42])澳大利亚墨尔本港口鲑鱼街的 E-Crete 预制桥面道路板(图 11-16)被维州管理局认定为最优质的混凝土道路板。用于该桥面道路板的产品需要较高的早期强度、低收缩率和长期耐久性。这个项目已经在 2009 年完成,被用来作为结构应用中的示范性试验项目。

图 11-16 澳大利亚墨尔本港口鲑鱼街的 E-Crete™预制桥面道路板

11.9.4　桥梁挡土墙

这个项目是墨尔本天鹅街桥梁挡土墙的重建项目（图 11-17）2009 年的 E-Crete™ 被维州道路管理局选中应用于这个项目，该项目位置重要并且要求苛刻。挡土墙需要强度 40MPa 的结构混凝土[42]，所用混凝土具有较长坍落度保持时间，具有泵送能力，墙上还安装了检测设备监测钢筋的长期性能。

图 11-17　澳大利亚墨尔本天鹅街桥的 E-Crete™ 挡土墙

11.9.5　人行道

西部高速公路联盟使用 E-Crete™ 建造了墨尔本港布雷迪街承压 25MPa 的人行道（图 11-18）。这是在该类项目中的首次使用，维州道路管理局在 2010 年制定了 E-Crete™ 20MPa、25MPa 和 32MPa 等级的混凝土在普通混凝土铺路以及在人行道，路肩和排水管非结构用途中的规范[41]。这个项目涉及多方的参与和合作，包括维州道路管理局、墨尔本市议会、墨尔本市港务局和联盟合作伙伴。

图 11-18　澳大利亚墨尔本港布雷迪街 E-Crete™ 人行道

11.9.6 步行道和地面工程

汤马氏镇的娱乐和水上中心的工程（图11-19）在2010年竣工，标志着与市议会、设计方、施工方和建筑商合作关系达到了新的高度。这是从设计到施工使用E-Crete™的第一个样品工程项目。人行道和车道用一系列装饰性的25MPa级别的E-Crete™混凝土完成。

图1-19　E-Crete™路面和澳大利亚墨尔本的休闲水中心工程

11.9.7 预制板

墨尔本的梅尔顿图书馆有3500m²，两层楼，图书馆设计特别注重能源效率和可持续发展。E-Crete™预制板用于建筑物的外墙（图11-20），于2012年安装。每个面板长9m，由

图11-20　澳大利亚墨尔本E-Crete™预制板建造的图书馆

40 MPa 等级的 E-Crete™ 混凝土制造，设计的早期强度很高，骨料用的是天然的河卵石。

11.10 不同地区的铁路轨枕

另一个与市政基础设施工程相关的是铁路枕木的生产，E-Crete™ 混凝土在预制方面能够达到很高的强度。符合日本标准的预应力轨枕已经用碱硅酸盐激发粉煤灰在实验室制备出来了[43]。20 世纪 80 年代在波兰也开发了碱激发矿渣轨枕，强度可以达到 70MPa[44]，5 年的使用期内，性能与使用硅酸盐水泥制造的轨枕相当[45]。在西班牙的一个示范研究计划中[46,47]，使用碱氢氧化物激发粉煤灰制备预应力蒸养的轨枕得到了迅速发展，能够满足西班牙和欧洲规范要求。1988 年，预应力碱硅酸盐激发矿渣混凝土轨枕安装在俄罗斯圣彼得堡和莫斯科之间的铁路上，就在 Tchudovo 火车站附近，轨道长约 20m。该轨枕设计的强度为 45MPa，矿渣用量为 500kg/m³，水灰比为 0.3。2000 年，对这些轨枕现场检测发现工作性能良好[48]。混凝土的碳化速率小于 1mm/年，在使用期间混凝土强度增加到 82MPa，并且没有腐蚀、开裂、剥落或其他缺陷问题[1,48]，嵌入的钢筋也没有锈蚀现象，材料的微观结构在服役期间是稳定的[48]，而且都是在具有冻融和冲击载荷的环境下。

11.11 小结

本章描述了在欧洲、亚洲、北美和澳大利亚的处于服役期内的碱激发混凝土的性能。自 20 世纪 50 年代以来，在西欧，商业化生产已经开始了（尽管在大部分情况下这种生产后期中断），20 世纪 60 年代在苏联，20 世纪 80 年代在芬兰、中国和美国，产品已经被投入使用。通过这些工程示范，为我们提供了各种气候和服役条件下的 AAM 混凝土耐久性能。这些工程示范表明已经投入使用的碱激发混凝土能够满足要求，没有明显的碳化、冻融、机械或化学稳定性、耐酸性、钢筋防腐、碱-骨料反应或任何其他形式的裂化问题。一般，强度数据取自服役十年以上的工程或者强度已经明显高于初始设计强度要求的产品。值得注意的是，并不是所有与 AAM 混凝土相关的商业运作都实现了市场化。在 AAM 混凝土应用中，需水量的敏感性、养护条件和工作性方面要比硅酸盐水泥混凝土的应用更具挑战性。然而，越来越多的证据表明在市政基础设施工程中，碱激发混凝土的可用性、耐久性和市场占有率将会逐渐扩大。

参考文献

[1] Shi C., Craven P. V., Roy D. M.. Alkali-Activated Cements and Concretes. Taylor & Francis, Abingdon, 2006.

[2] Broke O. A.. Experience of exploitation of the alkaline cement concretes. In: Krivenko P. V. (ed.) Proceedings of the Second International Conference on Alkaline Cements and Concretes, 1999: 657-684. ORANTA, Kiev.

[3] Craven P. V.. Alkaline cements: from research to application. In: Lukey G. C. (ed.) Geopolymers 2002. Turn Potential into Profit, Melbourne, Australia. CD-ROM Proceedings. Siloxo Pty. Ltd. 2002.

[4] Xu H., Provis J. L., Van Deventer J. S. J., Craven P. V.. Characterization of aged slag concretes. ACI Mater. J, 2008, 105 (2): 131-139.

[5] Deja J.. Carbonation aspects of alkali activated slag mortars and concretes. Silic. Indus, 2002, 67 (1): 37-42.

[6] Ilyin V. P.. Durability of materials based on slag-alkaline binders. In: Craven P. V. (ed.) Proceedings of the First International Conference on Alkaline Cements and Concretes, Kiev, Ukraine. 1994, 2: 789-836. VIPOL Stock Company.

[7] Ozyildirim C.. A Field Investigation of Concrete Overlays Containing Latex, Silica Fume, or Pyrament Cement. Virginia Transportation Research Council, Charlottesville, 1996.

[8] Brouwers H.. The work of Powers and Brownyard revisited: Part 1. Cem. Concr. Res, 2004, 34 (9): 1697-1716.

[9] Volovikov A., Kosenko S.. Experience from production and application of slag alkaline cements and concretes. In: Ertl Z. (ed.) Proceedings of the International Conference on Alkali Activated Materials - Research, Production and Utilization, Prague, Czech Republic, 2007: 737-744. Česká rozvojová agentura.

[10] Małolepszy J.. The hydration and the properties of alkali activated slag cementitious materials. Ceramika, 1989, 53: 7-125.

[11] Pavía S., Condren D.. Study of the durability of OPC versus GGBS concrete on exposure to silage effluent. J. Mater. Civil Eng, 2008, 20 (4): 313-320.

[12] De Belie N., Verselder H. J., De Blaere B., Van Nieuwenburg D., Verschoore R.: Influence of the cement type on the resistance of concrete to feed acids. Cem. Concr. Res, 1996, 26 (11): 1717-1725.

[13] Purdon A. O.. The action of alkalis on blast-furnace slag. J. Soc. Chem. Ind.-Trans. Commun, 1940, 59: 191-202.

[14] Vanooteghem M.. Duurzaamheid van beton met alkali-geactiveerde slak uit de jaren 50-Het Purdocement. M. Ing. thesis, Universiteit Gent, 2011.

[15] Davidovits J.. Process for the fabrication of sintered panels and panels resulting from the application of this process. Comppler: French, 2204999 and 2246382 [P]. 1973.

[16] Davidovits J.. Mineral polymers and methods of making them. Compiler: U. S, 4349386 [P], 1982.

[17] Davidovits J.. 30 years of successes and failures in geopolymer applications. Market trends and potential breakthroughs. In: Lukey G. C. (ed.) Geopolymers 2002. Turn Potential into Profit, Melbourne, Australia. CD-ROM Proceedings. Siloxo Pty. Ltd. 2002.

[18] Davidovits J.. Geopolymer Chemistry and Applications. Institut Géopolymère, Saint-Quentin, 2008.

[19] Davidovits J., Sawyer J. L.. Early high-strength mineral polymer. Compiler: U. S, 4509985[P]. 1985.

[20] Husbands T. B., Malone P. G., Wakeley L. D.. Performance of Concretes Proportioned with Pyrament Blended Cement, U. S. Army Corps of Engineers Construction Productivity Advancement Research Program, Report CPAR-SL-94-2, 1994.

[21] Zia P., Leming M. L., Ahmad S. H., Schemmel J. J., Elliott R. P.. Mechanical Behavior of High Performance Concretes, 1993, 2: Production of High Performance Concrete. SHRP-C-362. Strategic Highway Research Program, National Research Council, Washington, DC.

[22] Malone P. G., Randall C. J., Kirkpatrick T.. Potential Applications of Alkali-Activated Aluminosilicate Binders in Military Operations, Geotechnical Laboratory, Department of the Army, GL-85-15, 1985.

[23] Jones K.. Special Cements for Fast Track Concrete, Final Report MLR-87-4, Iowa Department of Transportation, Highway Division, 1988.

[24] Yu H. T., Mallela J., Darter M. I.. Highway Concrete Technology Development and Testing Volume IV: Field Evaluation of SHRP C-206 Test Sites (Early Opening of Full-Depth Pavement Repairs), Federal Highways Administration, 2006.

[25] Ozyildirim C.. A Field Investigation of Concrete Patches Containing Pyrament Blended Cement. Virginia Department of Transportation, Charlottesville, 1994.

[26] Czarnecki B., Day R. L.. Service life predictions for new and rehabilitated concrete bridge structures. In: Biondini F., Frangopol D. M. (eds.) Proceedings of the International Symposium on Life-cycle Civil Engineering, IALCCE'08, 2008: 311-16. CRC Press, Varenna.

[27] Geopolymer Institute. PYRAMENT cement good for heavy traffic after 25 years. 2011.
http://www.geopolymer.org/news/pyrament-cement-good-for-heavy-traffi c-afer-25-years.

[28] Zia P., Ahmad S. H., Leming M. L., Schemmel J. J., Elliott R. P.. Mechanical Behavior of High Performance Concretes, 1993, 3: Very High Early Strength Concrete. SHRP-C-363. Strategic Highway Research Program, National Research Council, Washington, DC.

[29] U. S. Army Corps of Engineers. Natural Weathering Exposure Station Treat Island Test Results-CPAR-High-Performance Blended Cement System. 1999. http://www.wes.army.mil/SL/TREAT_ISL/Programs/cparHighPerformance6/data6.html.

[30] Muszynski L. C.. Corrosion protection of reinforcing steel using Pyrament blended cement concrete. In: Swamy R. N. (ed.) Blended Cements in Construction, 1991: 442-454. Elsevier, Barking.

[31] Wheat H. G.. Corrosion behavior of steel in concrete made with Pyrament® blended cement. Cem. Concr. Res, 1992, 22: 103-111.

[32] Rodriguez-Gomez J., Nazarian S.. Laboratory Investigation of Delamination and Debonding of Thin-Bonded Overlays Due to Vehicular Vibration, RR1920-1, Univer-

sity of Texas at El Paso/Texas Department of Transportation, 1992.

[33] Forss B.. Process for producing a binder for slurry, mortar, and concrete. Compiler: U. S, 4306912[P]. 1982.

[34] Talling B.. Effect of curing conditions on alkali-activated slags. In: Malhotra V. M. (ed.) 3rd International Conference on Fly Ash, Silica Fume, Slag and Natural Pozzolans in Concrete, ACI SP114, Trondheim, Norway. 1989, 2: 1485-1500. American Concrete Institute.

[35] Talling B.. Geopolymers give fire safety to cruise ships. In: Lukey G. C. (ed.) Geopolymers 2002. Turn Potential into Profit, Melbourne, Australia. CD-ROM Proceedings. Siloxo Pty. Ltd. Detroit, MI, 2002.

[36] Buchwald A.. ASCEM® cement-a contribution towards conserving primary resources and reducing the output of CO_2. Cem. Int, 2012, 10 (5): 86-97.

[37] Lamers F. J. M., Schuur H. M. L., Saraber A. J., Braam J.. Production and application of a use-ful slag from inorganic waste products with a smelting process. In: Goumans J. J. J. M., VanerSloot H. A., Aalbers T. G. (eds.) Studies in Environmental Science 48: Waste Materials inConstruction, 1991: 513-522. Elsevier.

[38] van Deventer J. S. J., Provis J. L., Duxson P.. Technical and commercial progress in the adoption of geopolymer cement. Miner. Eng, 2012, 29: 89-104. Amsterdam, Netherlands.

[39] Van Deventer J. S. J., Brice D. G., Bernal S. A., Provis J. L.. Development, standardization an applications of alkali-activated concretes. In: ASTM Symposium on Geopolymers, San Diego CA. ASTM STP 1566. ASTM International, 2012.

[40] Wimpenny D., Duxson P., Cooper T., Provis J. L., Zeuschner R.. Fibre reinforced geopoly-mer concrete products for underground infrastructure. In: Concrete 2011, Perth, Australia. CD-ROM proceedings. Concrete Institute of Australia, 2011.

[41] VicRoads. VicRoads Standard Specifications. In: Section 703-General Concrete Paving, VicRoads, Melbourne, 2010.

[42] VicRoads. VicRoads Standard Specifications. In: Section 620-Precast Concrete Units, VicRoads, Melbourne, 2009.

[43] Uehara M.. New concrete with low environmental load using the geopolymer method. Quart. Rep. RTRI, 2010, 51 (1): 1-7.

[44] Małolepszy J., Deja J., Brylicki W.. Industrial application of slag alkaline concretes. In: Craven P. V. (ed.) Proceedings of the First International Conference on Alkaline Cements and Concretes, Kiev, Ukraine. 1994, 2: 989-1001. VIPOL Stock Company.

[45] Deja J., Brylicki W., Małolepszy J.. Anti-filtration screens based on alkali-activated slag binders. In: Ertl Z. (ed.) Proceedings of the International Conference on Alkali Activated Materials-Research, Production and Utilization, Prague, Czech Republic, 2007: 163-184. Česká rozvojová agentura.

[46] Fernández-Jiménez A., Palomo A., Rivulet D.. Alkali activation of industrial by-products to develop new Earth-friendly cements. In: 11th International Conference on Non-conventional Materials And Technologies (NOCMAT 2009), Bath, UK. CD-ROM proceedings, 2009.

[47] Palomo A., Fernández-Jiménez A., López-Hombrados C., Lleyda J. L.. Railway sleepers made of alkali activated fly ash concrete. Rev. Ing. Constr, 2007, 22 (2): 75-80.

[48] Poletayev A.. Assessment of durability of railway sleepers based on slag-alkaline binders.

第 12 章 碱激发材料的其他潜在应用

Susan A. Bernal, Pavel V. Krivenko, JohnL. Provis, Francisca Puertas, William D. A. Rickard, CaijunShi and Arie van Riessen

12.1 引言

本章重点讨论碱激发胶凝材料和混凝土各种潜在的商业化应用方向（除了作为大规模民用基础设施建筑以外）。在除本章各节所指出的工程化应用案例以外，其他碱激发材料还没有实现大规模应用。本章列出的每个领域都有中试或规模化示范项目，并且每一项在科学和技术方面都有发展的空间。除本章具体讨论的应用外，碱激发材料在其他特别的应用领域也有商业和学术上的发展，比如一种瓷砖填缝剂已经推向市场[1]，它具有自清洁性能，还有碱激发偏高岭土胶凝材料作为缓释药物的载体[2,3]。毫无疑问这些应用确实有商业前景，但它们是相当专业的应用。因此，本章的重点是更广泛地研究和开发，而不是提供特定产品的详细分析。下面将讨论的领域包括轻质材料、油井水泥、防火材料和纤维增强复合材料。

12.2 轻质碱激发材料

12.2.1 发泡胶凝材料和蒸压加气混凝土

轻质碱激发胶凝材料的制备方法很多，广泛用于轻质硅酸盐水泥混凝土生产的方法之一是水热法，这也是蒸压加气混凝土的生产方式；在美国、欧洲和苏联将这种方法应用于碱激发体系都有成功案例[4-9]。大多数用于生产发泡碱激发胶凝材料的方法，都是在胶凝材料呈浆体状态下，通过氧化胶凝材料多种组分中的任何一种，来产生氢气或氧气。金属铝粉[10-13]，直接加入[13]或作为硅粉组分[14-17]的金属硅、过氧化氢[4,18,19]、过氧化钠[20]和过硼酸钠[4,19]等已经被用作发泡剂。通过分解次氯酸钙产生氯气的方法也已被提出[21]，但产生的氯气对材料的耐久性的长期影响尚不明确。还可以通过使用合适的表面活性剂来实现发泡[22,23]，该方法已经实际应用于生产轻质碱激发 BFS 板，板中加入纤维状外加剂来起到隔声的功能[24]。

1978 年，俄罗斯的 Berezovo 使用混合碱金属氢氧化物废液作为蒸压发泡碱激发 BFS 混凝土的中试生产的激发剂，持续了许多年[8]。在基辅，后来还开发了偏高岭土和粉煤灰碱激发蒸汽加压混凝土[6,7]，图 12-1 是作为该研究计划的结果已经开发并投入使用的产品之一的示例。

许多作者还利用局部聚合硅（铝）酸盐凝胶在升温时发泡，开发泡沫碱激发材料[25-28]。这种材料在被动防火应用中是有价值的[19,29]，因为它是吸热的，并且还生产了不可燃的泡沫材料来填充空间。AAM 的高温性能将在本章的第 12.4 节中更详细地讨论。

图 12-1 使用铝金属粉末发泡的防火碱激发粉煤灰砌块,用作马弗炉内衬
(a) 胶凝材料结构(可见孔径为几十至几百微米,材料密度为 $600kg/m^3$);(b) 在实验室炉中测试的材料;(c 和 d) 用该材料制成的尺寸为 40cm×40cm×4cm 的砖,在高达 700℃ 的炉中工作 6 个月之后的样品(来自参考文献 [7] 的图片,版权 Springer)

12.2.2 轻骨料碱激发混凝土

轻骨料也能与碱激发胶凝材料很好地结合制备混凝土,其中液体[5,30]和固体[31,32]碱都可用作激发剂。轻骨料碱硅酸盐激发的 BFS 混凝土,强度高达 90MPa[33];骨料本身的强度远低于这个水平,但胶凝材料能够产生很高的强度。碱激发混凝土的冻融性能也随着轻骨料的加入而改善[8]。可使用的轻骨料(或者确切使用的是多孔陶瓷生产中的填料)包括粉煤灰空心微珠[34]和发泡聚苯乙烯[35,36]。预饱和轻骨料(膨胀黏土)也已成功地作为内养护的外加剂用于碱激发 BFS 胶凝材料,以减少自收缩和干缩的影响[37]。

铝硅酸盐轻骨料(来自黏土或珍珠岩)的废料可以提供相结合的活性氧化铝和二氧化硅,与碱激发反应制备胶凝材料[18,38,39]。发生这种反应会让大家想到当相同材料的较粗颗粒用作骨料时,可能存在碱激发的化学反应,这还需要做进一步的研究。

还值得注意的是,碱激发的粉煤灰(AAFA)既可以制备混凝土的轻骨料,也可以制备胶凝材料[40]。这种骨料在实际混凝土体系中的作用尚不明确,这是由于碱激发粉煤灰的吸湿性质将有可能导致混凝土的需水量出现问题,这种情况在使用再生骨料时经常出现。无论是作为玻璃相组分,还是硅酸盐水泥混凝土的额外碱源,都需要考虑这种骨料的潜在碱反应活性。

12.3 油井水泥

在苏联,碳酸钠激发 BFS 水泥被开发用作油井水泥,应用在深达 3500m、温度高达 80℃、压力超过 60MPa 油井下[8]。碱激发矿渣水泥也用于地下硬化工程,特别是在盐矿和硫矿的井眼密封以及欧洲的水下密封中[41-43]。AAMs 也被开发用于 CO_2 储存和封存地下固井[44]。钻井液和泥浆还可以与碱激发 BFS 混合用于固井的水泥浆[45,46],也有使用碱硅酸盐激发的粉煤灰/BFS 复合胶凝材料用于固井[47]。本应用的碱激发 BFS 配方最初由 Shell Oil Co. 于 1991 年开发并首次应用于墨西哥湾。同时类似的混合物材料在中国[48]和巴西[49]已成功使用。壳牌产品,名为"Slag-Mix",20 世纪 90 年代在墨西哥湾的几个项目中被用作完全替代硅酸盐水泥基油井水泥[45]。据报告表明,在这项应用中,技术、环境效益和经济效益都优于硅酸盐水泥[46],它作为"通用液"起到钻井液和固井的作用[50]。一些其他石油公司[51]的研究工作表明,在一些情况下碱激发 BFS 浆料在固化后可能会开裂和变化,这在一定程度上限制了它在整个行业的广泛使用。但是,人们还是在持续关注这类材料,Schlumberger 的专利申请涉及在油井[52]和 CO_2 隔离井[53]中使用碱激发水泥。

由 Brookhaven 国家实验室和 Halliburton[54,55]合作开发的硅酸钠激发的 BFS 和 BFS/粉煤灰水泥用于高温下含有浓 H_2SO_4 和溶解的 CO_2 的地热井。它们能够产生大于 80MPa 的抗压强度和极低的水渗透能力,并且在混合物中掺入的粉煤灰也被认为有助于耐酸性[55]。在高压釜中形成的凝胶相主要是部分结晶的 C-S-H 相,包括托勃莫来石相。较高温度(300℃)下 BFS 凝胶会过度结晶,随着温度升高需要掺入更多的粉煤灰来稳定体系[55]。

这确实是一个商业化、大宗应用的好领域,也为这种非传统胶凝材料批量化使用提供了机会。最近与 2010 年墨西哥湾"深水地平线"油井水泥失控事件可能预示着在油田应用中选择油井水泥时表现出相对保守的态度,但与此同时,推动碳捕获与储存技术和地热能将提振人们开发特种水泥的兴趣,来挑战极其恶劣的地下环境。

12.4 高温性能

本节概述了碱激发材料的耐热性能。这些材料,特别是被称为"地质聚合物"的低钙体系,与硅酸盐水泥基材料相比,在许多情况下已显示出具有优异的耐高温性能。从工程应用角度来说,这引起了人们对该技术的强烈兴趣。实际上,自 20 世纪 70 年代以来,法国 Geopolymer 品牌和系列产品推动着地质聚合物的商业化,也是从那时起,Davidovits 及其合作者集中获得了大量专利[19,56]。考虑到关于碱激发 BFS 胶凝材料在高温下的性能数据有限,本节中的大部分讨论将基于对地质聚合物体系的讨论。低钙 AAM 体系,包括地质聚合物,也可以被称为通过热处理生产特种陶瓷的前驱体材料[57-59]。

碱激发铝硅酸盐胶凝材料的无机框架结构使其具有优异的热稳定性[60]。这种特性使得这些材料和含有 AAM 胶凝材料的复合材料能够用于高温领域中,例如炉内衬、耐热涂层、隔热和墙板。值得注意的一个特别应用是,使用碱激发胶凝材料生产游船的防火板和内饰板,Renotech Oy 在芬兰[61]已经做了十多年。

当用作高温绝热材料时,需要注意以下性能。当用作基体涂层时,具有低热膨胀性或兼容的膨胀系数、低热导率、低剥离量和高熔点。此外,耐热材料必须保持物相相对稳定,形貌很少变化。

本章对高温下 OPC 的热性能只做简单的比较分析,读者可以参考 Mendes[62,63] 和 Hertz[64] 的工作,以获得关于这个议题更详细的概述。

12.4.1 热膨胀性能

任何胶凝材料用于高温领域时都必须考虑其热膨胀性能或收缩性能。加热过程的收缩或膨胀会引起应力变化,从而削弱或损坏基体的结构。对于涂层,尺寸变化导致涂层从基底开裂或剥落。AAM 的热膨胀/收缩率可以使用热膨胀计原位测量[65-67],还可以直接用"推杆"收缩测量仪测量或使用远程激光测量仪来测量尺寸变化。

由于地质聚合物的结构是无定形凝胶结构,所以其热膨胀是各向同性的,但是由于组成和温度的局部变化,可能发生不均匀膨胀,产生更大的热应力导致开裂或剥落。大多数地质聚合物材料常见的热膨胀特性列于表 12-1。该表由 Duxson 等人首先提出[68],Rickard 等人加以扩展[69],按温度范围划分了不同区域,在表 12-1 和图 12-2 中做了详细描述。根据样品组成和测试条件,对应于每个区域的精确温度范围有些变化。还应注意,不是所有的地质聚合物的热性能都将反映于表 12-1 和图 12-2 中,这还取决于微观结构和纳米结构的变化。

表 12-1 地质聚合物的热膨胀性能

区域	温度范围(℃)	描述	影响	因素
Ⅰ	0~150	受阻脱水	轻微膨胀	样品杨氏模量;加热速率
Ⅱ	100~300	自由水损失	显著收缩	含水量;加热速率
Ⅲ	250~600	脱羟基	轻微收缩	羟基丰度(化学结合水)
Ⅳ	550~900	黏性烧结致密化	显著收缩	残留水分;硅铝比
Ⅴ	高于致密化温度	凝胶膨胀和/或结晶;由于开裂而膨胀	适度膨胀	凝胶组成和微观结构;杂质含量和类型
Ⅵ	高于致密化温度	进一步致密化	大幅收缩	凝胶组成

像大多数固体材料一样,低钙 AAM 在受热时会膨胀(区域Ⅰ)。低钙 AAM 通常都含有孔隙中的吸附水和化学结合水。在加热时,水分会逐渐损失,导致基体总体收缩。在 100℃以上,脱水放慢,主要是固体凝胶的膨胀导致基体膨胀。随着温度升高,脱水速率也增加,基体的膨胀是胶凝材料的膨胀和含水孔的收缩(区域Ⅱ)之和。在大多数低钙 AAM 样品中,区域Ⅱ总体表现为收缩,其收缩率与样品的含水量成正比。

然而,在含水量很低的样品和含有外加剂(如蛭石)的样品中,在该区域中可以存在净膨胀[70]。其他反应,例如(凝胶和二次相)结晶、氧化(仅二次相)、烧结和熔融,可能影响高温下的热膨胀性能。二次相被定义为除了硅铝酸盐凝胶之外的相,通常是未反应的前驱体,例如偏高岭土,或者存在于粉煤灰中的石英、莫来石和氧化铁。

脱水收缩的程度取决于材料的含水量。例如,Rickard 等人[69]发现,预养护的含水量(质量分数)为 15.2%($w/c=0.2$)的碱硅酸盐激发粉煤灰样块,在 100~300℃之间有 2% 的脱水收缩(图 12-2)。脱水收缩的特性(例如起始温度和加热时间)取决于地质聚合物的

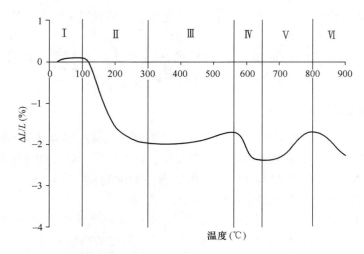

图 12-2　碱硅酸盐激发粉煤灰的热膨胀，显示了温度范围的划分
（表 12-1）。Si/Al = 2.3，w/c = 0.2[69]

结构和测量期间的加热速率。Duxson 等人[68]提出碱激发偏高岭土的脱水收缩与样品的杨氏模量成正比。具有较高杨氏模量的胶凝材料可以承受在脱水期间产生的更大的毛细管应力，因此初始收缩的起始温度提高，脱水速率由水从结构中扩散的速率控制。因此，孔结构对脱水速率也有很强的影响。Duxson 等[68]发现增加加热速率则增加了碱激发偏高岭土地质聚合物中脱水收缩的开始温度和持续时间。

脱羟基化发生在 250～400℃之间，伴随有小的质量损失。在区域 Ⅲ 中发生的热收缩（通常发生在 300～600℃之间）是由于随着羟基的释放导致的凝胶的物理收缩[68]。然而，该区域中的少量收缩可以被固相（凝胶和二次相）膨胀掩盖，如图 12-2 所示。在碱激发粉煤灰中通常这样，是由于二次相（例如石英、莫来石和赤铁矿）的浓度相对较高。

第二个主要的收缩发生在 550～900℃之间（区域Ⅳ），由于样品的致密化、凝胶烧结和黏性流动填充材料中的孔隙收缩。Rahier 等人[71]提出，该区域的收缩是凝胶的玻璃化转变温度（T_g）的指示。Duxson 等人[68]发现，随着 Si 与 Al 比的增加，致密化的开始温度降低。

超过致密区（区域 Ⅴ）在文献中的热膨胀性能众说不一。Rickard 等人[69]和 Rahier 等人[71]检测到的是热膨胀，Duxson 等人[68]和 Dombrowski 等人[72]检测到的是急剧的热收缩，Barbosa 和 MacKenzie[60]观察到的是样品尺寸稳定。关于该区域热膨胀性能出现的差异可能是由于组成的差异和二次相的存在。Provis 等人[65]发现，该区域的热膨胀性能与碱激发粉煤灰体系中的液固比成正比，特征参数（起始温度和总膨胀性能）随固体和液体前驱体的反应活性和配合比设计变化而变化[65,73]。膨胀是由于硅酸盐凝胶部分解聚，随着液固比的增加而增加。

研究还发现结晶是区域 Ⅴ 出现膨胀的原因。长石类相，如霞长石（K^+激发）、白榴石（K^+激发）和霞石（Na^+激发）在高温下从非晶态地质聚合物中结晶[60,74-76]。Bell 等人[58,59]证明了 Cs^+ 和 K^+ 激发的地质聚合物可以分别用作形成陶瓷相铯榴石和白榴石的前驱体。Barbosa 和 MacKenzie[60]观察到非化学计量的地质聚合物由于存在未结合的碱阳离子而会生成更多长石。

影响区域Ⅴ中热膨胀的其他因素是裂纹的产生和孔隙率的增加。Rickard等人[69]还发现，该区域的热膨胀性能取决于样品尺寸，较大的样品表现出较大的热膨胀性能。这可能是因为大尺寸样品中心和表面之间的较大温差引起的开裂。

热膨胀的最终特征区域是区域Ⅵ，该区域温度范围宽，通常样品会快速收缩，也是材料的破坏点。Subaer和Duxson都对区域Ⅵ做了研究，两人在此区域的收缩幅度相似，但Subaer[77]报道的急剧收缩导致材料碎裂，而Duxson等人[74]观察到收缩较慢。在该区域中收缩的原因是由于以下一种或多种原因：连续致密化（类似于区域Ⅳ）；在区域Ⅴ中形成的结晶相的破坏或进一步相变；样品孔结构的塌陷。

12.4.2 高温应用

碱激发材料的耐热性能使得它们被考虑应用到一些高温环境，特别是用于防火。已有大量报道，AAM相比OPC具有更好的防火性能，AAM置于火中碎裂现象明显减少，且能保持很好的机械强度[78]。耐火制品可以应用到隧道窑衬里、高层建筑、电梯门和海工结构件/涂层[79]。海工结构件可以在施工阶段喷涂（作为喷浆混凝土）或预制防火涂层，保护金属梁免受热变形的影响。纯防火混凝土结构件，可以免除由涂层和结构之间的不同热膨胀引起的分层现象，防火效果更好。用防火混凝土建造的隧道比使用传统材料建造的隧道更安全。在过去几十年，有不少隧道出现灾难性火灾，这有助于各种防火混凝土和非常规钢筋混凝土的研发。

地质聚合物复合材料由于其耐高温和低密度特性而被试用于飞机领域[80]。这项技术仍处于起步阶段，但它已经显示出推广的潜质。地质聚合物复合材料也被用作一级方程式赛车排气管的绝热材料[56]。特定的地质聚合物配比也适用于防火工程，其低成本、耐高温的特点明显优于其他可用材料[81-83]。低含水量和高纯的地质聚合物适用于工业耐火材料，最高可耐1200℃的高温。

对于防火工程应用，有两种不同类型的产品：一类用作结构部件（隧道、墙壁等），另一类是用作隔离结构钢梁或其他部件的涂层。第一种需要在高温下具有高的抗压强度，使结构不受损害，而第二种需要对基底具有高粘附性，且质量轻。在涂料应用中，重要的是耐磨性，而不是机械强度。

12.4.3 涂料

Goode对高层建筑中结构钢的防火保护做了全面深入的综述[84]。图12-2显示，对于典型的地质聚合物，加热将导致收缩。先前已经说明地质聚合物涂层暴露在高温下是坚固的，但是由于与基底不匹配的热膨胀性能（与钢相比，地质聚合物的热膨胀系数通常是正），将会出现分层现象。因此，需要对铝硅酸盐结构进行改性，使其具有类似于钢的热膨胀系数，复合材料在加热时具有更好的兼容性，而不分层。Temuujin等人[85]制备了系列不同Si/Al和水胶（w/b）比的偏高岭土基地质聚合物，研究发现Si/Al=2.5和w/b=0.74时，热膨胀系数是正值，可以与钢的膨胀系数匹配（图12-3）。此外，该化合物暴露于高温之后在钢质基材上的粘附性能更好了。将该地质聚合物煅烧至1000℃会出现铝硅酸钠结晶，冷却至室温后结晶相对涂层并没有有害影响。注意到在水中72h后，质量损失约8%，这表明存在残留的可溶性硅酸钠。然而，这些整体性能表明它们非常适合作为耐热涂层使用。

Temuujin 等人[29]也使用 F 级粉煤灰制备耐热的碱激发铝硅酸盐涂层。相比于偏高岭土，粉煤灰制备这种涂料的需水量较少，并且在 Si/Al=3.5 和 w/b=0.25 可制备性能优良的涂料（图 12-4）。

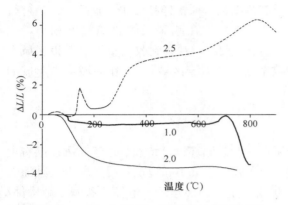

图 12-3　w/b=0.74 时，不同 Si/Al 比例下的偏高岭土基 AAM 样品的热膨胀特性
（数据来自文献 [85]）

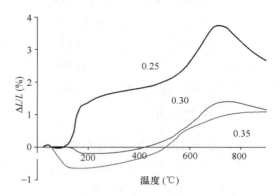

图 12-4　Si/Al=3.5 时，不同 w/b 比下粉煤灰基地质聚合物的热膨胀特性
（数据来自文献 [29]）

12.4.4　耐高温结构件

目前，耐高温碱激发材料的大部分研究集中在工程应用上，例如利用 AAM 优异的工程性能和耐高温性能制备结构件。Rickard 等人[86]使用澳大利亚三种不同化学组成的粉煤灰制备了一系列碱硅酸盐激发胶凝材料，将样品置于 1000℃后强度还显著增加（表 12-2）。并不是所有的地质聚合物在烧结后强度都会增加，这主要因为一些粉煤灰的氧化铁含量相对较高。在 1000℃下，样品的形态发生显著变化。未反应的粉煤灰颗粒和铝硅酸盐凝胶的烧结导致在所有样品中更均匀和微结构连接更好。但这并不能解释 Collie 粉煤灰样品中的强度损失的原因。通常情况下，烧结会在一定程度上提高基体的强度，但氧化铁的结晶会引起基体开裂，导致强度下降。

表 12-2　三种澳大利亚粉煤灰制成的碱硅酸盐激发材料的抗压强度

粉煤灰种类	氧化铁含量（wt%）	Si/Al	28d 抗压强度（MPa）	1000℃后的抗压强度（MPa）	室温强度百分比
Collie	13.2	2.0	128 (9)	24 (9)	19
		2.5	53 (10)	15 (4)	29
		3.0	29 (3)	—	—
Eraring	4.03	2.0	31 (2)	78 (11)	249
		2.5	33 (8)	132 (19)	396
		3.0	28 (5)	126 (20)	457
Tarong	0.64	2.0	26 (2)	13 (8)	49
		2.5	26 (4)	73 (17)	277
		3.0	25 (2)	99 (24)	396

注："—"的样品强度太低，无法进行测试[86]；括号中的值是最终数字的偏差值。

第 12 章 碱激发材料的其他潜在应用

除了氧化铁，粉煤灰其他一些特性也会明显影响地质聚合物的热性能。粉煤灰粒度、形态和结晶相的存在将极大地影响所得胶凝材料的特性。更细的粉煤灰粒度可以提高地质聚合物的抗压强度。球状粉煤灰颗粒的需水量较低（浆体工作性较好），地质聚合物在高温下收缩会降低。粉煤灰中游离石英颗粒会降低工作性，高温下可能会引起膨胀开裂。

Kong 等人[87-89]将偏高岭土基和粉煤灰基碱激发胶凝材料分别置于高温环境下，偏高岭土基样品的强度降低，而粉煤灰基样品的强度增加。压汞法测试偏高岭土基 AAM 主要存在大量介孔（2~50nm），而粉煤灰基 AAM 主要是微孔（<2nm）。孔尺寸和合理的孔分布使粉煤灰基胶凝材料保持更高的强度，这主要是在加热期间水分逸出而不损坏结构的能力有所不同。Bakharev[90]对碱硅酸盐激发和 NaOH 激发的粉煤灰地质聚合物进行了类似的实验，结果表明，高温烧结会引起孔隙率变化，直接影响煅烧样品的抗压强度变化。Bernal 等人[91]研究了碱硅酸盐激发的偏高岭土和偏高岭土/BFS 复合地质聚合物，发现加入 20% BFS 样品在 800℃ 下进行烧结后出现更高的强度，而 1000℃ 下强度并没有增长，单独使用偏高岭土的样品主要是凝胶致密化过程，强度也没有提高。

Provis 等[65,66]研究了硅酸钠激发粉煤灰体系的强度和热性能之间的关系。在 700~800℃，胶凝材料表现出最高的强度和较小的膨胀性，这主要与凝胶结构中高硅相的膨胀有关，高硅相不存在或过量的样品强度都会降低。XRD 测试显示存在沸石及类似相，如八面沸石、水方钠石和 Na-菱沸石。这些相在地质聚合物受热时都有自己的特性，如脱水温度、热膨胀性能和熔点等。

Kovalchuk 和 Krivenko[79]控制粉煤灰体系中的 Si 含量（$SiO_2/Al_2O_3=2\sim8$），来设计含有沸石的胶凝材料。当这些地质聚合物经高温煅烧后，产生霞石、方英石和钠长石等晶体。这些产物是很好的耐火材料。Dombrowski 等人[72]证明向碱激发胶凝材料中加入氢氧化钙可提高强度，减少收缩；在添加 8wt% $Ca(OH)_2$ 的样品中，在 800℃ 会形成霞石，在 1000℃ 会形成长石。这与 Bernal 等人[91]关于 800℃ 的偏高岭土/BFS 体系的产物类似。

大多数碱激发胶凝材料都是复合材料，因为它们包含石英、莫来石、氧化铁、玻璃态火山灰或矿渣，和/或其他前驱体颗粒以及碱激发凝胶。这种复杂的微结构使材料在恶劣环境条件（如高温）下能稳定存在。也有研究把蛭石掺入地质聚合物进行改性，制备的产物的耐热性能明显提高。Zuda 及其同事[30,92]将蛭石和绝缘陶瓷掺入矿渣中，碱激发混合物制备性能优良的轻质复合材料。当加热到 800℃ 时，它们的复合材料的强度降低到室温值的 35%，但此后强度增加，在 1200℃ 时，比室温时提高 30%。Lin 等人[93]还通过向 KOH 激发的偏高岭土中加入 $\alpha\text{-}Al_2O_3$ 制造了地质聚合物复合材料。$\alpha\text{-}Al_2O_3$ 填料掺入后，降低了收缩率，孔隙率相应提高，抗弯强度也提高了。随着 $\alpha\text{-}Al_2O_3$ 掺量增加，晶化起始温度也提高了。1400℃ 时，胶凝材料完全结晶成白榴石。Kong 和 Sanjayan[89]将粉煤灰地质聚合物加热到 800℃，基体的强度增加了 53%，而含骨料样品的强度降低了 65%。含骨料样品的强度降低归因于热膨胀不匹配：在 800℃ 下，含骨料地聚物总体积膨胀了约 2%，而地质聚合物本身收缩了 1.6%。如果这种材料用作防火工程的建筑材料，不同组分的热膨胀不匹配就会出现问题，这仍然需要做进一步深入研究。Lyon 等人[94]也将玻纤增韧地质聚合物放置于高热流（50kW/m²）中，高温处理后的强度是初始强度的 67%。

在 2002 年出版的地质聚合物书籍中，Davidovits[56]介绍了他过去 30 年的工程经验和商业化工作，其中有大量耐高温的复合地质聚合物的开发工作，包括碳纤维和其他多种体系。

这部分工作都可以通过专利文献查阅,其中大部分胶凝材料化合物都是以摩尔比来设计的。最近本书又出了新版,这里就不做详细叙述了[19]。事实上,商业化产品能够很好地推广都是大量专利经过几十年转化才实现的。

上述大量研究都有一些局限,那就是经高温处理后的样品都是在室温下进行强度测试和微观结构评估的。为了克服这个限制,Pan 和 Sanjayan[95]直接测量水玻璃激发粉煤灰试样在高温下的原位应力-应变行为。他们观察到,与初始室温强度相比,样品强度在52℃下增加了几乎两倍,但是超过520℃,玻璃化转变过程导致强度突然损失。Fernández-Jiménez 等人[96]还对碱激发粉煤灰试样进行了原位压缩和光谱强度分析,发现在800℃以上明显软化,尽管水玻璃激发的粉煤灰强度在400~600℃之间显示出增加,超过600℃之后则下降。这些研究表明,必须进行更多的动态和原位测量以确保强度-温度特性被正确地解释。

12.4.5 防火性能

上述耐高温性能的研究主要集中在 AAMs 的开发、设计和表征上。然而,为了将这些材料应用到建筑和公共设施中,需要进行标准的防火测试。在实际中,材料不仅要置于高温环境,还要以一定速率逐渐升温。火的温度随时间变化趋势要根据产生火情的地点和类型变化。

构成"标准"的火灾有很多种说法。室内火灾发生的典型顺序可以用房间的平均空气温度来表示。图12-5给出了这种火灾发生的三个阶段:

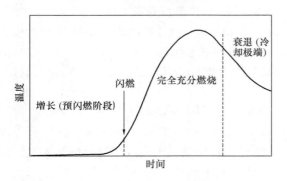

图 12-5 典型房间的时间-温度曲线,基于文献[97]

(1)产生火或前闪燃阶段,其中平均温度低,火焰位于其原点附近。

(2)完全着火或后闪燃阶段,室内的所有可燃物品都被卷入,火焰似乎充满整个空间。

(3)熄灭或冷却期。

建筑和结构件通常需要抵挡火灾。为此,有必要采用标准的火灾曲线,以便有一个共同的基准来比较建筑构件的不同选择。最常采用的火焰曲线在 ISO 834[98]中已给出,而 ASTM E119[99]火焰曲线也是常用的,与 ISO 曲线略有不同。ISO 834 曲线基于纤维的火灾,并且也被澳大利亚标准(AS 1530.4)、挪威(Nordtest NT Fire 046)标准和欧洲规范(EN 1991-1-2:2002)[100]采用。标准(纤维)火的时间-温度曲线如图12-6所示。标准火曲线旨在从闪燃阶段开始模拟图12-5的温度-时间曲线。火灾的前闪燃阶段通常被忽略,因为它对建筑构件的影响不大。对于许

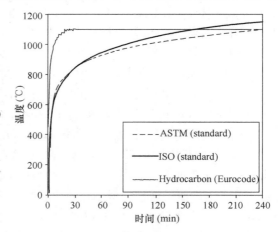

图 12-6 标准火灾的温度-时间关系图
(ASTM E119,ISO 834,欧洲规范
EN1991-1-2)

多材料，可以通过材料暴露的最高温度来预测在火灾中的性能。然而，脆性材料，如混凝土，会受到不对称加热产生的热梯度影响。影响热梯度的一个重要参数就是在初始阶段火焰的升温速率。当与纤维火焰建模的标准火焰相比时，烃类化合物的火灾初始阶段升温更快。烃类化合物发生火灾的可能性更大，例如道路和铁路隧道、海上和石化工业。遇到这类火灾应该从设计角度考虑建筑构件的防火能力。

欧洲规范（EN1991-1-2）为此提供了一条曲线，如图12-6所示。烃类火灾对混凝土材料特别有害，因为快速升温引起大斜率的温度梯度，混凝土孔中的蒸汽压累积可导致混凝土炸裂。但是，对于钢材就不同，其暴露的实际温度和持续时间是影响火灾响应的主要因素，但对脆性材料，升温速率是关键因素。

如果想真实地评估材料的防火性能，就需要按服役构件相同比例制备大尺寸结构件。但是进行这种测试的设备很少，其试件尺寸和测试的复杂性意味着这些测试会非常昂贵。Vilches等人使用一系列设备测试缩小比例的试件[101]，这要比大尺寸构件测试费用更低，更容易实现。

硅酸盐水泥混凝土，特别是高强混凝土，在火灾中极易剥落[62-64]。高强混凝土通常被定义为抗压强度在50MPa以上的混凝土，并且自20世纪80年代后期已经广泛用于建筑中。在火灾中，高强混凝土的剥落是从表面爆炸，片状脱落。高强混凝土剥落风险更高是因为它比正常强度混凝土渗透性更低，脆性更高[102]。当混凝土暴露于快速升温环境时，例如在烃类火焰中，剥落的风险进一步加剧。碱硅酸盐激发粉煤灰混凝土和高强OPC基混凝土的比较试验表明，AAM混凝土在高温下比OPC具有显著的优势[78,103]。结果表明[78]，低钙AAM胶凝材料的多孔性促进了在高温下蒸汽压的释放，与具有类似初始抗压强度的OPC混凝土相比，大大减少了剥落。

12.4.6 热性能小结

前面主要介绍偏高岭土和粉煤灰基AAM的耐热性能，碱激发矿渣的耐热性能相对少一些。下面的12.6节中会重点讨论纤维对纤维增强碱激发矿渣高温性能的影响。已发表的关于低钙AAM材料的文章主要集中在耐热性能的（主要）加热后强度分析上。高温条件下的热导率测量文献相对较少。光了解胶凝材料高温下的性能还不够，还需要知道受热期间热如何传导到钢筋或基底，所述材料的全部性能才能被确定。此外，需要在一定温度范围内对物相组成进行原位检测，来与膨胀测试结果对比分析，得到关联曲线，这方面其实已经做了不少有意义的工作[66,67]。但还需要评估在加热周期期间孔尺寸和熔点的变化。这将是具有挑战的工作。利用同步加速器X射线成像技术还是可以实现的。最后也是最关键的就是为了商业化应用，需要进行大规模的火灾测试，使用标准的防火等级来标定地质聚合物产品。没有标准的防火等级，很难使地质聚合物产品被消防领域接受。

12.5 废弃物的稳定化/固化

12.5.1 简介

一般来说，固体/稳定（S/S）是将废弃物与胶凝材料结合，形成单一固体，以减少有害成分释放到环境中。通过物理和/或化学手段将废物组分结合，并将危险废物转化为环境可接受的废弃物，再处置或在某些情况下的增值、使用[104-108]。水泥和相关胶凝材料固化污染物有两种机理：一是通过污染物和胶凝材料相之间相互反应的化学固化；二是污染物物理吸附到固化的胶凝材料的孔结构内表面上，和/或在低渗透率阶段内的物理封存，以降低暴露于潜在侵蚀性环境期间的可能性。上述固化机理与每种特定情况下普适性机制的内在联系根本上取决于胶凝材料和污染物（如有放射性存在的话）的化学性质以及整体产物的微观结构。因此，虽然没有哪种胶凝材料可以提供固化的最优方案，但是低渗透性材料通常固化效果更优。

目前固化应用中最常用的胶凝材料是硅酸盐水泥，为了降低成本，解决大体积构件的温升问题并且为了降低较大整体件内的温度升高[109]，在水泥中掺加 BFS 或粉煤灰。碱激发胶凝材料经过测试验证，在实际中用于固化各种固废，已有文献大量报道了使用碱激发胶凝材料处理有毒的[8,110-116]和放射性[8,111,113,117-119]固废，这里只做一个概述。考虑到一些独特的放射性固废的处理，这些将与其他一般有毒废物分开讨论。

通常，在碱激发胶凝材料中含阳离子固废比阴离子固化效果更好，含氧阴离子的过渡金属很难固化。在图 12-7 中给出了在 AAM 胶凝材料中通过固化处理元素的概述；被称为"结合"的元素是经处理以后完全无法迁移的元素，包括形成空间网络结构的材料；如 N、S、Cl 等元素也是固化的对象。固化其实是非常广义的概念，这里给出一些定义是为了下面更好地讨论。

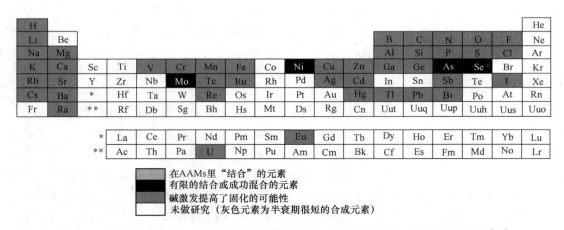

图 12-7　基于现有文献，在碱激发胶凝材料中结合（和/或通过固化处理）的元素[112]

12.5.2　危险固废的稳定化/固化

1. 碱激发 BFS 水泥

Pb、Cr、Cd 和 Zn 等元素掺入碱激发 BFS 浆料（单独掺入或作为工业固废的组分掺入）中的浸出率很低[120-122]。Cho 等人[121]使用 NaOH 和硅酸钠作为 BFS 的激发剂，两种激发剂固化以上元素的浸出率很低，而 Deja[120]使用 Na_2CO_3 和硅酸钠，发现 Na_2CO_3 能够形成重金属的碳酸盐沉淀，但也会导致硅酸盐激发样品可能出现微裂纹。另外，沸石掺入碱激发 BFS 中可以降低 Pb、Cr 的浸出率[123]。

铬是一种广泛传播的环境污染物，一些矿物加工和冶金固废中含量较高。六价铬的毒性很大，还能在环境中迁移。Cr（Ⅲ）毒性较小，流动性较小，$Cr(OH)_3$是相对难溶的，可与含Ca氢氧化物混合[124]。矿渣中的硫化物给孔溶液提供了很好的氧化还原环境，可将Cr（Ⅵ）转化为Cr（Ⅲ）减少迁移[120,122]。这种氧化还原过程也会把Fe（Ⅲ）还原为Fe（Ⅱ），使其容易移动[122]，但是Fe（Ⅱ）的毒性远低于Cr（Ⅵ），问题不大。Ahmed和Buenfeld[122]发现，镍和钼对Eh值相对不敏感，但由于碱激发BFS胶凝材料中较高的pH值，使迁移的可能性增加。被硫氧化形成可溶性化合物的元素（例如As、Sb和Sn）不适合在碱激发BFS体系中固化[122]。文献数据总结在图12-7，根据图示，以上研究结果是正确的。酸性固废需要先中和再碱激发处理，否则可能导致激发剂的碱性失效，这意味着需要消耗更多碱，使成本上升[125]。

低浓度（掺入到0.5%）Zn对NaOH激发BFS胶凝材料的强度发展影响较小，但如果掺入同量的Zn，在硅酸盐水泥水化过程中会形成锌酸钙，缓凝严重，对固化非常有害[126]。当Zn在NaOH激发BFS胶凝材料中的掺量达到2%时，会出现缓凝现象，最终强度也会明显降低[127]。Zn掺量进一步提高后就无法固化了。使用同种胶凝材料固化2%掺量的Hg，性能也明显下降；低掺量下对固化或强度影响不大，元素基本不会迁移，而达到2%时，就会出现缓凝，强度和固化效果下降[128]。

根据文献报道，一般来说，重金属对碱激发BFS固化和强度发展的影响远小于对硅酸盐水泥的影响[8,113]。这可能是由于在大多数碱激发胶凝材料体系（特别是水玻璃作激发剂的体系）中，SiO_2的利用率提高，这也是碱激发胶凝材料中凝胶化学的优势，而硅酸盐水泥在水化过程中，水泥颗粒延缓溶解，降低了SiO_2的利用率。Shi等人[129]使用量热法研究OPC和碱激发BFS胶凝材料固化炼钢炉（EAF）飞灰（含有有毒金属的混合物）的早期反应，相比AAM体系，OPC体系（都没有使用大掺量来加入过多重金属离子）的凝结速率、放热过程和最终强度都受到很大影响。掺加45%的EAF飞灰、BFS、石灰、水玻璃和少量硅灰的固化体系表现出很好的固化性能。然而，实验室养护样品和现场养护样品的性能存在明显差异，这是因为两种样品的养护条件不同导致[130,131]。

2. 碱激发偏高岭土或粉煤灰胶凝材料

已有大量研究是关于碱激发粉煤灰或偏高岭土基胶凝材料应用于固化领域的；这些研究绝大多数是基于实验室的，而且性能数据都是优化遴选出来的（这里存在明显的"出版偏爱"，作者不愿报告负面的测试数据，只是医学和科学领域都众所周知的）。然而，现场的研究结果显示这类材料具有很好的固化性能，现场研究与实验室研究数据结合明确了固化和浸出机理（这就不同于一组给定的胶凝材料体系的单一变量得出的性能数据），这对固化领域是非常有用的。碱激发胶凝材料也已被用作高盐尾矿堆场的封盖材料[132]，使用少量的活性铝硅酸盐激发剂掺入大量的低活性尾矿中制备成可控低强度材料，封盖于未处理的尾矿表面，可以有效减少扬尘，防止重金属浸出。

铅可能是使用AAM固化的有害元素中研究最多的，但固化机理尚不明确。Van Jaarsveld等人的早期研究[114,133,134]表明，铅可能在铝硅酸盐基体内被化学键合，但其具体的化学结合形式还不明确。铅改变了胶凝材料的孔结构，在浸出试验中，大量的Pb^{2+}被固化在基体中，不能向外扩散。Palacios和Palomo[135,136]认为铅在NaOH激发粉煤灰胶凝材料中被固化为不溶性Pb_3SiO_5，Zhang等人[137]的研究结果也对上述结果做了佐证，而XRD的图谱

只有单个 Pb_3SiO_5 的衍射峰。Zhang 等人[137]明确提出 Pb^{2+} 在硅酸钠激发粉煤灰中固化肯定是化学固化，而不是简单的物理封存。如果铅仅在物理结合，则添加微溶的 $PbCrO_4$ 会比添加可溶的 $Pb(NO_3)_2$ 固化效果更好，而在 H_2SO_4 或 Na_2CO_3 浸出实验的结果并非如此，表明肯定是发生了某种形式的化学反应。

文献中有几篇报道详述了使用低 Ca 碱激发胶凝材料固化 Cr（Ⅵ）的难度[135,137-139]。有文献详细讨论了在碱激发粉煤灰反应过程中 Cr 会迁移出来[140]，但使用硅酸盐激发比在高 pH 值下碱激发，粉煤灰颗粒的铬迁移量更少。像碱激发 BFS 体系的还原环境一样，将 0.5% 的 S^{2-} 以 Na_2S 的形式加到碱硅酸盐激发粉煤灰胶凝材料中，可以提高固化性能[141]。部分还原的铬可能在铝硅酸盐凝胶生长过程中形成 $Cr(OH)_3$ 或类似碱性盐的微胶囊。但是，如果 Cr 以微溶的铬酸盐如 $PbCrO_4$ 的形式加入，在固化过程中就不能被还原，残留的 Cr（Ⅵ）就可以浸出[141]。这就需要在固化处理之前对废弃物做前处理，但在与胶凝材料混合之前，试图做还原可溶性 Cr 的预处理，效果适得其反。

砷是大宗固废中需要特别固化的元素，因为它毒性高、可溶，且通常存在于矿物加工和其他多种工业固废中。因为固化效果一般，所以水泥固化处理砷只有一些个案[142]。通常砷都是作为多组分混合固废中的一种元素，研究其掺入碱激发铝硅酸盐胶凝材料中的固化效果[110,118,143,144]。大量研究显示，As 固化效果不错，只有几篇报道[144-146]显示碱激发方式固化 As 污染的粉煤灰效果一般。Fernández-Jiménez 等人[147,148]研究了砷以 $NaAsO_2$ 形式加入固化基体的情况，能够更多地分析特定成分的化学作用。有文献研究，通过透射电子显微镜观察到砷与富铁粉煤灰颗粒结合，尽管当把 Fe_2O_3 颗粒直接添加到混合物中时未观察到类似现象。这在一定程度上表明砷可能吸附到了氢氧化铁的表面[149]，这应该是砷在碱激发粉煤灰基胶凝材料中的固化机理。

关于地质聚合物中锌固化的文献较少。Minaříková 和 Škvára[150]研究发现，在碱硅酸盐激发粉煤灰基体中添加 ZnO 会使最终抗压强度降低约 50%，但强度发展的相对速率并没有明显变化。鉴于稳定/固化样品的强度要求通常不是很高，强度减少 50% 问题并不严重。在该研究中，固化性能一般良好，但不均匀性突出，而且少于 5% 石膏掺入体系后使浸出降低了 10 倍。Fernández Pereira 等人[151]将碱激发粉煤灰胶凝材料用于富锌电炉粉尘的固化，也显示出相对良好的性能（>90%）。

镉有剧毒，是致癌物，在许多采矿和冶金固废中出现，是常用的电池材料。在硅酸盐水泥中固化效果较好，Cd 部分代替 Ca^{2+} 形成混合的 Ca/Cd-水化硅酸盐凝胶[152,153]。但镉在低钙碱激发胶凝材料中的固化效果并不好[150,154]。特别考虑到 Cd^{2+} 浸出性能的环境规定非常严格，因此需要高性能的碱激发胶凝材料。碱激发粉煤灰固化镉的效果与浸出条件直接相关[137]。在硫酸中做浸出实验，固化效果很差（90d 浸出测试中存在超过 30% 迁移）；相同时间内，去离子水的固化率达到 99.95%，而在饱和 Na_2CO_3 溶液中会完全浸出。这些结果归因于在地质聚合物体系中镉以独立的氢氧化物或类碱盐存在，加入少量钙，Cd^{2+} 与高钙凝胶无法形成固化体。在高 pH 值下，$Cd(OH)_2$ 的溶解度是低的，中性条件下，溶解度更低，这也就是固化优良的原因。然而，当暴露于强酸和/或浓酸时，Cd^{2+} 的溶解度明显提高。因此，在高钙胶凝材料体系（硅酸盐水泥或碱激发 BFS）中固化富镉固废效果更优。

已经使用碱激发粉煤灰或偏高岭土基胶凝材料成功固化的其他固废组分包括

第12章 碱激发材料的其他潜在应用

铜[114,115,133,134,155]、钴[110]和汞[156]。在添加或不添加铝硅酸盐原料的情况下，垃圾焚烧飞灰（含有一些 Ca、Si 和 Al，可以参与基体反应）与碱硅酸盐的复合都可以固化有害元素。高氯化物和硫酸盐存在会使固化难度加大，这就需要预先洗涤[157-160]。

12.5.3 放射性固废的稳定化/固化

放射性固废的固化是碱激发技术在研究和开发方面另一个重要领域。大多数使用水泥和类似胶凝材料的工作集中在低/中放射性固废（L/ILW）固化，陶瓷或玻璃用于高放射性固废固化。L/ILW 的主要放射性组分是 ^{137}Cs 和 ^{90}Sr，大多数实验室研究通过使用非放射性模拟物（Cs 和 Sr 的稳定同位素）来避免对实验室工作人员的危害。这些研究能够复制放射性同位素的性质，但无法研究诸如放射性氢气产生或辐射诱导胶凝损伤的问题，在研究环境下是很难实现的。碱激发固废在某些情况下通过硬化后的热处理脱水以减少辐射[161,162]，在热处理过程中，有时可能会出现大尺寸试块开裂的问题，但也不是所有样块都会出现这类问题。在一些情况下，混合核废料会含有活性金属（特别是用于包裹核废料的 Al），会被高 pH 值下的 AAM（或 OPC）胶凝材料腐蚀，导致氢气产生，可能会使固化基体开裂；在这种情况下需要特殊的低碱胶凝材料[163]。在设计和建造地下核废料储存库时，通常使用膨胀黏土作外层防护，目的是防止地下水进入保护层，但黏土也可能会与高浓度区域的碱反应，因此在这种条件下优选低碱胶凝材料体系[164]。

在大多数 L/ILW 放射性固废中，铯通常被认为是最难以固化的放射性核元素，因为它本身结合弱，且很容易从许多常用的固化体系中逃逸[165]。然而，它较容易跟铝硅酸盐和其他类沸石物相结合[166]。当与硅酸盐水泥复合时，锶不太成问题，因为它通过置换到 Ca^{2+} 位点相对容易地结合到 C-S-H 相中[167]，但是鉴于两种同位素 ^{137}Cs 和 ^{90}Sr 通常作为铀裂变产物（铀裂变产生比锶更多的铯），并且要经过数十年至数世纪的时间，在辐射发射的裂变产物中的主要贡献者，它们才是主要关注的组分。

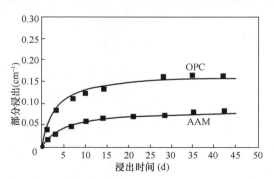

图 12-8 在 25℃时，硬化硅酸盐水泥和碱激发 BFS 水泥浆中 Cs^+ 的浸出（数据来自文献[171]）

一些实验室研究已经证实，从碱激发 BFS 基体浸出到去离子水中的铯比从硅酸盐水泥中的浸出量低得多[113,168-170]。图 12-8 显示了 Cs^+ 从硅酸盐水泥和含有 0.5%$CsNO_3$ 的碱激发 BFS 基体浸出量对比，在 25℃、湿养护 28d，OPC 比 AAM 浸出率高很多。这是因为 AAM 胶凝材料中 C-S-H 中具有不同的孔结构和较低的 C/S 比[171]。如在 12.5.2.1 节所述的 Pb 和 Cr 固化情况中，由于 C-(A)-S-H 和沸石前驱体具有优良的吸附性质，可以把沸石或偏高岭土部分替代 BFS 以减少 Cs^+ 和 Sr^{2+} 从硬化基体中浸出[170,172]。在俄罗斯还开发了氧化镁-矿渣碱激发胶凝材料用于处理放射性铯、钴和钌的危废，在宏观物理性能和浸出性能方面都表现优良[173]。

由于 Cs^+ 与类沸石（或原沸石）凝胶结构结合较好，碱激发偏高岭土和粉煤灰胶凝材料已经被广泛研究用于危废固化[117,161,162,174-178]。Chervonnyi 和 Chervonnaya[179]研究了来自受损切尔诺贝利反应堆周围区域木灰中的 ^{137}Cs 和 ^{90}Sr，发现硅酸盐水泥固化，木灰与热活化膨润土基地质聚合物固化后,浸出性能提高了 20 倍。在由俄罗斯 Kurchatov 研究所开发的名为 EKOR 的胶凝材料体系(名为"硅基地质聚合物",Eurotech 商业推广的)作为一个密封降尘剂,用作石棺来保护受损的反应堆堆芯结构的一部分[180,181]。ALLDECO 公司开发的碱激发黏土基胶凝材料也已被开发并应用于原捷克斯洛伐克共和国的放射性固废的固化中[182]。同样,德国(处理含铀和镭的物流)矿山运营商与法国 Cordi-Géopolymère 合作开展了类似的工作[118,183]。

在低 Ca 碱激发胶凝材料中的 Cs^+ 和 Sr^{2+} 的固化机理有所不同,这是由两种离子的不同电荷状态决定的。如上所述,Sr^{2+} 可以在 C-S-H 凝胶的形成中代替钙,当这些凝胶不存在时(即存在很少或没有钙的情况下),锶似乎完全存在于基体的孔溶液中,如果材料暴露于大气中则非常容易沉淀为碳酸盐[162,184]。$SrCO_3$ 是非常难溶的,因此这确实提供了相当有效的固化手段,如图 12-9 所示。铯不会形成这样的沉淀;相反,作为碱金属阳离子,它或多或少直接取代碱金属铝硅酸盐凝胶中的钠或钾,并且在胶凝材料结构中形成与这些其他碱相同的结构形式[59,177,185-187],这在第 4 章已经讨论。然而,当以总浓度标定时,Cs 的浸出远比含有 Cs^+ 和 Sr^{2+} 两种阳离子的胶凝材料中 Na 的浸出慢得多(图 12-9),这主要是因为 Cs^+ 作为一个极化强、半径最大的碱金属离子,与硅酸盐反应性更强[188],大的阳离子很难穿过受限的孔道。

通过水热法(高达 200℃),用大掺量粉煤灰/OPC 复合胶凝材料固化高碱低放射性固废溶液,会生成沸石,固化效果很好,其中危废溶液的碱性起到了激发作用[189,190]。术语"水化陶瓷"通常是指黏土在水热条件下碱激发生成一系列水化材料,可用于处理含钠固废,产物主要是沸石(废液中富含含氧阴离子时会生成钙霞石和方钠石)和碱铝硅酸盐凝胶[191-194]。原料中还可以掺入蛭石来增强 ^{137}Cs 的固化,也可以加入少量 Na_2S 作为氧化还原剂沉淀重金属。美国能源部用水化陶瓷固废来处理爱达荷州、华盛顿州和佐治亚州的高碱和富钠固废,也用于硼硅酸盐玻璃废弃物的玻璃化。美国能源部的能源实验室还在持续其研发(使用水热法和低温养护产品)[195]。

在放射性或核燃料循环相关的危废处理领域中,碱激发的其他应用有离子交换树脂的固化[125],^{152}Eu、^{60}Co 和 ^{59}Fe[196]、^{99}Tc[197,198]和 ^{129}I[198]的放射性同位素的固化。这确实是碱激

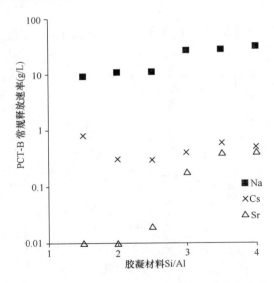

图 12-9 固化了 Na、Cs 和 Sr 的碱硅酸盐激发偏高岭土胶凝材料在去离子水中的标准浸出速率(PCT-B 试验),随 Si/Al 变化的曲线;胶凝材料中作为的 1% Cs 以 CsOH 加入,1% Sr 以 $Sr(OH)_2$ 的加入(来自文献 [161])

发材料一个很好的应用领域,扩展了其用途。钙硅酸盐体系、碱铝硅酸盐体系,或两种凝胶体系的复合,其性能都优于纯硅酸盐体系,特别是为了防止辐射分解氢气而给体系除水(通过中温热处理)时,上述碱激发体系更具优势。高碱的碱激发胶凝体系可能在有些情况下存在问题,这就需要仔细调控孔隙率和渗透性以确保长期固化性能,但可以证明,碱激发体系

是未来放射性固废处理的重要方式,也是未来几十年更清洁的核废料循环处理的一个选择方式。

12.6 纤维增韧

纤维增韧用于 AAM 中,与 OPC 体系同出一辙:通过提高拉伸强度和断裂韧性来提高材料性能,以便提高最终产品的延展性和耐久性。更好的延展性减少了开裂和裂化的延展,也有利于材料的体积稳定性。特别是材料处于不同载荷和高温时,减少徐变和收缩的有害影响。

纤维增韧方面使用了很多材料,如碳化硅[199]、聚丙烯[200-202]、玻璃[203]、碳[204-209]、玄武岩[210-212]、PVA[213]、硅灰石[214]和钢[215-217]等。在一些情况下,经纤维增韧的试样抗折强度和弹性模量都有提高,自干燥收缩减少,室温下尺寸稳定性更好。

在碱硅酸盐激发 BFS 砂浆中加入聚丙烯、玻璃或碳纤维对干燥收缩(在 50%的相对湿度下)的影响如图 12-10 所示。研究发现,聚丙烯和耐碱(AR)玻璃纤维可以控制收缩,控制量减少高达 35%。这证明只要这些纤维在高 pH 值下保持稳定,很多短纤维可以用于提高 AAM 基体的尺寸稳定性。Sakulich[218]认为,这类复合材料可以应用到大规模的基础设施项目上,因为它们没有各向异性现象,可以用标准混料机制备。

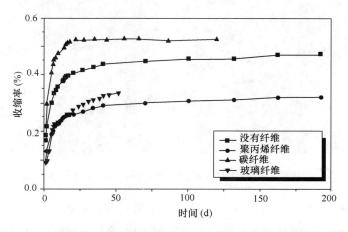

图 12-10　纤维增韧碱激发 BFS 试块的干燥收缩与掺入纤维类型的
关系(数据由 F. Puertas 提供)

Puertas 等人[200]使用聚丙烯纤维(体积在 0 和 1%之间)增韧三种碱激发复合试块,研究了聚丙烯纤维对试块的机械性能和耐久性的影响。这三种试块分别是水玻璃(4gNa$_2$O/100g BFS)激发,室温养护的 BFS 试块;用 8M NaOH 激发,85℃养护 24h 的 F 类粉煤灰试块;用 8M NaOH 激发,室温养护的,50%粉煤灰和 50%BFS 的试块。排除碱激发基体本身特性,这些复合材料强度发展与纤维掺加与否和掺量直接相关。可以看出(表 12-3)纤维增韧复合材料的抗折强度和弹性模量均较低,这是因为与空白碱激发样块相比,纤维的机械强度较低,且掺入纤维降低了浆体的工作性。Zhang 等人[201]用聚丙烯纤维增韧的硅酸盐激发粉煤灰/偏高岭土混合物也获得类似结果。

表 12-3　具有和不具有聚丙烯纤维（1%，v/v）的砂浆的机械参数[200]

胶凝材料	纤维（%）	弯曲强度（MPa）	弹性模量（MPa）	偏移（mm）
BFS	0	7.36	4860	0.1277
	1	5.91	3896	0.1361
粉煤灰	0	5.79	4441	0.1071
	1	4.79	3660	0.1084
粉煤灰/BFS	0	4.80	4906	0.0852
	1	4.66	3810	0.1068
OPC	0	7.76	5679	0.1136
	1	7.61	6137	0.1051

另一方面，在碱激发 BFS 试块中掺入玻璃纤维（以胶凝材料质量分数为 0.22% 掺入）提高了抗折强度，而没有改变抗压强度[203]。虽然玻璃纤维增韧试块在高温下性能较优，但韧性和耐冲击性均未改善。这是因为在高温下熔融的无水渣和玻璃纤维填充了孔道，导致强度恢复到初始机械强度的 50% 以上。另外注意到，由于 pH 值较高，基体中的玻璃纤维表面有劣化现象（图 12-11）。

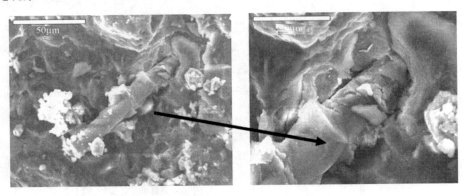

图 12-11　玻璃纤维增韧的 BFS 试块的扫描电镜图，其中显示了劣化的玻璃纤维[203]

用碳纤维[204]和纳米管[219]增韧的 AAM 也有研究，并没有改善试验样品的机械强度。但是，含 1% 碳纤维改善了样块的耐腐蚀性。还观察到碳纳米管而不是石墨提高了地质聚合物的导电性。Bernal 等人[205]确定、使用预处理（通过去除聚合物涂层）碳纤维，提高掺量，可以改善 AAS 砂浆的抗折强度、弹性模量和断裂韧性。这些结果之间的差异主要归因于各自研究使用纤维的长度和用量不同。

Thaumaturgo 及其同事[210,214]研究了用玄武岩纤维和硅灰石微纤维增韧 AAM 的断裂韧性。研究表明，劈裂拉伸强度和弯曲强度随着增韧材料掺量的增加而增加。这说明两种纤维与基体的结合都很好。

在碱激发 BFS 混凝土中加入高性能钢纤维，机械强度明显提升[215,216]。在混凝土中掺入大量钢纤维，弯曲强度和韧性都有提高（图 12-12）。分析结果显示，AAM 混凝土和钢纤维界面相互作用，在单轴拉伸载荷下表现出临界的滑动硬化响应。其控制机理是高摩擦剪切位移的典型过峰阻力之后产生了高韧性。这与 Penteado Dias 和 Thaumaturgo[210]以及 Silva 和

Thaumaturgo[214]报道的用玄武岩和硅灰石纤维增韧的结果一致。钢纤维增韧混凝土还表现出吸水性、吸附性和水渗透性的降低,这有助于增强混凝土的耐久性。

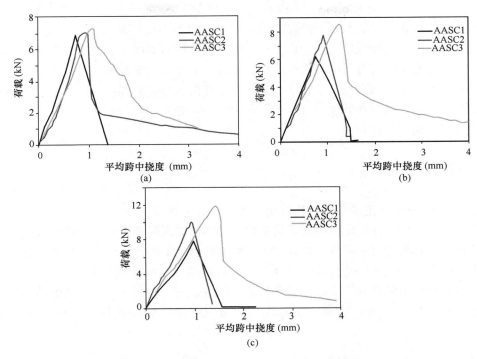

图 12-12　碱硅酸盐激发 BFS 混凝土固化后的弯曲载荷-挠度曲线,钢纤维掺量分别为
0(AAS1)、400kg/m³(AAS2)和 1200kg/m³(AAS3)
(a) 7d,(b) 14d 和 (c) 28d(来自文献 [215])

还有人使用连续纤维增韧 AAM 基体。Lyon 等人[94]开发了一种铝硅酸钾和硅酸钾激发偏高岭土与碳纤维复合的材料,He 等人[207,208]使用碳纤维板,Zhao 等人[217]使用不锈钢网。这些研究突出了纤维/基体界面作为控制复合材料强度和耐久性因素的重要性,偶尔也有性能降低现象。用连续纤维增韧的 AAM 体系的弯曲强度明显提高,这也与制备方法(通常是在真空下热处理)、纤维的类型和掺量有关。

Sakulich[218]做了纤维增韧 AAM 复合材料的"绿色"认知度的比较分析,综述了材料延展性改善的各种方法,图 12-13 是由文献分析简化绘制的示意图。2D 纤维增韧复合材料表现出各向异性的机械性能,需要对加工技术进行改进,而具有随机取向和不连续纤维的 AAM 复合材料可与高性能纤维增韧水泥复合材料相提并论。

可以确定,当用于高温领域时,纤维对

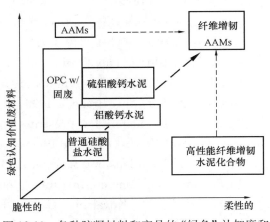

图 12-13　各种胶凝材料和产品的"绿色"认知度和延展性之间的关系(来自文献 [218])

AAM 的性能有显著贡献。基于 AAM 的复合材料已经被用于不同领域，如耐火[94,220,221]，玻璃工业[222]的耐火材料、绝热材料[18,81,223]和其他需要在高温时具有良好性能的领域[7,79]。根据 Papakonstantinou 等人[224]，几个因素使 AAM 材料成为纤维增韧体系中的理想基材：(1) AAM 在高于 1000℃的温度下是稳定的，并且掺入适当比例的纤维增加了体积稳定性，可降低热收缩，保持机械强度；(2) 由于 AAM 是接近环境温度下制备的，多种纤维都可掺入复合材料。这些优势与基材的耐火性能结合使这类复合材料更具实用价值。

纤维或网布增韧的碱激发砂浆和混凝土具有很好的机械性能和耐久性，适合于很多应用领域（建筑、交通、航空和其他）。而且还需要进一步加强 AAM 的研究，使其更好地应用于高性能纤维增韧复合材料中。

12.7 小结

除了在第 11 章中讨论的市政工程应用之外，还有一些碱激发材料可以在市政和工程材料领域做商业化推广。任何特定的胶凝材料配方都不可能同时具有所有的优良性能，但是肯定可以通过调整碱激发材料配方，使其用于轻质材料、地下固井水泥、耐高温领域、危险或放射性固废的固化，也可以结合纤维增韧，提高抗折强度、抗拉强度和/或收缩性能。尽管在这些领域材料的用量远低于市政基础设施建筑材料的用量，但一些高附加值的利用方式还是可以进行商业化推广的，特别是政府或商业组织在固废处理或环境修复方面存在较大商机。

参考文献

[1] BASF SE. PCI technical data sheet 254，PCI Geofug. 2011. http：//www.pci-augsburg.eu/en/ products/product-information/g.html.

[2] F orsgren J.，Pedersen C.，Strømme M.，Engqvist H.．Synthetic geopolymers for controlled delivery of oxycodone：adjustable and nanostructured porosity enables tunable and sustained drug release. PLoS，ONE，2011，6（3）：e17759.

[3] J ämstorp E.，Forsgren J.，Bredenberg S.，Engqvist H.，Strømme M.．Mechanically strong geopolymers offer new possibilities in treatment of chronic pain. J. Control. Release，2010，146(3)：370-377.

[4] Liefke E.．Industrial applications of foamed geopolymers In：Davidovits J.，Davidovits R.，James C.（eds.）Proceedings of Géopolymère '99-Second International Conference，Saint-Quentin，France，1999，1：189-199.

[5] Talling B.，Brandstetr J.．Present state and future of alkali-activated slag concretes. In：Malhotra V. M.（ed.）3rd International Conference on Fly Ash，Silica Fume，Slag and Natural Pozzolans in Concrete，ACI SP114，Trondheim，Norway. 1989，2：1519-1546. American Concrete Institute. Detroit，MI.

[6] Krivenko P. V.，Kovalchuk G. Y.．Heat-resistant fly ash based geocements. In：Lukey G. C.（ed.）Geopolymers 2002. Turn Potential into Profit，Melbourne，Australia. CD-ROM Proceedings. Siloxo Pty. Ltd，2002.

[7] Krivenko P. V., Kovalchuk G. Y.. Directed synthesis of alkaline aluminosilicate minerals in a geocement matrix. J. Mater. Sci, 2007, 42(9): 2944-2952.

[8] Shi C., Krivenko P. V., Roy D. M.. Alkali-Activated Cements and Concretes. Taylor & Francis, Abingdon, 2006.

[9] Grutzeck M. W., Kwan S., DiCola M.. Zeolite formation in alkali-activated cementitious systems. Cem. Concr. Res, 2004, 34(6): 949-955.

[10] Brooks R., Bahadory M., Tovia F., Rostami H.. Properties of alkali-activated fly ash: high performance to lightweight. Int. J. Sustain. Eng, 2010, 3(3): 211-218.

[11] Arellano Aguilar R., Burciaga Díaz O., Escalante García J. I.. Lightweight concretes of activated metakaolin-fly ash binders, with blast furnace slag aggregates. Constr. Build. Mater, 2010, 24(7): 1166-1175.

[12] Helferich R. L.. Lightweight hydrogel-bound aggregate shapes and process for producing same. Compiler U. S, 4963515[P]. 1990.

[13] Bell J. L., Kriven W. M.. Preparation of ceramic foams from metakaolin-based geopolymer gels. Ceram. Eng. Sci. Proc, 2008, 29(10): 97-111.

[14] Prud'homme E., Michaud P., Joussein E., Clacens J. M., Rossignol S.. Role of alkaline cations and water content on geomaterial foams: monitoring during formation. J. Non-Cryst. Solids, 2011, 357(4): 1270-1278.

[15] Prud'homme E., Michaud P., Joussein E., Peyratout C., Smith A., Arrii-Clacens S., Clacens J. M., Rossignol S.. Silica fume as porogent agent in geo-materials at low temperature. J. Eur. Ceram. Soc, 2010, 30(7): 1641-1648.

[16] Prud'homme E., Michaud P., Joussein E., Peyratout C., Smith A., Rossignol S.. In situ inorganic foams prepared from various clays at low temperature. Appl. Clay Sci, 2011, 51(1-2): 15-22.

[17] Henon J., Alzina A., Absi J., Smith D. S., Rossignol S.. Porosity control of cold consolidated geomaterial foam: temperature effect. Ceram. Int, 2012, 38(1): 77-84.

[18] Vaou V., Panias D.. Thermal insulating foamy geopolymers from perlite. Miner. Eng, 2010, 23(14): 1146-1151.

[19] Davidovits J.. Geopolymer Chemistry and Applications. Institut Géopolymère, Saint-Quentin, 2008.

[20] Bean D. L., Malone P. G.. Alkali-activated glassy silicate foamed concrete. Compiler: U. S, 5605570[P]. 1995.

[21] Birch G. D.. Cellular cementitious Composition. Compiler: U. S, 8167994 B2[P]. 2012.

[22] Zhao Y., Ye J., Lu X., Liu M., Lin Y., Gong W., Ning G.. Preparation of sintered foam materials by alkali-activated coal fly ash. J. Hazard. Mater. 2010, 174(1-3): 108-112.

[23] Laney B. E., Williams F. T., Rutherford R. L., Bailey D. T.. Advanced geopolymer

composites. Compiler U. S, 5244726[P]. 1993.

[24] Chislitskaya H.. Acoustic properties of slag alkaline foamed concretes. In: Krivenko P. V. (ed.) Proceedings of the First International Conference on Alkaline Cements and Concretes, Kiev, Ukraine. 1994, 2: 971-979. VIPOL Stock Company.

[25] Buchwald A., Oesterheld R., Hilbig H.. Incorporation of aluminate into silicate gels and its effect on the foamability and water resistance. J. Am. Ceram. Soc, 2010, 93 (10): 3370-3376.

[26] Fletcher R. A., MacKenzie K. J. D., Nicholson C. L., Shimada S.. The composition range of aluminosilicate geopolymers. J. Eur. Ceram. Soc, 2005, 25(9): 1471-1477.

[27] Pushkareva E., Guziy S., Sukhanevich M., Borisova A.. Influence of inorganic modifiers on structure, properties and durability of bloating geocement compositions. In: Ertl Z. (ed.) Proceedings of the International Conference on Alkali Activated Materials - Research, Production and Utilization, Prague, Czech Republic, 2007: 581-592. Česká rozvojová agentura.

[28] Sukhanevich M. V., Guzii S. G.. The effect of technological factors on properties of alkali aluminosilicate systems used for preparation of fireproof coatings. Refract. Ind. Ceram, 2004, 45(3): 217-219.

[29] Temuujin J., Minjigmaa A., Rickard W., Lee M., Williams I., Van Riessen A.. Fly ash based geopolymer thin coatings on metal substrates and its thermal evaluation. J. Hazard. Mater, 2010, 180 (1-3): 748-752.

[30] Zuda L., Drchalová J., Rovnaník P., Bayer P., Keršner Z., Černý R.. Alkali-activated a luminosilicate composite with heat-resistant lightweight aggregates exposed to high temperatures: mechanical and water transport properties. Cem. Concr. Compos, 2010, 32(2): 157-163.

[31] Yang K. H., Song J. K., Lee J. S.. Properties of alkali-activated mortar and concrete using lightweight aggregates. Mater. Struct, 2010, 43(3): 403-416.

[32] Yang K.-H., Mun J.-H., Sim J.-I., Song J.-K.. Effect of water content on the properties of lightweight alkali-activated slag concrete. J. Mater. Civil Eng, 2011, 23(6): 886-894.

[33] Tulaganov A. A.. Structure formation and properties of the high-strength alkaline lightweight concretes. In: Krivenko P. V. (ed.) Proceedings of the Second International Conference on Alkaline Cements and Concretes, Kiev, Ukraine, 1999: 168-184. ORANTA.

[34] Simonovic M.. Flexural properties and characterization of geopolymer based sandwich composite structures at room and elevated temperature. Ph. D. thesis, Rutgers, The State University of New Jersey, 2007.

[35] Wu H.-C., Sun P.. New building materials from fly ash-based lightweight inorganic

polymer. Constr. Build. Mater, 2007, 21: 211-217.

[36] Mallicoat S. W., Sarin P., Kriven W. M.. Novel, alkali-bonded, ceramic filtration membranes. Ceram. Eng. Sci. Proc, 2005, 26(8): 37-44.

[37] Sakulich A. R., Bentz D. P.. Mitigation of autogenous shrinkage in alkali activated slag mortars by internal curing. Mater. Struct, 2013, 46: 1355-1367.

[38] Soares P., Pinto A. T., Ferreira V. M., Labrincha J. A.. Geopolymerization of lightweight aggregate waste. Mater. Constr, 2008, 59(291): 23-34.

[39] Vance E. R., Perera D. S., Imperia P., Cassidy D. J., Davis J., Gourley J. T.. Perlite waste as a precursor for geopolymer formation. J. Aust. Ceram. Soc, 2009, 45(1): 44-49.

[40] Jo B.-W., Park S.-K., Park J.-B.. Properties of concrete made with alkali-activated fly ash lightweight aggregate (AFLA). Cem. Concr. Compos, 2007, 29(2): 128-135.

[41] Małolepszy J., Deja J., Brylicki W.. Industrial application of slag alkaline concretes. In: Krivenko P. V. (ed.) Proceedings of the First International Conference on Alkaline Cements and Concretes, Kiev, Ukraine. 1994, 2: 989-1001. VIPOL Stock Company.

[42] Brylicki W., Małolepszy J., Stryczek S.. Industrial scale application of the alkali activated slag cementitious materials in the injection sealing works. In: Goumans J. J. J. M., Van der Sloot H. A., Aalbers T. G. (eds.) Environmental Aspects of Construction with Waste Materials, Maastricht, Netherlands, 1994: 841-849. Elsevier. Amsterdam, Netherlands.

[43] Deja J., Brylicki W., Małolepszy J.. Anti-filtration screens based on alkali-activated slag binders. In: Ertl Z. (ed.) Proceedings of the International Conference on Alkali Activated Materials - Research, Production and Utilization, Prague, Czech Republic, 2007: 163-184. Česká rozvojová agentura.

[44] Nasvi M. C. M., Ranjith P. G., Sanjayan J.. The permeability of geopolymer at down-hole stress conditions: application for carbon dioxide sequestration wells. Appl. Energy, 2013, 102: 1391-1398.

[45] Javanmardi K., Flodberg K. D., Nahm J. J.. Mud to cement technology proven in offshore drilling project. Oil Gas J, 1993, 91(7): 49-57.

[46] Nahm J. J., Javanmardi K., Cowan K. M., Hale A. H.. Slag mix mud conversion cementing technology: reduction of mud disposal volumes and management of rig-site drilling wastes. J. Petroleum Sci. Eng, 1994, 11(1): 3-12.

[47] Ruiz-Santaquiteria C., Fernández-Jiménez A., Palomo A.. Rheological properties of alkali activated cement for oil well linings. In: Shi C., Yu Z., Khayat K. H., Yan P. (eds.) 2nd International Symposium on Design, Performance and Use of Self Consolidating Concrete, Beijing, China, 2009: 878-891. RILEM. Bagneux, France.

[48] Wu D., Peiyan A., Huang B.. Slag/mud mixtures improve cementing operations in China. Oil Gas J, 1996, 94(52): 95-100.

[49] Silva M. G. P., Miranda C. R., D'Almeida A. R., Campos G., Bezerra M. T. A.. Slag cementing versus conventional cementing: comparative bond results. In: 5th Latin American and Caribbean Petroleum Engineering Conference and Exhibition, Rio de Janeiro, Brazil. Paper SPE39005, 1997.

[50] Nahm J. J., Romero R. N., Hale A. A., Keedy C. R., Wyant R. E., Briggs B. R., Smith T. R., Lombardi M. A.. Universal fluids improve cementing. World Oil, 1994, 215 (11)67-72.

[51] Benge O. G., Webster W. W.. Blast furnace slag slurries may have limits for oil field use. Oil Gas J. 1994, 92(29): 41-49.

[52] Barlet-Gouedard V., Porcherie O., Pershikova E.. Pumpable geopolymer formulation for oilfield application. Compiler: World, WO/2009/103480[P]. 2009.

[53] Barlet-Gouedard V., Zusatz-Ayache C. M., Porcherle O.. Geopolymer composition and application for carbon dioxide storage. Compiler U. S, 7846250[P]. 2010.

[54] Sugama T., Brothers L. E.. Sodium-silicate-activated slag for acid-resistant geothermal well cements. Adv. Cem. Res, 2004, 16(2): 77-87.

[55] Sugama T., Brothers L. E., Van de Putte T. R.. Acid-resistant cements for geothermal wells: sodium silicate activated slag/fly ash blends. Adv. Cem. Res, 2005, 17(2): 65-75.

[56] Davidovits J.. 30 years of successes and failures in geopolymer applications. Market trends and potential breakthroughs. In: Lukey G. C. (ed.) Geopolymers 2002. Turn Potential into Profit., Melbourne, Australia. CD-ROM Proceedings. Siloxo Pty. Ltd. 2002.

[57] Xie N., Bell J. L., Kriven W. M.. Fabrication of structural leucite glass-ceramics from potassium- based geopolymer precursors. J. Am. Ceram. Soc, 2010, 93(9): 2644-2649.

[58] Bell J. L., Driemeyer P. E., Kriven W. M.. Formation of ceramics from metakaolin-based geopolymers. Part II: K-based geopolymer. J. Am. Ceram. Soc, 2009, 92(3): 607-615.

[59] Bell J. L., Driemeyer P. E., Kriven W. M.. Formation of ceramics from metakaolin-based geopolymers: Part I - Cs-based geopolymer. J. Am. Ceram. Soc, 2009, 92(1): 1-8.

[60] Barbosa V. F. F., MacKenzie K. J. D.. Thermal behaviour of inorganic geopolymers and composites derived from sodium polysialate. Mater. Res. Bull, 2003, 38(2): 319-331.

[61] Talling B.. Geopolymers give fire safety to cruise ships. In: Lukey G. C., (ed.) Geopolymers 2002. Turn Potential into Profit., Melbourne, Australia. CD-ROM Proceedings. Siloxo Pty. Ltd. 2002.

[62] Mendes A., Sanjayan J., Collins F.. Phase transformations and mechanical strength of OPC/ slag pastes submitted to high temperatures. Mater. Struct, 2008, 41(2): 345-350.

[63] Mendes A., Sanjayan J., Collins F.. Long-term progressive deterioration following fire exposure of OPC versus slag blended cement pastes. Mater. Struct, 2009, 42(1): 95-101.

[64] Hertz K. D.. Concrete strength for fire safety design. Mag. Concr. Res, 2005, 57(8): 445-453.

[65] Provis J. L., Yong C. Z., Duxson P., Van Deventer J. S. J.. Correlating mechanical and thermal properties of sodium silicate-fly ash geopolymers. Colloids Surf. A, 2009, 336(1-3): 57-63.

[66] Provis J. L., Harrex R. M., Bernal S. A., Duxson P., Van Deventer J. S. J.. Dilatometry of geopolymers as a means of selecting desirable fly ash sources. J. Non-Cryst. Solids, 2012, 358(16): 1930-1937.

[67] Rickard W. D. A., Temuujin J., Van Riessen A.. Thermal analysis of geopolymer pastes synthesised from five fly ashes of variable composition. J. Non-Cryst. Solids, 2012, 358(15): 1830-1839.

[68] Duxson P., Lukey G. C., Van Deventer J. S. J.. Physical evolution of Na-geopolymer derived from metakaolin up to 1000 C. J. Mater. Sci, 2007, 42(9): 3044-3054.

[69] Rickard W. D. A., Van Riessen A., Walls P.. Thermal character of geopolymers synthesized from class F fly ash containing high concentrations of iron and α-quartz. Int. J. Appl. Ceram. Technol, 2010, 7(1): 81-88.

[70] Temuujin J., Rickard W., Lee M., Vvan Riessen A.. Preparation and thermal properties of fire resistant metakaolin-based geopolymer-type coatings. J. Non-Cryst. Solids, 2011, 357(5): 1399-1404.

[71] Rahier H., Wastiels J., Biesemans M., Willem R., Van Assche G., Van Mele B.. Reaction mechanism, kinetics and high temperature transformations of geopolymers. J. Mater. Sci, 2007, 42(9): 2982-2996.

[72] Dombrowski K., Buchwald A., Weil M.. The influence of calcium content on the structure and thermal performance of fly ash based geopolymers. J. Mater. Sci, 2007, 42(9): 3033-3043.

[73] Provis J. L., Duxson P., Harrex R. M., Yong C. Z., Van Deventer J. S. J.. Valorisation of fly ashes by geopolymerisation. Global NEST J, 2009, 11(2): 147-154.

[74] Duxson P., Lukey G. C., Van Deventer J. S. J.. The thermal evolution of metakaolin geopolymers: part 2 - phase stability and structural development. J. Non-Cryst. Solids, 2007, 353(22-23): 2186-2200.

[75] White C. E., Provis J. L., Proffen T., Van Deventer J. S. J.. The effects of temperature on the local structure of metakaolin-based geopolymer binder: a neutron pair distribution function investigation. J. Am. Ceram. Soc, 2010, 93(10): 3486-3492.

[76] Duxson P., Lukey G. C., Van Deventer J. S. J.. Evolution of gel structure during thermal processing of Na-geopolymer gels. Langmuir, 2006, 22(21): 8750-8757.

[77] Subaer. Influence of aggregate on the microstructure of geopolymer. Ph. D. thesis, Curtin University of Technology, 2005.

[78] Zhao R., Sanjayan J. G.. Geopolymer and Portland cement concretes in simulated fire. Mag. Concr. Res, 2011, 63(3): 163-173.

[79] Kovalchuk G., Krivenko P. V.. Producing fire- and heat-resistant geopolymers. In: Provis J. L., Van Deventer J. S. J. (eds.) Geopolymers: Structures, Processing, Properties and Industrial Applications, 2009: 229-268. Woodhead, Cambridge.

[80] Giancaspro J., Balaguru P. N., Lyon R. E.. Use of inorganic polymer to improve the fire response of balsa sandwich structures. J. Mater. Civil Eng, 2006, 18(3): 390-397.

[81] Comrie D. C., Kriven W. M.. Composite cold ceramic geopolymer in a refractory application. Ceram. Trans, 2003, 153: 211-225.

[82] Fullston D., Sagoe-Crentsil K.. Small footprint aluminosilicate matrix - refractory hybrid materials. J. Aust. Ceram. Soc, 2009, 45(2): 69-74.

[83] Medri V., Fabbri S., Ruffini A., Dedecek J., Vaccari A.. SiC-based refractory paints prepared with alkali aluminosilicate binders. J. Eur. Ceram. Soc, 2011, 31(12): 2155-2165.

[84] Goode M. G.. NIST GCR 04-872, Fire protection of structural steel in high-rise buildings, National Institute of Standards and Technology, 2004.

[85] Temuujin J., Minjigmaa A., Rickard W., Lee M., Williams I., Van Riessen A.. Preparation of metakaolin based geopolymer coatings on metal substrates as thermal barriers. Appl. Clay. Sci, 2009, 46(3): 265-270.

[86] Rickard W. D. A., Williams R., Temuujin J., Van Riessen A.. Assessing the suitability of three Australian fly ashes as an aluminosilicate source for geopolymers in high temperature applications. Mater. Sci. Eng. A, 2011, 528: 3390-3397.

[87] Kong D. L. Y., Sanjayan J. G.. Damage behaviour of geopolymer composites exposed to elevated temperatures. Cem. Concr. Compos, 2008. 30(10): 986-991.

[88] Kong D. L. Y., Sanjayan J. G., Sagoe-Crentsil K.. Comparative performance of geopolymers made with metakaolin and fly ash after exposure to elevated temperatures. Cem. Concr. Res, 2007, 37: 1583-1589.

[89] Kong D. L. Y., Sanjayan J. G., Sagoe-Crentsil K.. Factors affecting the performance of metakaolin geopolymers exposed to elevated temperatures. J. Mater. Sci, 2008, 43: 824-831.

[90] Bakharev T.. Thermal behaviour of geopolymers prepared using class F fly ash and elevated temperature curing. Cem. Concr. Res, 2006, 36: 1134-1147.

[91] Bernal S. A., Rodríguez E. D., Mejía de Gutierrez R., Gordillo M., Provis J. L.. Mechanical and thermal characterisation of geopolymers based on silicate-activated

metakaolin/slag blends. J. Mater. Sci, 2011, 46 (16): 5477-5486.

[92] Zuda L., Černý R.. Measurement of linear thermal expansion coefficient of alkali-activated aluminosilicate composites up to 1000 C. Cem. Concr. Compos, 2009, 31 (4): 263-267.

[93] Lin T. S., Jia D. C., He P. G., Wang M. R.. Thermo-mechanical and microstructural characterization of geopolymers with $\alpha\text{-}Al_2O_3$ particle filler. Int. J. Thermophys, 2009, 30(5): 1568-1577.

[94] Lyon R. E., Balaguru P. N., Foden A., Sorathia U., Davidovits J., Davidovics M.. Fire- resistant aluminosilicate composites. Fire Mater, 1997, 21(2): 67-73.

[95] Pan Z., Sanjayan J. G.. Stress-strain behaviour and abrupt loss of stiffness of geopolymer at elevated temperatures. Cem. Concr. Compos, 2010, 32 (9): 657-664.

[96] Fernández-Jiménez A., Pastor J. Y., Martín A., Palomo A.. High-temperature resistance in alkali-activated cement. J. Am. Ceram. Soc, 2010, 93(10): 3411-3417.

[97] Institution of Engineers: Fire engineering for building structures and safety, working party on fire engineering, The National Committee on Structural Engineering, Institution of Engineers, 1989.

[98] ISO. Fire resistance tests - elements of building construction - Part 1: general requirements, 1999.

[99] ASTM International. Standard test methods for fire tests of building construction materials (ASTM E119-12a). West Conshohocken, 2005.

[100] European Committee for Standardization: Eurocode 1, Actions on structures - Part 1-2: General actions - actions on structures exposed to fire (EN 1991-1-2). Brussels, Belgium, 2002.

[101] Vilches L. F., Fernández-Pereira C., Olivares del Valle J., Vale J.. Recycling potential of coal fly ash and titanium waste as new fireproof products. Chem. Eng. J, 2003, 95: 155-161.

[102] Hertz K. D.. Limits of spalling of fire-exposed concrete. Fire Saf. J, 2003, 38(2): 103-116(2003)

[103] Van Riessen A., Rickard W., Sanjayan J.. Thermal properties of geopolymers. In: Provis J. L., Van Deventer J. S. J. (eds.) Geopolymers: Structures, Processing, Properties and Industrial Applications, 2009: 317-344. Woodhead, Cambridge, 2009.

[104] Conner J. R., Hoeffner S. L.. A critical review of stabilization/solidification technology. Crit. Rev. Environ. Sci. Technol, 1998, 28(4): 397-462.

[105] Glasser F. P.. Fundamental aspects of cement solidification and stabilisation. J. Hazard. Mater, 1997, 52(2-3): 151-170.

[106] Malviya R., Chaudhary R.. Factors affecting hazardous waste solidification/stabili-

[107] Glasser F. P.. Progress in the immobilization of radioactive wastes in cement. Cem. Concr. Res, 1992, 22(2-3): 201-216.

[108] Glasser F. P.. Mineralogical aspects of cement in radioactive waste disposal. Miner. Mag, 2001, 65(5): 621-633.

[109] Milestone N. B.. Reactions in cement encapsulated nuclear wastes: need for toolbox of different cement types. Adv. Appl. Ceram, 2006, 105(1): 13-20.

[110] Comrie D. C., Paterson J. H., Ritcey D. J.. Geopolymer technologies in toxic waste management. In: Davidovits J., Orlinski J. (eds.) Proceedings of Geopolymer '88 - First European Conference on Soft Mineralurgy, Compeigne, France. 1988, 1: 107-123. Universite de Technologie de Compeigne.

[111] Davidovits J., Comrie D. C.. Long term durability of hazardous toxic and nuclear waste disposals. In: Davidovits J., Orlinski J. (eds.) Proceedings of Geopolymer '88 - First European Conference on Soft Mineralurgy, Compeigne, France. 1988, 1: 125-134. Universite de Technologie de Compeigne.

[112] Provis J. L.. Immobilization of toxic waste in geopolymers. In: Provis J. L., Van Deventer J. S. J. (eds.) Geopolymers: Structure, Processing, Properties and Industrial Applications, 2009: 423-442. Woodhead, Cambridge.

[113] Shi C., Fernández-Jiménez A.. Stabilization/solidification of hazardous and radioactive wastes with alkali-activated cements. J. Hazard. Mater, 2006, B137(3): 1656-1663.

[114] Van Jaarsveld J. G. S., Van Deventer J. S. J., Lorenzen L.. The potential use of geopolymeric materials to immobilise toxic metals. 1. Theory and applications. Miner. Eng, 1997, 10(7): 659-669.

[115] Van Jaarsveld J. G. S., Van Deventer J. S. J., Schwartzman A.. The potential use of geopolymeric materials to immobilise toxic metals: Part II. Material and leaching characteristics. Miner. Eng, 1999, 12(1): 75-91.

[116] Van Deventer J. S. J., Provis J. L., Feng D., Duxson P.. The role of mineral processing in the development of cement with low carbon emissions. In: XXV International Mineral Processing Congress (IMPC), Brisbane, Australia, 2010: 2771-2781. AusIMM. Melbourne, Australia.

[117] Vance E. R., Perera D. S.. Geopolymers for nuclear waste immobilisation. In: Provis J. L., Van Deventer J. S. J. (eds.) Geopolymers: Structure, Processing, Properties and Industrial Applications, 2009: 403-422. Woodhead, Cambridge.

[118] Hermann E., Kunze C., Gatzweiler R., Kießig G., Davidovits J.. Solidification of various radioactive residues by Géopolymère® with special emphasis on long-term stability. In: Davidovits J., Davidovits R., James C. (eds.) Proceedings of Géopolymère '99 - Second International Conference, Saint-Quentin, France. 1991, 1: 211-228.

[119] Krivenko P. V., Skurchinskaya J. V., Lavrinenko L. V., Starkov O. V., Konovalov E. E.. Physico-chemical bases of radioactive wastes-Immobilisation in a mineral-like solidified stone. In: Krivenko P. V. (ed.) Proceedings of the First International Conference on Alkaline Cements and Concretes, Kiev, Ukraine. 1994, 1: 1095-1106. VIPOL Stock Company.

[120] Deja J.. Immobilization of Cr^{6+}, Cd^{2+}, Zn^{2+} and Pb^{2+} in alkali-activated slag binders. Cem. Concr. Res, 2002, 32(12): 1971-1979.

[121] Cho J. W., Ioku K., Goto S.. Effect of Pb-II and Cr-VI ions on the hydration of slag alkaline cement and the immobilization of these heavy metal ions. Adv. Cem. Res, 1999, 11(3): 111-118.

[122] Ahmed Y. H., Buenfeld N. R.. An investigation of ground granulated blastfurnace slag as a toxic waste solidification/stabilization reagent. Environ. Eng. Sci, 1997, 14(2): 113-132.

[123] Zhang D., Hou H., He X., Liu W.. Research on stabilization of heavy metals by alkali activated slag cement. Urban Environ. Urban Ecol, 2006, 19(4): 44-46.

[124] Omotoso O. E., Ivey D. G., Mikula R.. Quantitative X-ray diffraction analysis of chromium(III) doped tricalcium silicate pastes. Cem. Concr. Res, 1996, 26(9): 1369-1379.

[125] Ipatti A.. Solidification of ion-exchange resins with alkali-activated blast-furnace slag. Cem. Concr. Res, 1992. 22(2-3): 281-286.

[126] Fernández Olmo I., Chacon E., Irabien A.. Influence of lead, zinc, iron (III) and chromium (III) oxides on the setting time and strength development of Portland cement. Cem. Concr. Res, 2001, 31(8): 1213-1219.

[127] Qian G., Sun D. D., Tay J. H.. Characterization of mercury- and zinc-doped alkali-activated slag matrix: Part II. Zinc. Cem. Concr. Res, 2003, 33(8): 1257-1262.

[128] Qian G., Sun D. D., Tay J. H.. Characterization of mercury- and zinc-doped alkali-activated slag matrix: Part I. Mercury. Cem. Concr. Res, 2003, 33(8): 1251-1256.

[129] Shi C., Stegemann J., Caldwell R.. An examination of interference in waste solidification through measurement of heat signature. Waste Manag, 1998, 17(4): 249-255.

[130] Caldwell R. J., Stegemann J. A., Shi C.. Effect of curing on field-solidified waste properties. Part 1: physical properties. Waste Manag. Res, 1999, 17(1): 37-43.

[131] Caldwell R. J., Stegemann J. A., Shi C.. Effect of curing on field-solidified waste properties. Part 2: chemical properties. Waste Manag. Res, 1999, 17(1): 44-49.

[132] Van Jaarsveld J. G. S., Lukey G. C., Van Deventer J. S. J., Graham A.. The stabilisation of mine tailings by reactive geopolymerisation. In: MINPREX 2000 - International Congress on Mineral Processing and Extractive Metallurgy, Melbourne, Australia, 2000: 363-371. Australasian Institute of Mining and Metallurgy.

[133] Van Jaarsveld J. G. S., Van Deventer J. S. J., Lorenzen L.. Factors affecting the immobilization of metals in geopolymerized fly ash. Metall. Mater. Trans, 1998, B. 29(1): 283-291.

[134] Van Jaarsveld J. G. S., Van Deventer J. S. J.. The effect of metal contaminants on the formation and properties of waste-based geopolymers. Cem. Concr. Res, 1999, 29(8): 1189-1200.

[135] Palomo A., Palacios M.. Alkali-activated cementitious materials: alternative matrices for the immobilisation of hazardous wastes - part II. Stabilisation of chromium and lead. Cem. Concr. Res, 2003, 33(2): 289-295.

[136] Palacios M., Palomo A.. Alkali-activated fly ash matrices for lead immobilisation: a comparison of different leaching tests. Adv. Cem. Res, 2004, 16(4): 137-144.

[137] Zhang J., Provis J. L., Feng D., Van Deventer J. S. J.. Geopolymers for immobilization of Cr^{6+}, Cd^{2+}, and Pb^{2+}. J. Hazard. Mater, 2008, 157(2-3): 587-598.

[138] Bankowski P., Zou L., Hodges R.. Reduction of metal leaching in brown coal fly ash using geopolymers. J. Hazard. Mater, 2004, B114(1-3): 59-67.

[139] Luna Galiano Y., Salihoglu G., Fernández Pereira C., Vale Parapar J.. Study on the immobilization of Cr(VI) and Cr(III) in geopolymers based on coal combustion fly ash. In: 2011 World of Coal Ash Conference, Denver, CO. CD-ROM proceedings. ACAA/CAER. Lexington, KY, 2011.

[140] Provis J. L., Rose V., Bernal S. A., Van Deventer J. S. J.. High resolution nanoprobe X-ray fluorescence characterization of heterogeneous calcium and heavy metal distributions in alkali activated fly ash. Langmuir, 2009, 25(19): 11897-11904.

[141] Zhang J., Provis J. L., Feng D., Van Deventer J. S. J.. The role of sulfide in the immobilization of Cr(VI) in fly ash geopolymers. Cem. Concr. Res, 2008, 38(5): 681-688.

[142] Henke K. R.. Waste treatment and remediation technologies for arsenic. In: Henke K. R. (ed.) Arsenic: Environmental Chemistry, Health Threats and Waste Treatment, 2009: 351-430. Wiley, Chichester, 2009.

[143] Bankowski P., Zou L., Hodges R.. Using inorganic polymer to reduce leach rates of metals from brown coal fly ash. Miner. Eng, 2004, 17(2): 159-166.

[144] Álvarez-Ayuso E., Querol X., Plana F., Alastuey A., Moreno, N., Izquierdo M., Font O., Moreno T., Diez S., Vázquez E., Barra, M.. Environmental, physical and structural characterisation of geopolymer matrixes synthesised from coal (co-)combustion fly ashes. J. Hazard. Mater, 2008, 154(1-3): 175-183.

[145] Škvára F., Kopecký L., Šmilauer V., Bittnar, Z.. Material and structural characterization of alkali activated low-calcium brown coal fly ash. J. Hazard. Mater, 2009, 168(2-3): 711-720.

[146] Sansui O., Tempest B., Ogunro V., Gergely J., Daniels J.. Effect of hydroxy ion on immobilization of oxyanions forming trace elements from fly ash-based geopoly-

mer concrete. In: World of Coal Ash 2009, Lexington, KY. CD-ROM proceedings, 2009.

[147] Fernández-Jiménez A. M., Lachowski E. E., Palomo A., Macphee D. E.. Microstructural characterisation of alkali-activated PFA matrices for waste immobilisation. Cem. Concr. Compos, 2004, 26(8): 1001-1006.

[148] Fernández-Jiménez A., Palomo A., Macphee D. E., Lachowski E. E.. Fixing arsenic in alkali-activated cementitious matrices. J. Am. Ceram. Soc, 2005, 88(5): 1122-1126.

[149] Sherman D. M., Randall S. R.. Surface complexation of arsenic(Ⅴ) to iron(Ⅲ) (hydr)oxides: structural mechanism from abinition molecular geometries and EXAFS spectroscopy. Geochim. Cosmochim. Acta, 2003, 67(22): 4223-4230.

[150] Minaříková M., Škvára F.. Fixation of heavy metals in geopolymeric materials based on brown coal fly ash. Ceram. Silik, 2006, 50(4): 200-207.

[151] Fernández Pereira C., Luna Y., Querol X., Antenucci D., Vale J.. Waste stabilization/ solidification of an electric arc furnace dust using fly ash-based geopolymers. Fuel, 2009, 88(7): 1185-1193.

[152] Díez J. M., Madrid J., Macías A.. Characterization of cement-stabilized Cd wastes. Cem. Concr. Res, 1997, 27(3): 337-343.

[153] Pomiès M.-P., Lequeux N., Boch P.. Speciation of cadmium in cement: part I. Cd^{2+} uptake by C-S-H. Cem. Concr. Res, 2001, 31(4): 563-569.

[154] Xu J. Z., Zhou Y. L., Chang Q., Qu H. Q.. Study on the factors of affecting the immobilization of heavy metals in fly ash-based geopolymers. Mater. Lett, 2006, 60(6): 820-822.

[155] Phair J. W., Van Deventer J. S. J., Smith J. D.. Effect of Al source and alkali activation on Pb and Cu immobilization in fly-ash based "geopolymers". Appl. Geochem, 2004, 19(3): 423-434.

[156] Donatello S., Fernández-Jiménez A., Palomo A.. Alkaline activation as a procedure for the transformation of fly ash into new materials. Part Ⅱ - An assessment of mercury immobilisation. In: 2011 World of Coal Ash Conference, Denver, CO. CD-ROM proceedings. ACAA/ CAER, 2011.

[157] Zheng L., Wang W., Shi Y.. The effects of alkaline dosage and Si/Al ratio on the immobilization of heavy metals in municipal solid waste incineration fly ash-based geopolymer. Chemosphere, 2010, 79(6): 665-671.

[158] Zheng L., Wang C., Wang W., Shi Y., Gao X.. Immobilization of MSWI fly ash through geopolymerization: effects of water-wash. Waste Manag, 2011, 31(2): 311-317.

[159] Lancellotti I., Kamseu E., Michelazzi M., Barbieri L., Corradi A., Leonelli C.. Chemical stability of geopolymers containing municipal solid waste incinerator fly ash. Waste Manag, 2010, 30(4): 673-679.

[160] Luna Galiano Y., Fernández Pereira C., Vale J.. Stabilization/solidification of a municipal solid waste incineration residue using fly ash-based geopolymers. J. Hazard. Mater, 2011, 185(1): 373-381.

[161] Aly Z., Vance E. R., Perera D. S., Hanna J. V., Griffith C. S., Davis J., Durce D.. Aqueous leachability of metakaolin-based geopolymers with molar ratios of Si/Al=1.5~4. J. Nucl. Mater, 2008, 378(2): 172-179.

[162] Blackford M. G., Hanna J. V., Pike K. J., Vance E. R., Perera D. S.. Transmission electron microscopy and nuclear magnetic resonance studies of geopolymers for radioactive waste immobilization. J. Am. Ceram. Soc, 2007, 90(4): 1193-1199.

[163] Bai Y., Collier N. C., Milestone N. B., Yang C. H.. The potential for using slags activated with near neutral salts as immobilisation matrices for nuclear wastes containing reactive metals. J. Nucl. Mater, 2011, 413(3): 183-192.

[164] Cau Dit Comes C., Courtois S., Nectoux D., Leclercq S., Bourbon X.. Formulating a low- alkalinity, high-resistance and low-heat concrete for radioactive waste repositories. Cem. Concr. Res, 2006, 36: 2152-2163.

[165] El-Kamash A. M., El-Dakroury A. M., Aly H. F.. Leaching kinetics of 137C s and 60 Co radionuclides fixed in cement and cement-based materials. Cem. Concr. Res, 2002, 32(11): 1797-1803.

[166] Hoyle S. L., Grutzeck M. W.. Incorporation of cesium by hydrating calcium aluminosilicates. J. Am. Ceram. Soc, 1989, 72(10): 1938-1947.

[167] Tits J., Wieland E., Müller C. J., Landesman C., Bradbury M. H.. Strontium binding by calcium silicate hydrates. J. Colloid Interf. Sci, 2006, 300(1): 78-87.

[168] Wu X., Yen S., Shen X., Tang M., Yang L.. Alkali-activated slag cement based radioactive waste forms. Cem. Concr. Res, 1991, 21(1): 16-20.

[169] Shi C., Day R. L.. Alkali-slag cements for the solidification of radioactive wastes. In: Gillam T. M., Wiles C. C. (eds.) Stabilization and solidification of hazardous, radioactive, and mixed wastes, 1996: 163-173. ASTM STP 1240, West Conshohocken.

[170] Shen X., Yan S., Wu X., Tang M., Yang L.. Immobilization of stimulated high level wastes into AASC waste form. Cem. Concr. Res, 1994, 24(1): 133-138.

[171] Shi C., Shen X., Wu X., Tang M.. Immobilization of radioactive wastes with Portland and alkali-slag cement pastes. Il Cemento, 1994, 91(2): 97-108.

[172] Qian G., Sun D. D., Tay J. H.. New aluminium-rich alkali slag matrix with clay minerals for immobilizing simulated radioactive Sr and Cs waste. J. Nucl. Mater, 2001, 299(3): 199-204.

[173] Zosin A. P., Priimak T. I., Avsaragov K. B.. Geopolymer materials based on magnesia-iron slags for normalization and storage of radioactive wastes. At. Energy, 1998, 85(1): 510-514.

[174] Fernández-Jiménez A., Macphee D. E., Lachowski E. E., Palomo A.. Immobiliza-

tion of cesium in alkaline activated fly ash matrix. J. Nucl. Mater, 2005, 346(2-3): 185-193.

[175] Perera D. S., Vance E. R., Aly Z., Davis J., Nicholson C. L.. Immobilization of Cs and Sr in geopolymers with Si/Al~2. Ceram. Trans, 2006, 176: 91-96.

[176] Chen S., Wu M. Q., Zhang S. R.. Mineral phases and properties of alkali-activated metakaolin- slag hydroceramics for a disposal of simulated highly-alkaline wastes. J. Nucl. Mater, 2010, 402(2-3): 173-178.

[177] Berger S., Frizon F., Joussot-Dubien C.. Formulation of caesium based and caesium containing geopolymers. Adv. Appl. Ceram, 2009, 108(7): 412-417.

[178] Khalil M. Y., Merz E.. Immobilization of intermediate-level wastes in geopolymers. J. Nucl. Mater, 1994, 211(2): 141-148.

[179] Chervonnyi A. D., Chervonnaya N. A.. Geopolymeric agent for immobilization of radioactive ashes after biomass burning. Radiochemistry, 2003, 45(2): 182-188.

[180] Strzlecki D.. Geopolymer succeeds at Chernobyl field test. Pollut. Eng, 2001, 33(10): 36.

[181] Childress P.. The use of EKOR TM to stabilize fuel-containing material at Chernobyl. In: WM'01 Conference, Tucson, AZ. CD-ROM proceedings, 2001.

[182] Majersky D.. Removal and solidification of the high contaminated sludges into the aluminosilicate matrix SIAL during decommissioning activities. In: CEG Workshop on Methods and Techniques for Radioactive Waste Management Applicable for Remediation of Isolated Nuclear Sites, Petten. IAEA, 2004.

[183] Kunze C., Hermann E., Griebel I., Kießig G., Dullies F., Schreiter M.. Entwicklung und praxiseinsatz eines hocheffizienten selektiven sorbens für radium. Wasser-Abwasser, 2002, 143(7-8): 572-577.

[184] Provis J. L., White C. E., Gehman J. D., Vlachos D. G.. Modeling silica nanoparticle dissolution in TPAOH-TEOS-H_2O solutions. J. Phys. Chem, 2008, C112(38): 14769-14775.

[185] Berger S., Frizon F., Fournel, V., Cau-dit-Comes C.. Immobilization of cesium in geopolymeric matrix: a formulation study. In: Beaudoin J. J. (ed.) 12th International Congress on the Chemistry of Cement, Montreal, Canada. CD-ROM. National Research Council of Canada, Ottawa, 2007.

[186] He P., Jia D., Wang M., Zhou Y.. Effect of cesium substitution on the thermal evolution and ceramics formation of potassium-based geopolymer. Ceram. Int, 2010, 36(8): 2395-2400.

[187] Bell J. L., Sarin P., Provis J. L., Haggerty R. P., Driemeyer P. E., Chupas P. J., Van Deventer J. S. J., Kriven W. M.. Atomic structure of a cesium aluminosilicate geopolymer: a pair distribution function study. Chem. Mater, 2008, 20(14): 4768-4776.

[188] McCormick A. V., Bell A. T., Radke C. J.. Evidence from alkali-metal NMR spec-

troscopy for ion pairing in alkaline silicate solutions. J. Phys. Chem, 1989, 93(5): 1733-1737.

[189] Brough A. R., Katz A., Sun G. K., Struble L. J., Kirkpatrick R. J., Young J. F.. Adiabatically cured, alkali-activated cement-based wasteforms containing high levels of fly ash: formation of zeolites and Al-substituted C-S-H. Cem. Concr. Res, 2001, 31(10): 1437-1447.

[190] Brough A. R., Katz A., Bakharev T., Sun G.-K., Kirkpatrick R. J., Struble L. J., Young J. F.. Microstructural aspects of zeolite formation in alkali activated cements containing high levels of fly ash. In: Diamond S., Mindess S., Glasser F. P., Roberts L. W., Skalny J. P., Wakely L. D. (eds.) Microstructure of Cement-Based Systems/Bonding and Interfaces in Cementitious Materials. Materials Research Society Symposium Proceedings 370, 1995: 197-208. Materials Research Society, Pittsburgh, PA.

[191] Siemer D. D.. Hydroceramics, a "new" cementitious waste form material for US defense- type reprocessing waste. Mater. Res. Innov, 2002, 6(3): 96-104.

[192] Bao Y., Kwan S., Siemer D. D., Grutzeck M. W.. Binders for radioactive waste forms made from pretreated calcined sodium bearing waste. J. Mater. Sci, 2003, 39(2): 481-488.

[193] Bao Y., Grutzeck M. W., Jantzen C. M.. Preparation and properties of hydroceramic waste forms made with simulated Hanford low-activity waste. J. Am. Ceram. Soc, 2005, 88(12): 3287- 3302.

[194] Olanrewaju J.. Hydrothermal transformation and dissolution of hydroceramic waste forms for the INEEL calcined high-level nuclear waste. Ph. D. thesis, Pennsylvania State University, 2002.

[195] Cozzi A. D., Bannochie C. J., Burket P. R., Crawford C. L., Jantzen C. M.. Immobilization of radioactive waste in fly ash based geopolymers. In: 2011 World of Coal Ash Conference, Denver, CO. CD-ROM proceedings. ACAA/CAER, 2011.

[196] Hanzlíček T., Steinerova M., Straka P.. Radioactive metal isotopes stabilized in a geopolymer matrix: determination of a leaching extract by a radiotracer method. J. Am. Ceram. Soc, 2006, 89(11): 3541-3543.

[197] Pierce E. M., Cantrell K. J., Westsik J. H., Parker K. E., Um W., Valenta M. M., Serne R. J.. Report PNNL-19505: Secondary waste form screening test results - cast stone and alkali alumino-silicate geopolymer, 2010.

[198] Gong W., Lutze W., Pegg I. L.. Low-temperature solidification of radioactive and hazardous wastes. Compiler: U. S, 7855313[P]. 2010.

[199] Davidovits J.. Geopolymer chemistry and properties. In: Davidovits J., Orlinski J. (eds.) Proceedings of Geopolymer '88 - First European Conference on Soft Mineralurgy, Compeigne, France. 1988, 1: 25-48. Universite de Technologie de Compeigne.

[200] Puertas F., Amat T., Fernández-Jiménez A., Vázquez T.. Mechanical and durable behaviour of alkaline cement mortars reinforced with polypropylene fibres. Cem. Concr. Res, 2003, 33(12): 2031-2036.

[201] Zhang Z.-H., Yao X., Zhu H.-J., Hua S.-D., Chen Y.. Preparation and mechanical properties of polypropylene fiber reinforced calcined kaolin-fly ash based geopolymer. J. Cent. South Univ. Technol, 2009, 16: 49-52.

[202] Wimpenny D., Duxson P., Cooper T., Provis J. L., Zeuschner R.. Fibre reinforced geopolymer concrete products for underground infrastructure. In: Concrete 2011, Perth, Australia. CD-ROM proceedings. Concrete Institute of Australia, 2011.

[203] Puertas F., Gil-Maroto A., Palacios M., Amat T.. Alkali-activated slag mortars reinforced with AR glassfibre. Performance and properties. Mater. Constr, 2006, 56(283): 79-90.

[204] Alcaide J. S., Alcocel E. G., Puertas F., Lapuente R., Garces P.. Carbon fibre-reinforced, alkali-activated slag mortars. Mater. Constr, 2007, 57(288): 33-48.

[205] Bernal S., Esguerra J., Galindo J., Mejía de Gutiérrez R., Rodríguez E., Gordillo M., Delvasto S.. Morteros geopolimericos reforzados con fibras de carbono basados en un sistema binario de un subproducto industrial. Rev. Lat. Metal. Mater, 2009, S1(2): 587-592.

[206] Tran D. H., Kroisová D., Louda P., Bortnovsky O., Bezucha P.. Effect of curing temperature on flexural properties of silica-based geopolymer-carbon reinforced composite. J. Achiev. Mater. Manuf. Eng, 2009, 37(2): 492-495.

[207] He P., Jia D., Lin T., Wang M., Zhou Y.. Effects of high-temperature heat treatment on the mechanical properties of unidirectional carbon fiber reinforced geopolymer composites. Ceram. Int, 2010, 36(4): 1447-1453.

[208] He P., Jia D., Wang M., Zhou Y.. Improvement of high-temperature mechanical properties of heat treated Cf/geopolymer composites by sol-SiO_2 impregnation. J. Eur. Ceram. Soc, 2010, 30(15): 3053-3061.

[209] Lin T., Jia D., He P., Wang M., Liang D.. Effects of fiber length on mechanical properties and fracture behavior of short carbon fiber reinforced geopolymer matrix composites. Mater. Sci. Eng. A, 2008, 497(1-2): 181-185.

[210] Penteado Dias D., Thaumaturgo C.. Fracture toughness of geopolymeric concretes reinforced with basalt fibers. Cem. Concr. Compos, 2005, 27(1): 49-54.

[211] Li W., Xu J.. Mechanical properties of basalt fiber reinforced geopolymeric concrete under impact loading. Mater. Sci. Eng. A, 2009, 505: 178-186.

[212] Li W., Xu J.. Impact characterization of basalt fiber reinforced geopolymeric concrete using a 100-mm-diameter split Hopkinson pressure bar. Mater. Sci. Eng. A, 2009, 513-514: 145-153.

[213] Zhang Y., Wei S., Li Z.. Impact behavior and microstructural characteristics of

PVA fiber reinforced fly ash-geopolymer boards prepared by extrusion technique. J. Mater. Sci, 2006, 41: 2787-2794.

[214] Silva F. J., Thaumaturgo C.. Fibre reinforcement and fracture response in geopolymeric mortars. Fatigue Fract. Eng. Mater. Struct, 2003, 26(2): 167-172.

[215] Bernal S., De Gutierrez R., Delvasto S., Rodriguez E.. Performance of an alkali-activated slag concrete reinforced with steel fibers. Constr. Build. Mater, 2010, 24(2): 208-214.

[216] Bernal S., Mejía de Gutierrez R., Rodriguez E., Delvasto S., Puertas F.. Mechanical behaviour of steel fibre-reinforced alkali activated slag concrete. Mater. Constr, 2009, 59(293): 53-62.

[217] Zhao Q., Nair B. G., Rahimian T., Balaguru P.. Novel geopolymer based composites with enhanced ductility. J. Mater. Sci, 2007, 42(9): 3131-3137.

[218] Sakulich A. R.. Reinforced geopolymer composites for enhanced material greenness and durability. Sustain. Cities Soc, 2011, 1: 195-210.

[219] MacKenzie K. J. D., Bolton M. J.. Electrical and mechanical properties of aluminosilicate inorganic polymer composites with carbon nanotubes. J. Mater. Sci, 2009, 44: 2851-2857.

[220] Hussain M., Varley R. J., Cheng Y. B., Simon G. P.. Investigation of thermal and fire performance of novel hybrid geopolymer composites. J. Mater. Sci, 2004, 39(14): 4721-4726.

[221] Hammell J. A., Balaguru P. N., Lyon R. E.. Strength retention of fire resistant aluminosilicate-carbon composites under wet-dry conditions. Compos, 2000, B. 31(2): 107-111.

[222] Kriven W. M., Bell J. L., Gordon M.. Geopolymer refractories for the glass manufacturing industry. Ceram. Eng. Sci. Proc, 2004, 25(1): 57-79.

[223] Perera D. S., Trautman R. L.. Geopolymers with the potential for use as refractory castables. Adv. Technol. Mater. Mater. Proc, 2005, 7(2): 187-190.

[224] Papakonstantinou C. G., Balaguru P., Lyon R. E.. Comparative study of high temperature composites. Compos, 2001, B32(8): 637-649.

第 13 章 碱激发技术的总结与展望

David G. Brice, Lesley S. C. Ko, John L. Provis and
Jannie S. J. van Deventer

13.1 技术委员会的成果总结

经过对碱激发胶凝材料和混凝土的标准化讨论，RILEM TC 224-AAM 已经建立了概念性框架，形成了标志性成果。TC 所有成员达成一致共识，书中提到的材料配方（第 7 章作了介绍）和测试方法（第 8、第 9 和第 10 章作了介绍）对于全球范围内的碱激发混凝土制品的性能表征是必不可少的。但是也需要相对谨慎地推广碱激发材料，避免出现"下一代高铝水泥"应用到实际环境或一些体系里不适应的问题。因此，非常有必要严谨、慎重地推出碱激发材料的相关标准，来限制低品质 AAM 制品或用在不适当领域的产品，损害该项技术在全球的声誉。采用严格的性能测试指标（甚至是超过预期性能水平的更加苛刻的指标）和坚实的科学基础，才能支撑这类材料的相关测试方法，这也是 TC 小组花费大量精力想要实现的目标。

随着对全球气候变化的不断关注，公众和消费者们更倾向于使用"绿色"产品。在有碳税的相关国家市场，碱激发胶凝材料替代纯硅酸盐水泥胶凝材料是一个不错的选择。相对于其他胶凝化学体系，人们逐渐认识到水泥混合材对工业 CO_2 减排是有限的，而碱激发胶凝体系可以直接实现大量 CO_2 减排。除了传统的水泥体系，其实还有各种各样的胶凝体系具有高性能、环境友好和市场应用潜力。但是，水泥的替代产品在全产业链应用中会受到多种因素制约：（1）有效长期的耐久性数据；（2）适合的监管标准（这也需要 AAMs 技术成熟度的权威认证）；（3）材料设计、产品生产、质量控制、应用的产业化经验；（4）原材料稳定的供应链。

碱激发材料确实面临这些挑战，但对于许多竞争技术而言，它的服役期限更长（第 11 章和第 12 章中作了介绍），还利用了大量工业废弃物。像粉煤灰和矿渣这些含玻璃相的铝硅酸盐固体废弃物，完全可以用于大规模 AAMs 的工业生产。火山岩灰、天然火山灰和煅烧黏土也可以作为 AAMs 的原料，只是价格偏高，大宗工业化供应相对受限。

AAMs 的 R&D 与商业化发展是密不可分的，前沿科学研究与工程开发应同步进行，才能有效提高 AAM 混凝土的性能。要把 AAM 的配合比设计、流变性、反应过程、凝胶化学和胶凝材料微观结构的基本原理充分应用于这类材料在生产中的工作性、施工性和耐久性上。但实际上，简单的研究和在科技期刊上发表一些论文，对于推动 AAMs 的商业化是远远不够的。新材料的市场化推动需要开创性的研究，甚至有时要对约定俗成的习惯、现有标准和规程进行挑战，才能引领全球的可持续发展。

本书前边所有章节的重点总结为以下几点。

13.1.1 凝胶形成的反应机理

（1）低钙和高钙原料的AAM凝胶形成机理的研究取得了许多不错的进展，凝胶化学反应过程与前驱体中有效的钙含量直接相关（特别是基于硅酸盐链状和网状结构的凝胶）。近期，使用先进的核磁共振和同步加速技术来分析凝胶纳米结构，可以很好地理解AAM体系的物相演变。

（2）许多不同的原材料可以用于生产AAM产品。尽管固体废弃物原料存在材质均一性问题，研究人员还是在原材料性能与碱激发反应的相关性方面取得很多进展。

（3）AAM混凝土比OPC混凝土孔隙更多，这就不光需要理解凝胶的纳米结构，还需要理解水在这些胶凝材料的纳米结构和微米结构中的作用。在测试孔隙率和渗透性的同时，通过显微和层析成相技术充分分析材料的微观结构才能理解这些问题。

13.1.2 耐久性

（1）从市政和基础设施对废弃物固化的特种应用看，AAM混凝土在很多领域都表现出很好的性能。

（2）如果AAMs暴露在低湿度环境下养护，凝胶在早龄期不能与水化的结合水很好地结合，就会出现干燥问题，因此养护对于AAMs非常重要。在实际生产中，早期养护和产品静置干燥过程，AAMs都有可能出现收缩和表面微裂问题。

（3）许多传统的耐久性测试干燥的方法对于AAM凝胶的稳定性都会造成不利影响，最终会影响测试的结果。

（4）AAMs具有很好的耐Cl^-、耐酸、防火、抗侵蚀、耐硫酸盐性能。使用$MgSO_4$做硫酸盐侵蚀测试，SO_4^{2-}对AAMs的侵蚀要低于Mg^{2+}。

（5）在一些常规实验室测试中，AAMs的抗碳化性能有限，这与现场使用的情况有所区别，也不是什么大的问题，可能是在测试过程中的一些步骤存在偏差。

（6）AAM混凝土对钢筋的腐蚀性研究较少，人们也不能根据碱度来简单预测，所以这个领域的研究非常重要。

13.1.3 实用技术和设计思路

（1）尽管AAM的工程性能已有一些发表/未发表的研究报道，但人们对AAM的长期耐久性，特别是徐变行为，仍然缺乏深入的研究总结。环境对徐变性能的影响仍不明确，这也限制了AAM在结构构件领域的大量应用。

（2）纵观OPC和AAM化学胶凝体系的差别，在碱激发体系里，通用减水剂和其他有机外加剂都不能有效工作，这就需要开发新型外加剂。所以需要对AAM体系的表面化学现象进行深入的研究和理解。

13.2 碱激发的未来

碱激发材料的商业化之路与其他许多混凝土的替代胶凝材料一样，不仅有赖于技术的成熟，还有赖于经济和社会因素的成熟。标准是商业化的重要组成部分，但事实上只是商业化过程的一个小的组成部分（尽管许多研究者花费大量精力致力于标准研究）。图 13-1 给出了工业化生产全过程中 AAM 混凝土商业化进程中关键步骤的示意图。

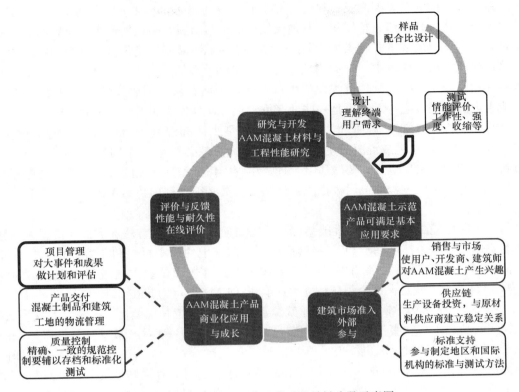

图 13-1　AAM 混凝土商业化进程关键步骤示意图

产品进入市场前进行中试，单独测试产品性能是必要环节。对于非结构混凝土，典型的评价指标包括体积密度、含气量、坍落度、凝结时间、强度（早期和服役期）和收缩等。对于高强和结构混凝土，还需要测试抗折强度或抗拉强度、耐酸或耐化学腐蚀、防火性能、抗渗性能、碳化速率、吸水性、徐变和钢筋防护等。中试的测试数据反馈用于工业生产和研发，也可用于进一步的基础和应用研究。

许多企业倾向于把 AAM 混凝土首先应用于低风险的工程领域，特别是一些性能指标不是严格限制的工程项目中，进而把新材料应用到高风险的项目上，满足监管部门、工程师和规范要求。这些企业更喜欢渐进地推动标准和商业化应用。目前碱激发材料产业化面临的主要挑战包括：

（1）原材料。数量（稳定的大批量供应）和质量（质量可控、均质）。在一个相对较长的时间周期内，铝硅酸盐原材料和碱激发剂都需要一个稳定、独立的供应链来实现建厂的投资回报。对于硅酸盐水泥来说，至少可以稳定供应原材料 50 年。

问题是如何得到主要以固体废弃物或工业副产物为主的 AAM 原材料，或者说，工厂必须接收不同来源的原材料，依据现有技术生产同一类产品。其实，几十年前乌克兰已经出台了一系列标准规程来实现这一目标。尽管在欧洲燃煤灰渣和冶金渣主要作为硅酸盐水泥的混合材，但在中国、印度这样的大市场，燃煤灰渣和冶金渣完全可以应用于碱激发材料的生产。

(2) 成本。假如 CO_2 税或其他污染收费在全球或一些地区强制执行的话，建材行业也会面临一系列同样的问题，碱激发材料就会非常有市场前景了。如果 CO_2 税成为传统 OPC 产品的主要成本，包括矿渣、粉煤灰、其他天然铝硅酸盐材料和碱激发剂等原材料的成本就会低于现有水泥熟料的成本。燃煤电厂向共燃和生物质燃料转变一定程度上会影响粉煤灰的质量，这也应该考虑在成本中。根据当地市场和工业生产情况，选择是单独建立 AAMs 的绿色生产基地，还是直接放到现有水泥的生产企业。当前石灰石采矿场的限制开采也是 OPC 生产成本上涨的因素之一，这也是一些地区推动 AAM 替代水泥形成优势的原因。

(3) 质量控制（QC）与质量认证（QA）。这是水泥和混凝土产品最重要的把控环节。像在 OPC 生产中，水泥或混凝土生产企业的生产人员都必须遵守 QC 和 QA 管理。AAM 的原材料入厂和产品出厂也都需要做质量管理。尽管不像硅酸盐水泥的 QC 和 QA 管理那么复杂，也没有熟料生产过程那道门槛把关，但是技术员必须深刻理解 AAM 原材料质量对产品质量的影响。

(4) 长期性能。在做标准的性能指标时，需要使用快速试验方法和数据评价。大部分耐久性的快速试验方法都是基于 OPC 材料设计的，而其中的胶凝材料和孔溶液化学过程并不一定都适用于 AAM 材料。科学家和研究者应该对这些快速试验方法进行合理、正确的修订。TC 224-AAM 就是在这个领域建立另一套 RILEM TC，为此在 2012 年建立了 TC 247-DTA 数据库。另一个经常被问及的问题就是暴露在真实环境和规模化生产制品的现场数据，以及这些产品在生产和服役期间的性能指标。第 2 章和第 11 章对碱激发胶凝材料体系的长期服役性能做了大量讨论，也给出了很多数据。事实上，AAM 具有很好的耐久性。即使在一些特殊环境下，AAM 也能表现出很好的性能指标，证明其有很好的耐久性。这就需要终端用户把 AAM 材料用到世界各个工程领域。往往，新材料的美好预期与规模化创新发展是有冲突的，这就需要通过大型工程来验证新材料。在一些地区（如日本、澳大利亚），政府和职能部门就这些材料长期服役示范给了特别许可，但是在许多国家，这仍是一大挑战。

(5) 标准化。许多地区市场，因为没有专用的标准和认证指标，新的水泥或混凝土产品面临市场准入问题（第 7 章有作讨论）。起草一项新标准并不容易，需要让大部分的利益相关单位参与到编委会，达成一致。这些利益相关单位包括生产企业、行业协会、研究院、政府部门、材料使用单位、学术界代表、教育机构、终端用户、认证机构等。这些单位或机构不仅关心产品的质量和服役性能，也关心产品的安全性、环保性，还要产品在各自领域具有竞争力，同时确保在各自领域满足法律要求。一旦供应商能把 AAM 商业优势传递给终端客户，也就突破了各方最终达成共识的技术壁垒。因此，非常有必要让各方参与者能够认识到 AAM 技术的商业前景和环境潜质。

(6) 客户认知。想要让终端用户接受，那就要与 OPC 做充分的对照，拿出有利的条件说服别人。这些条件可能是机遇，也可能是挑战，如经济效益、更优的性能（如强度、耐久性）、环保优势（如绿色标志、LEED 信用标志）。还需要对当地政府、其他政府部门、企业、项目开发商、建筑师等集中宣贯教育，强调 CO_2 减排的潜在好处，摒弃传统市场既得利益者的错误观点。对于产品最好的说服方式还是产品的性能，当然也需要建立一批对项目和技术忠实拥护的用户群。在市场上，"绿色优势"逐渐注入终端用户的思想，这是实现产品差异化的关键因素。这就有一些关键问题会被问及：社会是否愿意真正接受？我们是否要把环境因素（如温室气体排放）置于经济效益之前考虑？在此推动下，一些"绿标"产品才能被市场认可，一些基础类的生命全周期分析研究才有意义。假如能够真正倡导可持续发展，那就应该执行更加严格的环境评估，把流通使用过程中的评价指标和精确数据都用于全生命周期的研究。

13.3 结语

学界和工业界的先行者们都在共同努力，选择合适的 AAMs 示范于各种混凝土领域，并对 AAM 混凝土的长期性能进行评价。不同地区的用户都在致力于开发更多的非硅酸盐水泥体系的技术，AAMs 正是一类非常具有优势的理想材料。尽管 AAM 产品仍然面临很多挑战，但是产业化过程必须与开拓性研究并驾齐驱才能真正实现这项技术的大规模应用。

基础研究的目的还是为了提升 AAM 的性能，推动 AAM 的应用，包括化学外加剂的开发、耐久性分析、技术和产业化推动。

总结起来，碱激发胶凝材料在混凝土制品中的大规模应用需要实现以下几个目标：

（1）非结构材料的大尺度现场中试与应用。
（2）结构材料的先行先试。
（3）标准的国际合作与推动。
（4）质量研究——长期服役性能的分析与预测。

以上问题不局限于 AAMs，在非传统水泥和混凝土开发与商业化中都会碰到。这里仅作为 TC 小组关于 AAM 开发的技术研究报告的主要结论呈现给大家。